The Molecular Evolutionary Clock

Simon Y. W. Ho
Editor

The Molecular Evolutionary Clock

Theory and Practice

Springer

Editor
Simon Y. W. Ho
School of Life and Environmental Sciences
University of Sydney
Sydney, New South Wales, Australia

ISBN 978-3-030-60180-5   ISBN 978-3-030-60181-2 (eBook)
https://doi.org/10.1007/978-3-030-60181-2

This Springer imprint is published by the registered company Springer Nature Switzerland AG.
The registered company address is: Gewerbestrasse 11, 6330 Cham, Switzerland

Preface

Evolutionary biology has been transformed over the past few decades by the remarkable wealth of information offered by genomic data. Throughout this period, one enduring concept has been the molecular evolutionary clock, which proposes that rates of genetic change are relatively constant through time. Following its origins in the first half of the 1960s, the molecular clock has played an important role in molecular evolutionary theory while also serving as a valuable tool for inferring evolutionary timescales. It has sustained a range of challenges and criticisms over the years, undergoing considerable development and evolution in response. In the genomic era, the molecular clock remains an exceptionally useful means of placing a timescale on the tree of life.

The purpose of this book is to provide an overview of the molecular evolutionary clock, including its theory and applications. Despite its importance, the molecular clock has not been the focus of any scholarly books in the past two decades, although numerous review articles on this topic have been published. The chapters of this book are grouped into four sections. Part I deals with **Evolutionary Rates** and introduces the molecular clock, the principles of molecular evolution, spontaneous mutation rates, and the causes of evolutionary rate variation. Part II introduces **Molecular Dating**, including the principles of molecular dating, Bayesian molecular dating, and clock models for morphological evolution. Part III describes approaches for **Calibrating Molecular Clocks**, including calibrations from the fossil record, biogeographic calibrations, tip-dating, and total-evidence dating and the fossilized birth–death model. Part IV examines the growing field of **Phylogenomics** and describes rapid dating methods and phylogenomic dating. In Chap. 1, I briefly introduce the individual chapters in these four sections of the book.

I take this opportunity to thank all of the contributors to this book. The authors are leading experts in their respective fields of research and generously gave their time to contribute high-quality chapters to this project. The chapters also benefited from helpful evaluations by anonymous reviewers. I sincerely hope that this book provides a useful, informative, and comprehensive resource for students and researchers in molecular evolution and other fields.

Sydney, Australia — Simon Y. W. Ho

Contents

Part I

Evolutionary Rates

The Molecular Clock and Evolutionary Rates Across the Tree of Life

1

Simon Y. W. Ho

Abstract

The molecular evolutionary clock was proposed in the 1960s and has undergone considerable evolution over the past six decades. After arising from early studies of the amino acid sequences of proteins, the molecular clock became a point of contention between competing theories of molecular evolution. In this chapter, I describe the origins of the molecular clock hypothesis and the mixture of evidence that emerged throughout the 1970s and 1980s, including the discovery of departures from clocklike evolution in proteins and DNA. I review some of the broad patterns of evolutionary rate variation across the tree of life, including rates of spontaneous mutation and long-term evolution in viruses, bacteria, animals, and plants. With the remarkable growth of genomic data over the past two decades, the molecular clock is now primarily seen as a tool for reconstructing evolutionary timescales. In the final parts of this chapter, I summarize the key developments in molecular dating methods and describe how these approaches have been used to infer the timing of major evolutionary events.

S. Y. W. Ho (✉)
School of Life and Environmental Sciences, University of Sydney, Sydney, New South Wales, Australia
e-mail: simon.ho@sydney.edu.au

Keywords

Molecular clock · Neutral theory · Mutation rate · Evolutionary rate · Rate variation · Molecular dating · Tree of life

1.1 Introduction

The molecular evolutionary clock has had a profound influence on molecular evolutionary theory, while also providing an indispensable tool for inferring evolutionary rates and timescales. Starting from the simple premise that evolutionary change at the molecular level proceeds at a relatively constant rate, the molecular clock has undergone considerable evolution over the past six decades (Fig. 1.1). The history of research on the molecular clock has featured an extensive debate over molecular evolutionary theory, persistent challenges to its assumptions and predictions, and applications to questions about the timing of major biological events. Throughout this time, researchers have devoted substantial efforts to understand the causes of evolutionary rate variation across the tree of life, and to apply the principle of the molecular clock in methods for estimating evolutionary timescales. The molecular clock has now confirmed its important role in research in the life sciences, finding applications in such diverse fields as evolutionary biology, molecular ecology, archaeology, and epidemiology.

S. Y. W. Ho (ed.), *The Molecular Evolutionary Clock*, https://doi.org/10.1007/978-3-030-60181-2_1

The idea of a molecular clock emerged from studies of proteins in the mid-twentieth century, a time when new biochemical and genetic data were bringing important insights into evolutionary biology. In particular, efforts to determine the amino acid sequences of proteins were yielding valuable data sets that could inform evolutionary thinking. A series of innovative studies in the early 1960s gave rise to the molecular clock (Zuckerkandl and Pauling 1962, 1965; Margoliash 1963; Doolittle and Blombäck 1964), which soon grew to become an integral part of the neutral theory of molecular evolution (Kimura 1968, 1969). In the ensuing decades, the molecular clock played a central role in the debates between neutralists and selectionists, who supported opposing theories of molecular evolution (Ohta and Gillespie 1996). In the present genomic age, the molecular clock is perhaps most widely recognized as a tool for estimating the timing of evolutionary events (Bromham and Penny 2003).

This book provides an overview of the molecular evolutionary clock, including its theory and practice. It attempts to cover a huge field of research that cannot be satisfactorily summarized in an individual review article; nevertheless, this book can only be considered as an introductory text. Many of the chapters in this book focus on recent developments in this fast-moving field, including the latest endeavours to cope with genome-scale data sets and to combine molecular, phenotypic, and palaeontological data in a biologically meaningful way.

In this opening chapter, I describe the origins of the molecular clock and its evolution over the past six decades. I then provide an overview of the different forms of evolutionary rate variation across the tree of life, ranging from viruses and bacteria to eukaryotes. The chapter concludes with a description of how molecular clocks are used to infer evolutionary timescales, including a summary of some of the major applications of molecular dating. Throughout this chapter, I introduce the contents of the remaining chapters of the book.

1.2 The Molecular Clock Hypothesis

1.2.1 Origins of the Molecular Clock

The term 'molecular evolutionary clock' was proposed by Emile Zuckerkandl and Linus Pauling in 1965. Zuckerkandl had joined Pauling in the California Institute of Technology in late 1959 and the two worked on the sequencing and analysis of the haemoglobin protein (Morgan 1998). Less than a decade earlier, the first amino acid sequence of a protein, insulin, had been determined. Zuckerkandl and Pauling (1962) noted that the divergence in the amino acid sequence of haemoglobin increased over time with the evolutionary distance between species. They made the inspired assumption that a simple linear relationship existed between the two quantities.

Zuckerkandl and Pauling (1962) raised the possibility of using this clocklike property to develop a tool for estimating the timing of divergence between haemoglobin chains and between vertebrate species. Based on a palaeontological estimate of 100–160 million years (Myr) for the divergence between human and horse, they inferred an evolutionary rate of 1 amino acid substitution per 14.5 Myr (Fig. 1.2a). Their application of this rate to the amino acid sequences yielded estimates of the divergence times between haemoglobin chains, with the α chain splitting from the β and γ chains about 565–600 Myr ago in the late Precambrian. The divergences between the β chain and the γ and δ chains were estimated to have occurred much more recently, at 260 Myr ago in the Permian and 44 Myr ago in the Eocene, respectively.

In their analysis of haemoglobin, Zuckerkandl and Pauling (1962) also obtained an estimate of 11 Myr for the evolutionary split between gorilla and human (Fig. 1.2a). They noted that this estimate was at the lower end of the timing of 11–35 Myr ago suggested by the fossil record. Their estimate, and other molecular estimates of the hominid evolutionary timescale reported in the 1960s (Sarich and Wilson 1967a), were controversial because they were inconsistent with the

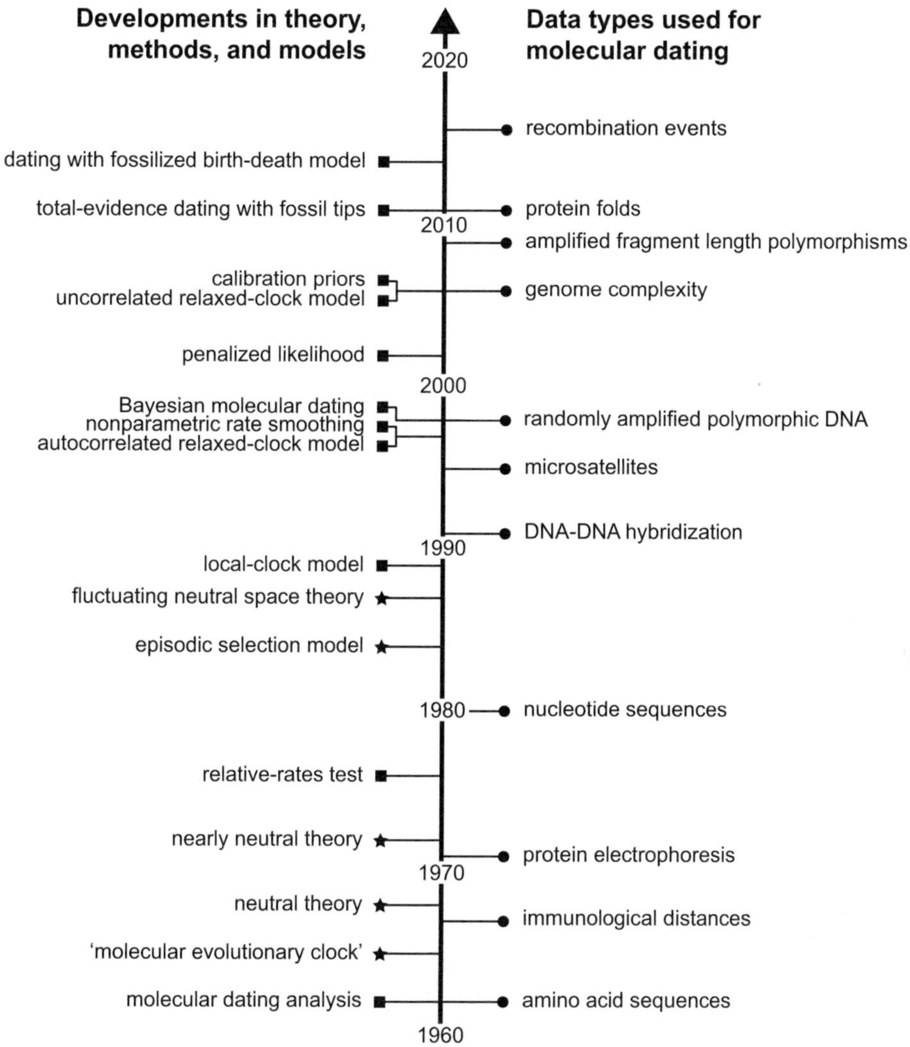

Fig. 1.1 Timeline of advances throughout the history of the molecular clock, beginning with its application to amino acid sequences (Zuckerkandl and Pauling 1962). The left side of the timeline lists some of the key developments in molecular evolutionary theory (stars) and in molecular dating methods and models of evolutionary rate variation (squares). The term 'molecular evolutionary clock' was introduced in 1965. Most of the developments listed here are referred to explicitly in the main text of this chapter. The right side of the timeline lists the first use of different data types for molecular dating (circles; for references, see Ho et al. 2016). Nucleotide sequences are now the most widely used type of genetic data in molecular dating analyses

prevailing notion of a large evolutionary distance between modern humans and the other great apes (Wilson et al. 1977). However, reports would soon emerge of constant evolutionary rates in the amino acid sequences of cytochrome *c* (Margoliash 1963) and fibrinopeptides (Fig. 1.2b; Doolittle and Blombäck 1964), lending support to the molecular clock hypothesis.

In addition to developing a tool for inferring evolutionary timescales, Zuckerkandl and Pauling (1962) foresaw some of the problems that would beset molecular clock analyses in subsequent decades. They referred to the problems posed by repeated substitutions at the same amino acid site (including back-mutations), the potentially confounding impacts of natural selection, and the influence of population size.

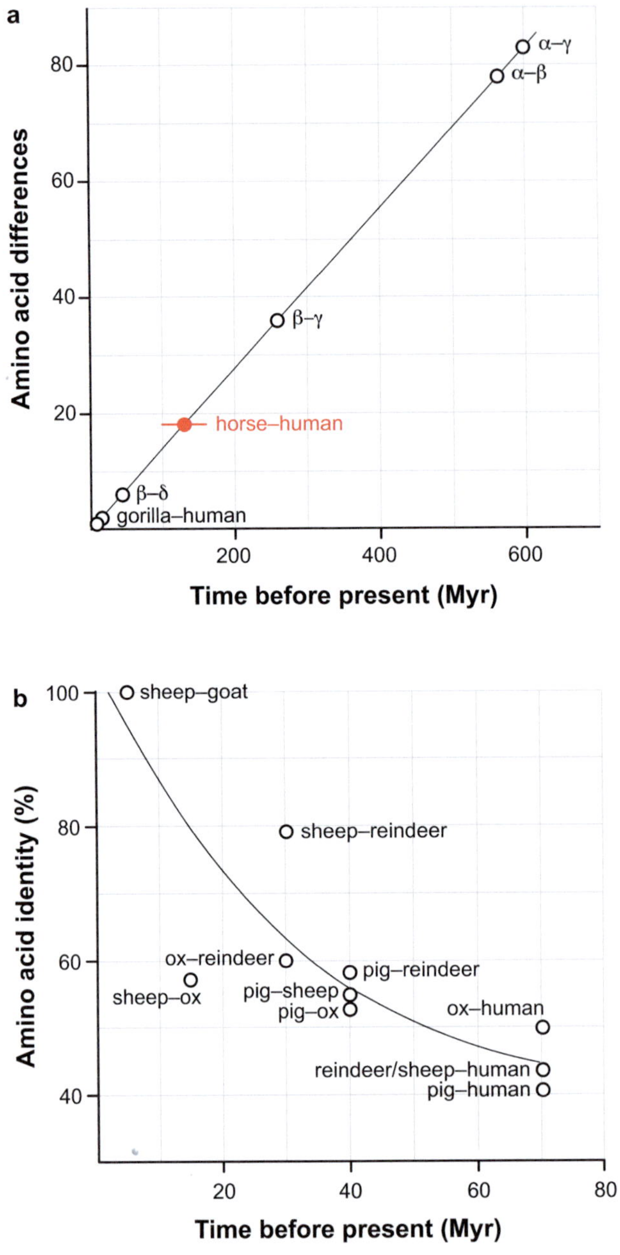

Fig. 1.2 (**a**) The earliest use of the molecular clock to infer evolutionary divergence times, based on amino acid sequences of haemoglobin (Zuckerkandl and Pauling 1962). The evolutionary rate was calibrated using a palaeontological estimate of the horse–human divergence at 100–160 million years ago. Assuming a constant rate of amino acid replacements, the divergence time of gorilla and human was estimated (represented by two data points corresponding to the divergences between the two α chains and between the two β chains), along with the divergence times of various pairs of haemoglobin chains (denoted by Greek letters). Data from Zuckerkandl and Pauling (1962). (**b**) Clocklike evolution in fibrinopeptides, based on pairwise comparisons of amino acid sequences in sheep, goat, reindeer, ox, pig, and human. Pairwise amino acid sequence identity (%) is plotted against the time of divergence estimated from the fossil record. Data from Doolittle and Blombäck (1964)

Their idea of the molecular clock acknowledged an important role for natural selection, although they later surmised that 'the changes that occur at a fairly regular over-all rate would be expected to be those that change the functional properties of the molecule relatively little' (p. 148, Zuckerkandl and Pauling 1965). This statement seemed to anticipate the close association that would soon form between the molecular clock and the neutral theory of molecular evolution (e.g., Kimura 1968; King and Jukes 1969; Wilson and Sarich 1969).

The neutral theory, put forward by Motoo Kimura in 1968, made the bold assertion that the majority of mutations are neutral. This contradicted the dominant view that such mutations are rare or transient (Fisher 1936; Mayr 1963), although the importance of neutral mutations in molecular evolution had been suggested earlier in the same decade (Freese 1962; Sueoka 1962). In Kimura's proposal, the term 'neutral' was not intended to suggest that the corresponding gene lacked function (e.g., Zuckerkandl 1978), but instead meant that the mutation conferred neither an advantage nor disadvantage to the organism and that the fate of the mutation would be governed by genetic drift. Although the molecular clock was influential in the development of the neutral theory (Takahata 2007), Kimura's case for the theory largely rested on estimates of enzyme variability from electrophoretic studies and rates of protein evolution inferred from analyses of amino acid sequences. He argued that these high evolutionary rates greatly exceeded the limits imposed by the 'cost of natural selection' (Haldane 1957), thus suggesting that many of the mutations must be neutral (Kimura 1968).

A significant consequence of the neutral theory is that the rate at which neutral mutations are fixed in the population (known as the 'substitution rate') is approximately equal to the rate at which the mutations are spontaneously generated (Kimura 1968). For this reason, the molecular clock was regarded as an additional source of evidence for the neutral theory (Kimura 1969, 1983). In Chap. 2, Soojin Yi provides an introduction to molecular evolution, including the neutral theory and its later developments, as well as some of the principles behind the molecular clock. She also explains the relationship between the mutation rate and substitution rate under the neutral theory.

The initial reactions to the proposal of the molecular clock were largely negative (e.g., Stebbins and Lewontin 1972), with criticisms being levelled by a number of eminent evolutionary biologists. For example, Ernst Mayr argued that 'evolution is too complex and too variable a process, connected with too many factors, for the time dependence of the evolutionary process at the molecular level to be a simple function' (p. 137, Zuckerkandl and Pauling 1965). At the time, the evolutionary biologist Morris Goodman was one of the few to recognize the potential applications of the clock (Morgan 1998). With further evidence for the constancy of molecular evolutionary rates, as well as growing appreciation of its great potential for reconstructing the timescale of evolution, the notion of a molecular clock endured. By the late 1970s, Allan Wilson et al. (1977) declared that the 'discovery of the evolutionary clock stands out as the most significant result of research in molecular evolution' (p. 577).

1.2.2 Decades of Evolution

The molecular clock was a prominent source of contention in the molecular evolutionary debates throughout the 1970s to 1990s, an era that also saw a shift in focus from protein sequences to DNA sequences (Fig. 1.1; Ohta and Gillespie 1996; Nei et al. 2010). In the early part of this period, there was growing evidence of a discrepancy between the evolutionary dynamics of 'silent' (synonymous or non-coding) and 'replacement' (nonsynonymous) changes in DNA. Replacement substitutions occurred at a constant rate per year, which was cited as support for the neutral theory (Kimura 1969). However, silent substitutions, which are expected to be under much lower selective constraint, appeared to occur at a constant rate per generation (Laird et al. 1969; Kohne 1970). There was evidence of a slowdown in evolutionary rates of both proteins and DNA in hominoids compared with other primates and mammals, particularly rodents (Goodman 1961; Kikuno et al. 1985; Wu and Li 1985), in accordance with the differences in generation times among these organisms.

Kimura (1983) later recognized that the neutral theory should predict a constant substitution rate per generation rather than per year, while admitting that evidence of the constancy of evolutionary change per unit time presented a 'difficult problem' (p. 246) for the theory. The different dynamics observed for silent and replacement substitutions were partly reconciled in the nearly neutral theory, developed by Tomoko Ohta (1972, 1973). The nearly neutral theory proposed that many mutations have a small impact on fitness and are mildly deleterious or mildly advantageous (see Chap. 2), and predicts a constant evolutionary rate per unit of time. However, this prediction relies on a negative correlation between population size and generation time, which was assumed but not explicitly demonstrated by Ohta (1972, 1973). In any case, as described by Gillespie (1991), Kimura 'quickly retreated from the [per-year constancy of mutation rates] when he adopted Ohta's mildly deleterious theory' (p. 274). Nevertheless, upon considering the evidence of a generation-time effect, Kimura (1987) noted that the departures from rate constancy across lineages were not as great as would be expected on the basis of differences in generation time.

A somewhat different challenge to the hypothesis of a molecular clock was that the occurrences of substitutions were often found to be more erratic than expected. Zuckerkandl and Pauling (1965) had suggested that amino acid substitutions occur stochastically, following a Poisson point process. Under this stochastic process, the variance in the number of substitutions per unit time is equal to the expected number of substitutions per unit time. The ratio of these quantities, known as the index of dispersion, provides a measure of the departure from a Poisson process; values exceeding 1 indicate overdispersion. Studies of proteins found that overdispersion was widespread among proteins (Ohta and Kimura 1971; Langley and Fitch 1974; Gillespie 1984, 1989), contradicting the expectations under the molecular clock. One attempt to explain this overdispersion within the framework of the neutral theory was based on a model of fluctuating neutral space (Takahata 1987), in which each neutral mutation changes the rate of neutral mutations. However, most explanations appealed to the effects of natural selection, with overdispersion being a potential outcome under some conditions of episodic, fluctuating, or negative selection (Gillespie 1984, 1993; Cutler 2000). There is now a body of evidence showing that some features of molecular and genomic evolution cannot be adequately explained by the neutral theory (e.g., Kreitman and Akashi 1995; Kern and Hahn 2018).

The molecular clock gradually moved away from its conspicuous role in the selectionist–neutralist debate and became increasingly appreciated for its practical applications in evolutionary biology. Although there is continued interest in the causes of evolutionary rate variation, the molecular clock is now most widely known as a tool for inferring evolutionary timescales. However, the utility of the molecular clock as a dating tool is potentially diminished by the presence of evolutionary rate variation. There have been considerable efforts to rescue the molecular clock from this quagmire, leading to major advances in molecular dating methods over the past two decades.

1.3 Evolutionary Rate Variation

1.3.1 Partitioning Variation in Rates

Evolutionary rate variation occurs in different modes and across a range of temporal, molecular, and biological scales. Early studies considered differences in rates across nucleotide or amino acid sites (site effects), across genes or loci (gene or locus effects; Fig. 1.3a), and across lineages (lineage effects; Fig. 1.3b). For a given gene, any overdispersion that remained after accounting for lineage effects was ascribed to residual effects (e.g., Langley and Fitch 1974;

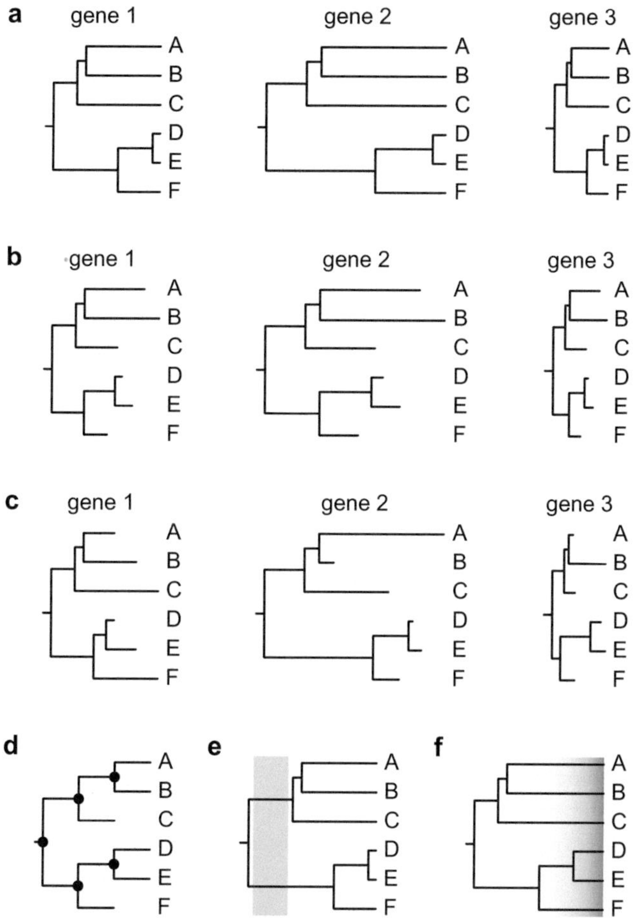

Fig. 1.3 Evolutionary rate variation depicted in phylogenetic trees with branch lengths proportional to the amount of genetic change. Each tree consists of six taxa, labelled A to F. (**a**) Gene effects lead to rate variation across genes, so that there are different total amounts of genetic change in the phylogenetic trees from genes 1, 2, and 3. (**b**) Lineage effects lead to rate variation across the branches of the tree, but this effect is shared by all of the genes. Accordingly, the branch lengths of the three phylogenetic trees share the same proportions. (**c**) Gene-by-lineage interactions lead to different patterns of among-lineage rate variation across the phylogenetic trees from the three genes. (**d**) Punctuated evolution proposes that bursts of genetic change occur at speciation events, denoted by circles. This leads to a pattern in which the length of the path between the root and any tip of the tree is approximately proportional to the number of speciation events along that path. (**e**) Epoch effects occur when there is a different evolutionary rate during a particular period of time, as indicated by the grey shaded area. (**f**) Time-dependent rates lead to a bias whereby evolutionary rates are higher when estimated over recent, short-term timescales than over longer periods of time

Gillespie 1991). In their comprehensive review of evolutionary rate variation in plants, Gaut et al. (2011) used an approach inspired by an analysis of variance that had been conducted 8 years earlier (Smith and Eyre-Walker 2003). Specifically, in addition to site effects, gene effects, and lineage effects, they considered the two- and three-way interactions among these three components: site-by-gene effects, site-by-lineage effects, gene-by-lineage effects, and site-by-gene-by-lineage effects. For molecular clocks, the most important of these effects are caused by gene-by-lineage interactions (Fig. 1.3c); these are analogous to residual effects (Gillespie 1991).

1.3.2 Site Effects and Gene Effects

Site effects can be caused by differences in selective constraints on individual nucleotides or amino acids and by heterogeneity in mutation rates (Hodgkinson and Eyre-Walker 2011). Functionally or structurally important sites tend to evolve more slowly than other sites, or might even be invariant to change, and such amino acid sites in cytochrome *c* and haemoglobin were discussed at length by Zuckerkandl and Pauling (1965). Differences in the proportions of such constrained sites were argued to be the main cause of rate variation across proteins under the neutral theory (King and Jukes 1969). In the nucleotide sequences of protein-coding genes, nonsynonymous mutations are more likely to be selected against than are synonymous mutations. The distinction between 'silent' and 'replacement' dynamics in DNA sequences was already well appreciated by the 1970s (e.g., King and Jukes 1969; Jukes and Kimura 1984), and its varying effects on rates at the three codon positions in protein-coding genes are now routinely taken into account in analyses of nucleotide sequences (e.g., Shapiro et al. 2006).

Mutation rates can also vary among nucleotides and according to the local context, with mutations at cytosine-guanine dinucleotides ('CpG') occurring at higher rates than at other dinucleotides partly because of the vulnerability of the cytosine to deamination (see Chap. 2). Studies of genomic data have revealed other forms of site effects, such as higher mutation rates in parts of the genome linked to insertions and deletions (Tian et al. 2008). In analyses of molecular sequence data, site effects are typically accommodated by modelling the site rates using a gamma distribution (Yang 1996). As with many models in biology, this approach aims to capture an important feature of sequence evolution without attempting to resolve the underlying mechanisms.

Gene effects were widely recognized during the development of the molecular clock (Fig. 1.3a), with evidence of evolutionary rate variation across haemoglobin, cytochrome *c*, fibrinopeptides, and other gene products (e.g., Zuckerkandl and Pauling 1965; Dickerson 1971). In her extensive survey of protein sequences, the pioneering biochemist Margaret Dayhoff (1978) found nearly 400-fold variation in evolutionary rates across proteins. Many of the causes of site effects also lead to rate variation across genes, so the two forms of variation are closely linked. However, the evolutionary rates of genes and proteins are most strongly correlated with their levels of expression (e.g., Rocha and Danchin 2004; Park et al. 2012) and not their functional importance (Wang and Zhang 2009). A negative relationship between the expression level of a protein and its evolutionary rate has been found across a wide range of organisms, including bacteria and eukaryotes, but the specific causes of this relationship remain unclear (Zhang and Yang 2015). In contrast with the rate variation across proteins, rates of synonymous substitutions show little variation across protein-coding genes in mammalian genomes (Kumar and Subramanian 2002).

On a broader scale, evolutionary rates can show substantial disparities between nuclear and organellar genomes. A widely recognized pattern in metazoans is that mutation rates are much higher in the mitochondrial genome than in the nuclear genome (Brown et al. 1979; Miyata et al. 1982). However, the ratio of mitochondrial to nuclear evolutionary rates has been found to be considerably greater in birds, reptiles, and other vertebrates than in insects and arachnids (Allio et al. 2017). The comparatively high information content of mitochondrial DNA ensured that it held a long reign as the preferred marker in studies of population genetics, molecular systematics, and phylogenetics in humans and other animals (Avise et al. 1987). The popularity of mitochondrial DNA declined with the advent of high-throughput sequencing technologies, which enabled nuclear genome data to be obtained efficiently and on large scales, and with growing concerns about excessive reliance on a single genetic marker.

In contrast with the trends observed in animals, elevated mutation rates are not seen in the mitochondrial genomes of other eukaryotes

(Baer et al. 2007). In plants, nuclear genomes evolve more rapidly than chloroplast genomes, which evolve more rapidly than mitochondrial genomes (Wolfe et al. 1987). This pattern is particularly pronounced in angiosperms, but less so in gymnosperms (Drouin et al. 2008). The reasons for the low evolutionary rates in the chloroplast and mitochondrial genomes of plants are not entirely clear, but might be related to DNA repair mechanisms (Christensen 2013). In plastid-bearing eukaryotes other than land plants, mitochondrial genomes have a higher evolutionary rate than plastid genomes (Smith 2015).

1.3.3 Lineage Effects and Gene-by-Lineage Interactions

Evidence of lineage effects emerged soon after the proposal of the molecular clock and continued to grow in the ensuing decades (Fig. 1.3b). The generation-time effect, as described in Sect. 1.2.2, appeared to be the most prominent form of evolutionary rate variation across lineages. The hominoid slowdown in evolutionary rates, first quantified by Goodman (1961), has been confirmed in genome-scale analyses of primates (Kim et al. 2006; Chintalapati and Moorjani 2020). A generation-time effect has now been found in a variety of organisms, including bacteria (Weller and Wu 2015), birds (Mooers and Harvey 1994), and invertebrates (Thomas et al. 2010), and broadly across animals (Allio et al. 2017). However, evolutionary rates appear to show a more complex relationship with generation time in plants, in which the germline is segregated at a late stage of their growth (Lanfear et al. 2013).

Lineage effects can be detected using a variety of methods. Sarich and Wilson (1967b, 1973) described a framework for comparing the relative rates between a pair of taxa, which was later developed into a statistical test (Fitch 1976; Wu and Li 1985). The relative-rates test has largely been superseded by methods that can test for among-lineage rate heterogeneity across an entire phylogenetic tree. These include the likelihood-ratio test, which can be used to compare a model in which the phylogeny is constrained to be ultrametric (all tips being equally distant from the root of the tree) against a model in which the branch lengths are unconstrained (Felsenstein 1981). The rapid increase in genetic data throughout the 1980s and 1990s led to an accumulation of evidence of evolutionary rate variation (Britten 1986; Drake et al. 1998). Some of the major patterns of rate variation across the tree of life are described in Sect. 1.4.

Gene-by-lineage interactions (Fig. 1.3c), which comprise the variation in evolutionary rates that are not accounted for by gene effects or lineage effects, represent an additional layer of complexity in patterns of rate variation (e.g., Gillespie 1989; Ayala 1997). These interactions have been found to be more prominent in nonsynonymous than synonymous rates in plant chloroplast genomes (Muse and Gaut 1997). Gene-by-lineage interactions appear to account for a small proportion of evolutionary rate heterogeneity in mitochondrial and nuclear genes from eutherian mammals (Smith and Eyre-Walker 2003), but are potentially important when large sets of genes are being analysed for the purposes of inferring evolutionary timescales. Variation across genes and across lineages are the dominant forms of genome-scale rate heterogeneity (Snir et al. 2012), although gene-by-lineage interactions have been detected in genomic data from eutherian mammals (Duchêne and Ho 2015) and flowering plants (Duchêne et al. 2016a). Further genomic analyses will allow the different forms of evolutionary rate variation to be characterized for other groups of organisms.

1.3.4 Other Forms of Evolutionary Rate Variation

The framework used in the previous section provides a helpful means of partitioning rate variation into its major components, allowing consideration of the biological and evolutionary drivers of rates of mutation and substitution (Fig. 1.3). Nevertheless, there are several important features of evolutionary rate variation that do not fit neatly into this classification. Here I describe three of these phenomena: punctuated evolution, epoch effects, and time-dependent rates. These forms of rate variation can pose substantial challenges for using molecular clocks to infer evolutionary timescales.

The punctuated equilibrium theory was put forward in an attempt to explain patterns in the fossil record, which appears to feature long periods of stasis punctuated by rapid bursts of morphological change (Eldredge and Gould 1972). Inspired by this theory, molecular evolutionary biologists have sought evidence of bursts of genetic change caused by founder effects at speciation events (Fig. 1.3d; Webster et al. 2003; Pagel et al. 2006). These can potentially be detected using a phylogenetic approach to analyse molecular sequence data, because the theory predicts that a measurable proportion of genetic change is correlated with the number of speciation events along any lineage in the evolutionary tree. However, tests of punctuated evolution have been seriously hindered by a problem known as the node-density effect, which produces patterns similar to those expected under punctuated molecular evolution (Fitch and Beintema 1990). Newly developed phylogenetic models of evolutionary rates might be able to shed further light on the occurrence of punctuated molecular evolution (Manceau et al. 2020).

Rates of molecular evolution can vary across time periods, leading to epoch effects (Fig. 1.3e; Lee and Ho 2016). For example, some external factors, such as environmental conditions, might raise evolutionary rates across an entire population or even an entire assemblage of organisms. One potential example is a several-fold increase in phenotypic and genomic evolutionary rates during the rapid diversification of metazoan phyla in the Cambrian, an event that is often referred to as the 'Cambrian explosion' (Lee et al. 2013). Epoch effects are particularly difficult to identify unless the period of evolutionary rate elevation can be bracketed by reliable age constraints from the fossil record. For example, epoch effects cannot be detected by a likelihood-ratio test for clocklike evolution, in which the null hypothesis is that all of the tips are the same distance from the root of the tree (Yang 2014).

The study of evolutionary rates has been hindered by a time-dependent bias, which causes rate estimates to scale negatively with the timeframe of their measurement (Fig. 1.3f). This pattern can be caused by various factors, including the effects of purifying selection and substitution saturation (Ho et al. 2011). On short timeframes, estimates of evolutionary rates can be inflated by the inclusion of deleterious mutations, which tend to be removed from the population by purifying selection over longer periods of time. Substitution saturation can cause underestimation of the amount of genetic change across longer evolutionary timescales, and this bias is exacerbated by model misspecification (Soubrier et al. 2012). The most striking disparities are seen when the short-term rate estimates from pedigrees and mutation-accumulation lines are compared with those inferred using phylogenetic analysis (e.g., Howell et al. 2003). There is evidence of a time-dependent pattern in evolutionary rate estimates from viruses (Duchêne et al., 2014; Aiewsakun and Katzourakis 2016), bacteria (Duchêne et al. 2016b; but see Gibson and Eyre-Walker 2019), and metazoan mitochondrial genomes (Molak and Ho 2015). The evidence for time-dependent biases in metazoan nuclear genomes has so far been limited, although spontaneous mutation rates appear to be greater than long-term evolutionary rates estimated using phylogenetic methods (with modern humans being at least one exception to this pattern; Scally 2016; Chintalapati and Moorjani 2020).

1.4 Evolutionary Rates Across the Tree of Life

1.4.1 Estimating Rates of Mutation and Evolution

Across the tree of life, evolutionary rates show striking variation and span multiple orders of magnitude. This variation can be considered at a range of biological scales: within individuals, between generations, between populations, among species, and across clades. Lying at one end of this spectrum are rates of spontaneous mutation, which have commonly been estimated by studying laboratory populations but are increasingly based on genome sequencing of closely related individuals or even of different tissues within the same individual. These rates have typically been difficult to estimate directly, because of the small numbers of mutations between generations and because studies often compare the genomes of somatic rather than germline cells. However, improvements in the efficiency and cost of genome sequencing have led to a stunning increase in studies of spontaneous mutation rates, even in multicellular eukaryotes that experience very few mutations per generation. In Chap. 3, Susanne Pfeifer presents an overview of the major approaches that have been used to estimate spontaneous mutation rates, along with a summary of the estimates that have been published so far. These studies have revealed considerable variation in mutation rates across species (Drake et al. 1998; Baer et al. 2007).

Given that most mutations have negative impacts on fitness, the question arises as to why mutation rates are nonzero (Sturtevant 1937). This can be understood in terms of the fitness costs of reducing mutation rates, because cellular and energetic resources are needed for proofreading and error correction (Kimura 1967). A nonzero mutation rate also provides genetic variation, allowing populations of organisms to adapt to changes in environmental conditions. These factors have led to the idea that mutation rates themselves are evolvable; the optimal mutation rate is expected to vary along the genome and across species (Baer et al. 2007). However, some have argued that mutation rates represent a balance between genetic drift and selection for reduced copying errors (Lynch 2010; Lynch et al. 2016). In Chap. 4, Lindell Bromham describes the current state of knowledge of the causes of rate variation across the tree of life, including the factors that affect rates of spontaneous mutation and the rates of fixation of these mutations (i.e., substitution rates).

In many phylogenetic studies using molecular clock models, evolutionary rates and timescales are jointly estimated. These analyses have produced a comprehensive picture of evolutionary rate variation across the diversity of life. In these cases, evolutionary rates are averaged along branches of the phylogeny, meaning that these estimates represent long-term quantities and are partly dependent on taxon sampling (Lanfear et al. 2010). Furthermore, they are somewhat removed from the underlying rates of spontaneous mutation because they have also been shaped by the effects of selection and drift. Some researchers have attempted to use rates estimated from noncoding or synonymous sites as an approximation of mutation rates. In any case, a more complete understanding of rate variation can be achieved by considering both spontaneous mutation rates and phylogenetic estimates of evolutionary rates.

1.4.2 Viruses and Bacteria

The genomes of viruses and bacteria show a remarkable range of mutation rates and evolutionary rates. Among viruses, rates broadly vary with the structure and composition of the genome. Viruses with single-stranded genomes evolve more rapidly than those with double-stranded genomes (Duffy et al. 2008; Sanjuán et al. 2010), although the reasons for this pattern remain unclear (Peck and Lauring 2018). RNA viruses copy their genomes using RNA-dependent RNA polymerases, which lack proofreading ability, so most of these viruses are unable to correct any copying errors that occur during genome replication. As a consequence, they generally have higher mutation rates than DNA viruses, especially double-stranded DNA viruses. There is also a negative correlation between genome size and evolutionary rate (Sanjuán et al. 2010), which is particularly noticeable in viruses but is also seen across a broad range of taxa (Drake 1991; Drake et al. 1998).

The most rapidly evolving viruses tend to be those with single-stranded RNA genomes, such as influenza virus, dengue virus, and coronaviruses. These viruses experience substitution rates as high as 10^{-3} substitutions per site per year (Duffy et al. 2008). At the other end of the spectrum, double-stranded DNA viruses, such as variola virus (which causes smallpox), can evolve at rates below 10^{-5} substitutions per site per year (Firth et al. 2010). For rapidly evolving viruses, evolutionary rates can be estimated using time-structured data sets in which genomes have been sampled at different points in time (Rambaut 2000; Drummond et al. 2001). In contrast, slowly evolving viruses, such as hepatitis B virus, might not undergo a sufficient amount of genetic change over such timeframes to permit any reliable inference of their substitution rate. In some of these cases, evolutionary rates can be estimated by assuming that viruses have codiverged with their hosts (e.g., Bernard 1994; Paraskevis et al. 2013). Virus-host codivergence appears to be more common in double-stranded DNA viruses than in RNA viruses (Geoghegan et al. 2017).

Bacteria have larger genomes than viruses and tend to evolve more slowly. Analyses of genomic data sets have revealed a wide variation in evolutionary rates among bacterial taxa (Duchêne et al. 2016b). The most rapidly evolving bacterial species, such as *Neisseria gonorrhoeae*, *Helicobacter pylori*, and *Enterococcus faecium*, experience nucleotide substitution rates of about 10^{-5} substitutions per site per year. In contrast, rates below 10^{-7} substitutions per site per year are seen in *Mycobacterium tuberculosis* and the plague bacterium *Yersinia pestis* (Duchêne et al. 2016b). The variation in evolutionary rates in bacteria has been ascribed to differences in generation time (Gibson and Eyre-Walker 2019), but attempts to resolve these patterns have been hindered by strong time-dependent biases in rate estimation (Rocha et al. 2006; Duchêne et al. 2016b). Nevertheless, a generation-time effect can be seen in the lower evolutionary rates of spore-forming bacteria compared with bacteria that do not form spores (Weller and Wu 2015).

1.4.3 Eukaryotes

Rates of molecular evolution in eukaryotes, particularly multicellular eukaryotes with long generation times, are generally lower than those of viruses and bacteria. Estimates of mutation rates in unicellular eukaryotes include 1.9×10^{-11} substitutions per site per generation for the protist *Paramecium tetraurelia* (Sung et al. 2012) and about 2×10^{-10} substitutions per site per generation for the yeasts *Saccharomyces cerevisiae* (Zhu et al. 2014) and *Schizosaccharomyces pombe* (Farlow et al. 2015). There have been relatively few estimates of spontaneous mutation rates in the nuclear genomes of animals, but these are growing rapidly with the application of high-throughput sequencing to pedigrees and parent-offspring trios (see Chap. 3).

Animal nuclear genomes evolve slowly, so per-generation mutation rates are difficult to estimate because of the confounding impacts of sequencing error. Inference of mutation rates is also complicated by rate differences between sexes and between the soma and germline. Analyses of genomes from pedigrees and parent-offspring trios have produced a range of estimates of the spontaneous mutation rate in modern humans, centred on a value of 5×10^{-10} mutations per site per year (Scally 2016). Spontaneous mutation rates have also been estimated for the nuclear genomes of the nematode worm *Caenorhabditis elegans*, the common fruit fly *Drosophila melanogaster*, Western honey bee *Apis mellifera*, collared flycatcher *Ficedula albicollis*, house mouse *Mus musculus*, and common chimpanzee *Pan troglodytes*, among other animal species (see Chap. 3; Smeds et al. 2016).

An alternative approach to estimating mutation rates has involved analyses of rates of synonymous substitutions and changes at third codon positions, which are under weaker selective constraints and so are believed to provide an approximation of mutation rates. These analyses have revealed that mitochondrial mutation rates vary considerably across birds and mammals (Nabholz et al. 2008, 2009) and invertebrates (Thomas et al. 2010). In contrast, studies of mitochondrial substitution rates in birds and mammals have identified a relative degree of constancy across lineages, with a mean rate of about 0.01 substitutions per site per Myr (Weir and Schluter 2008; but see Pereira and Baker 2006; Nguyen and Ho 2016). This has led to the notion of a 1% mitochondrial clock in birds and mammals. A similar 'universal' mitochondrial clock has been widely used in studies of invertebrates (Brower 1994; but see Papadopoulou et al. 2010).

Evolutionary rates show considerable heterogeneity across plant lineages, but a few general trends can be observed. The nuclear genomes of gymnosperms evolve at rates that are several times lower, on average, than those of angiosperms (De La Torre et al. 2017). This pattern can potentially be explained by the longer generations and large genomes of gymnosperms. Within flowering plants, there is evidence of a substantial increase in evolutionary rates in the early evolution of the grasses (Christin et al. 2014), whereas palms have evolved much more slowly (Gaut et al. 1992). Evolutionary rates are higher in annual plants than in perennial plants, a pattern that has been found in sequence analyses of the internal transcribed spacer of nuclear ribosomal DNA (Kay et al. 2006) and in larger sets of chloroplast and nuclear genes (Yue et al. 2010). Similarly, herbaceous flowering plants have higher rates of molecular evolution than woody plants with shrub or tree habits (Smith and Donoghue 2008). These patterns in rate variation between annual and perennial plants, and between herbaceous and woody plants, are believed to reflect broad differences in generation time.

Mutation rates in nuclear genomes have been estimated for a number of plant species, including thale cress *Arabidopsis thaliana* (Ossowski et al. 2010), common oak *Quercus robur* (Schmid-Siegert et al. 2017), Sitka spruce *Picea sitchensis* (Hanlon et al. 2019), and yellow box eucalypt *Eucalyptus melliodora* (Orr et al. 2020). Some of these studies were able to trace somatic mutations across the plant, such as along tree branches. For example, 300 mutations were identified along 90.1 metres of branch length in an individual tree of *Eucalyptus melliodora*, allowing the somatic mutation rate to be calculated at 2.75×10^{-9} mutations per nucleotide for each metre of tree branch (Orr et al. 2020). A detailed genomic analysis of eight plant species revealed evidence of higher per-year mutation rates in roots than in shoots in perennial plants, but such a pattern was not seen in annual plants (Wang et al. 2019). In addition, mutation rates were found to be higher in petals than in leaves. These studies have revealed the complexities of mutation rate variation in plants, while highlighting the difficulty in understanding the relationships of these rates to the long-term evolutionary rates in these taxa.

1.5 The Molecular Clock as a Tool for Inferring Timescales

1.5.1 Molecular Dating

In modern genetics and genomics, the molecular clock has its most prominent role as a tool for inferring evolutionary timescales. This application of the molecular clock is sometimes referred to as molecular clock dating, divergence-time estimation, or simply molecular dating. There is a rich history of development of molecular dating methods (Fig. 1.1), with much of the progress in this field being tied to advances in phylogenetic methods and computational power (Bromham and Penny 2003; Kumar 2005). In Chap. 5, Susana Magallón describes the principles behind molecular dating methods and the steps involved in using these methods to infer evolutionary timescales from molecular sequence data.

Research on molecular dating has led to the development of a range of phylogenetic dating methods and statistical models of evolutionary rates (Heath and Moore 2014; Ho and Duchêne 2014; Yang 2014; Kumar and Hedges 2016). These have included methods to cope with among-lineage rate variation, such as nonparametric rate smoothing (Sanderson 1997) and penalized likelihood (Sanderson 2002), as well as models of evolutionary rate variation across branches (Hasegawa et al. 1989; Thorne et al. 1998). Notably, much of the recent progress in molecular clocks has focused on phenomenological rather than mechanistic models, leaving these developments somewhat decoupled from the earlier theoretical context of the molecular clock.

Molecular dating was first performed using amino acid sequences (Zuckerkandl and Pauling 1962) and immunological comparisons by microcomplement fixation (Sarich and Wilson 1967a), but is now overwhelmingly based on the analysis of nucleotide sequences. The most important developments have been in the use of genome-scale data sets for inferring evolutionary timescales. Alongside these efforts, there have been various attempts to use other forms of genetic, genomic, and protein data for molecular dating (Fig. 1.1; Ho et al. 2016). For example, the timing of intraspecific events has been estimated using molecular clocks based on microsatellites (Goldstein et al. 1995), whereas deeper events have been dated using protein folds (Wang et al. 2011).

The application of Bayesian approaches to phylogenetic analysis has led to major developments in molecular dating (dos Reis et al. 2016; Bromham et al. 2018). In Chap. 6, Tianqi Zhu provides an introduction to the Bayesian framework for molecular dating, which permits the application of complex, parameter-rich models that would not be tractable using other methods. These include sophisticated models of evolutionary rate heterogeneity (clock models), models of lineage diversification (in the form of the tree prior), and various means of incorporating data from the fossil record (dos Reis et al. 2016; Bromham et al. 2018).

In Bayesian molecular dating, models of among-lineage rate variation have seen particularly active development. The most widely used are the relaxed-clock models, which allow a distinct rate of evolution along each branch of the phylogenetic tree. The earliest relaxed-clock models were inspired by the work of Gillespie (1991), who suggested that the substitution rate might evolve along lineages. Relaxed-clock models that allow such autocorrelation in the evolutionary rate were implemented in Bayesian dating methods in the late 1990s and subsequently expanded (e.g., Thorne et al. 1998; Kishino et al. 2001; Aris-Brosou and Yang 2002). Later work saw the appearance of relaxed-clock models that allow independent or uncorrelated rates across branches (e.g., Drummond et al. 2006; Rannala and Yang 2007).

The methods developed for molecular dating have also been applied, with some modifications, to analyses of morphological data. In Chap. 7, Michael Lee describes the use of phenotypic traits for estimating evolutionary timescales, focusing on the analysis of discrete morphological characters. The use of morphological clocks has produced useful insights into the evolution of birds and other groups of organisms (e.g., Polly 2001; Lee et al. 2014), although there continue to be various shortcomings that need to be addressed (Puttick et al. 2016). For example, questions persist about the strength of the association between molecular and morphological rates of evolution (Davies and Savolainen 2006; Seligmann 2010). Nevertheless, with continued advances in models of phenotypic evolution (e.g., Álvarez-Carretero et al. 2019), phylogenetic dating analyses of morphological characters present a promising avenue for further research.

Unless there is a priori information about the evolutionary rate, molecular dating methods need to calibrate the clock so that it gives date estimates measured in absolute time. The most widely used types of calibrating information are those based on palaeontological, geological, and biogeographic evidence. In Chap. 8, Jacqueline Nguyen and I describe the use of fossil evidence for calibration, which has a rich history of development and has fostered productive collaborations between geneticists and palaeontologists. In Chap. 9, Michael Landis explains how information from biogeography and palaeogeography can be used to calibrate the molecular clock, based on the timing of geological events such as the separation of landmasses.

Some phylogenetic methods have been extended to account for the inclusion of genomes and morphological data that have been sampled at distinct points in time. In Chap. 10, Sebastián Duchêne and David Duchêne describe the use of sampling times for calibration in analyses of rapidly evolving viruses and bacteria, and when analysing data sets containing ancient DNA sequences. Distinct sampling times are also a feature of morphological data sets that include fossil taxa. In Chap. 11, Alexandra Gavryushkina and Chi Zhang describe the analysis of combined morphological and molecular data, including the development of diversification models that explicitly include extinct species and fossil sampling (e.g., Ronquist et al. 2012; Heath et al. 2014).

The past two decades have seen remarkable growth in genomic data, which has been made possible by the development of high-throughput sequencing methods. This has provided a vast wealth of molecular sequence data for understanding molecular evolution at the genomic scale, but has also brought substantial challenges to molecular dating (Ho 2014; Tong et al. 2016). In Chap. 12, Qiqing Tao, Koichiro Tamura, and Sudhir Kumar review a range of methods that are designed to perform rapid molecular dating, allowing the analysis of data sets containing large numbers of sequences. In Chap. 13, Sandra Álvarez-Carretero and Mario dos Reis describe the application of Bayesian phylogenetic dating to genome-scale data sets, including some of the techniques that have been used to improve computational feasibility. These two closing chapters present a promising picture of how the molecular clock will retain its relevance and utility in the coming years.

1.5.2 Evolutionary Timescales

The molecular clock has been used extensively to reconstruct evolutionary timescales across the tree of life. Early studies focused on the divergence times of humans and related primates (Zuckerkandl and Pauling 1962; Sarich and Wilson 1967a), but often included other mammals (Margoliash 1963; Doolittle and Blombäck 1964). There continued to be a focus on the evolutionary rates and timescales of mammals, particularly eutherian mammals, primarily because of the availability of molecular data for this group of organisms. Developments in automated DNA sequencing in the late 1980s and early 1990s led to rapid growth in molecular sequence data, allowing a considerable expansion of the scope of molecular dating studies.

Molecular dating gained widespread attention in the 1990s when researchers began analysing large data sets to reconstruct the timescales of major evolutionary events. These studies often involved spectacular claims about the antiquity of major branches of the tree of life. These questions have held perennial interest, including the timing of the divergences among the kingdoms of life (e.g., Doolittle et al. 1996), the divergences among metazoan phyla (the 'Cambrian explosion'; e.g., Wray et al. 1996; dos Reis et al. 2015), the diversification of angiosperms (e.g., Martin et al. 1989; Magallón et al. 2015), and the radiations of eutherian mammals and modern birds (e.g., Hedges et al. 1996; Easteal 1999; Springer et al. 2003; dos Reis et al. 2013). The molecular date estimates for these events have often been at odds with the timescales supported by a literal reading of the palaeontological evidence, leading to deliberation about the relative merits of the fossil record and molecular clocks (Smith and Peterson 2002; Benton and Ayala 2003; Brochu et al. 2004). For example, many molecular estimates for the age of crown angiosperms have been greater than 200 Myr, whereas the oldest fossil evidence dates to about 136 Myr in the Early Cretaceous (Magallón et al. 2015). The debates over the discrepancies between molecular and fossil evidence identified some important shortcomings in molecular dating methods, which provided a strong impetus for methodological innovation. Improved modelling of evolutionary rate variation and use of fossil evidence has narrowed some of the gaps between molecular and palaeontological date estimates.

Molecular dating has been particularly valuable for understanding the evolutionary history and epidemiological dynamics of pathogens (Pybus and Rambaut 2009). Fine-scale sampling of pathogens, for example during contemporary virus outbreaks, can allow a detailed reconstruction of evolutionary rates, transmission dynamics, and phylogeographic spread (Pybus and Rambaut 2009). Over longer evolutionary timescales, molecular clocks can be used to determine when pathogens crossed species barriers and infected new hosts, and whether these pathogens continued to codiverge with the host populations.

One of the more surprising applications of molecular dating has been to estimate the ages of the biological samples from which genomic data have been obtained (Shapiro et al. 2011;

Moorjani et al. 2016). This approach can be used to estimate or validate the ages of any samples that have uncertain or contentious dates, such as those that are beyond the 50,000-year reach of radiocarbon dating or where the cost of direct radiometric dating is prohibitive. For example, a Bayesian dating analysis was used to estimate the age of a 400,000-year-old hominin sample from Sima de los Huesos in Spain (Meyer et al. 2014). Ancient hominin genomes have also been dated using a molecular clock based on the accumulation of recombination events over time (Moorjani et al. 2016).

Continued development of molecular clocks will allow evolutionary and demographic timescales to be resolved with increasing confidence. Some of the most promising areas of research include better techniques for incorporating fossil data, mechanistic models of evolutionary rate variation among lineages, and molecular dating methods that are able to process genome-scale data sets from large numbers of taxa. At the same time, these efforts will be substantially aided by advances in understanding of genomic evolution and other biological processes.

1.6 Concluding Remarks

This book is intended to provide an overview of the state of the art of molecular clocks, although the continual and rapid expansion of the field prevents a comprehensive treatment from being achievable. Nevertheless, I hope that this book provides a useful starting point for researchers and students interested in molecular evolutionary clocks. The field is likely to carry on developing at a great pace in response to the growth of genomic data. With international efforts to sequence the genomes of all vertebrates, invertebrates, and other eukaryotes, we will continue to make great strides towards placing a timescale on the tree of life.

References

Aiewsakun P, Katzourakis A (2016) Time-dependent rate phenomenon in viruses. J Virol 90:7184–7195

Allio R, Donega S, Galtier N, Nabholz B (2017) Large variation in the ratio of mitochondrial to nuclear mutation rate across animals: implications for genetic diversity and the use of mitochondrial DNA as a molecular marker. Mol Biol Evol 34:2762–2772

Álvarez-Carretero S, Goswami A, Yang Z, dos Reis M (2019) Bayesian estimation of species divergence times using correlated quantitative characters. Syst Biol 68:967–986

Aris-Brosou S, Yang Z (2002) Effects of models of rate evolution on estimation of divergence dates with special reference to the metazoan 18S ribosomal RNA phylogeny. Syst Biol 51:703–714

Avise JC, Arnold J, Ball RM, Bermingham E, Lamb T, Neigel JE, Reeb CA, Saunders NC (1987) Intraspecific phylogeography: the mitochondrial DNA bridge between population genetics and systematics. Annu Rev Ecol Syst 18:489–522

Ayala FJ (1997) Vagaries of the molecular clock. Proc Natl Acad Sci USA 94:7776–7783

Baer CF, Miyamoto MM, Denver DR (2007) Mutation rate variation in multicellular eukaryotes: causes and consequences. Nat Rev Genet 8:619–631

Benton MJ, Ayala FJ (2003) Dating the tree of life. Science 300:1698–1700

Bernard H-U (1994) Coevolution of papillomaviruses with human populations. Trends Microbiol 2:140–143

Britten RJ (1986) Rates of DNA sequence evolution differ between taxonomic groups. Science 231:1393–1398

Brochu CA, Sumrall CD, Theodor JM (2004) When clocks (and communities) collide: Estimating divergence times from molecules and the fossil record. J Paleontol 78:1–6

Bromham L, Penny D (2003) The modern molecular clock. Nat Rev Genet 4:216–224

Bromham L, Duchêne S, Hua X, Ritchie AM, Duchêne DA, Ho SYW (2018) Bayesian molecular dating: opening up the black box. Biol Rev 93:1165–1191

Brower AVZ (1994) Rapid morphological radiation and convergence among races of the butterfly *Heliconius erato* inferred from patterns of mitochondrial DNA evolution. Proc Natl Acad Sci USA 91:6491–6495

Brown WM, George M Jr, Wilson AC (1979) Rapid evolution of animal mitochondrial DNA. Proc Natl Acad Sci USA 76:1967–1971

Chintalapati M, Moorjani P (2020) Evolution of the mutation rate across primates. Curr Opin Genet Dev 62:58–64

Christensen AC (2013) Plant mitochondrial genome evolution can be explained by DNA repair mechanisms. Genome Biol Evol 5:1079–1086

Christin P-A, Spriggs E, Osborne CP, Strömberg CAE, Salamin N, Edwards EJ (2014) Molecular dating, evolutionary rates, and the age of the grasses. Syst Biol 63:153–165

Cutler D (2000) Understanding the overdispersed molecular clock. Genetics 154:1403–1417

Davies TJ, Savolainen V (2006) Neutral theory, phylogenies, and the relationship between phenotypic change and evolutionary rates. Evolution 60:476–483

Dayhoff MO (1978) Atlas of protein sequence and structure, vol 5, suppl 3. National Biomedical Research Foundation, Washington, DC

De La Torre AR, Li Z, Van de Peer Y, Ingvarsson PK (2017) Contrasting rates of molecular evolution and patterns of selection among gymnosperms and flowering plants. Mol Biol Evol 34:1363–1377

Dickerson RE (1971) The structure of cytochrome *c* and the rates of molecular evolution. J Mol Evol 1:26–45

Doolittle RF, Blombäck B (1964) Amino-acid sequence investigations of fibrinopeptides from various mammals: evolutionary implications. Nature 202:147–152

Doolittle RF, Feng D-F, Tsang S, Cho G, Little E (1996) Determining divergence times of the major kingdoms of living organisms with a protein clock. Science 271:470–477

dos Reis M, Donoghue PCJ, Yang Z (2013) Neither phylogenomic nor palaeontological data support a Palaeogene origin of placental mammals. Biol Lett 10:20131003

dos Reis M, Thawornwattana Y, Angelis K, Telford MJ, Donoghue PCJ, Yang Z (2015) Uncertainty in the timing of origin of animals and the limits of precision in molecular timescales. Curr Biol 25:2939–2950

dos Reis M, Donoghue PCJ, Yang Z (2016) Bayesian molecular clock dating of species divergences in the genomics era. Nat Rev Genet 17:71–80

Drake JW (1991) A constant rate of spontaneous mutation in DNA-based microbes. Proc Natl Acad Sci USA 88:7160–7164

Drake JW, Charlesworth B, Charlesworth D, Crow JF (1998) Rates of spontaneous mutation. Genetics 148:1667–1686

Drouin G, Daoud H, Xia J (2008) Relative rates of synonymous substitutions in the mitochondrial, chloroplast and nuclear genomes of seed plants. Mol Phylogenet Evol 49:827–831

Drummond AJ, Forsberg R, Rodrigo AG (2001) The inference of stepwise changes in substitution rates using serial sequence samples. Mol Biol Evol 18:1365–1371

Drummond AJ, Ho SYW, Phillips MJ, Rambaut A (2006) Relaxed phylogenetics and dating with confidence. PLOS Biol 4:e88

Duchêne S, Ho SYW (2015) Mammalian genome evolution is governed by multiple pacemakers. Bioinformatics 31:2061–2065

Duchêne S, Holmes EC, Ho SYW (2014) Analyses of evolutionary dynamics in viruses are hindered by a time-dependent bias in rate estimates. Proc R Soc B 281:20140732

Duchêne S, Foster CSP, Ho SYW (2016a) Estimating the number and assignment of clock models in analyses of multigene datasets. Bioinformatics 32:1281–1285

Duchêne S, Holt KE, Weill F-X, Le Hello S, Hawkey J, Edwards DJ, Fourment M, Holmes EC (2016b) Genome-scale rates of evolutionary change in bacteria. Microb Genom 2:e000094

Duffy S, Shackelton LA, Holmes EC (2008) Rates of evolutionary change in viruses: patterns and determinants. Nat Rev Genet 9:267–276

Easteal S (1999) Molecular evidence for the early divergence of placental mammals. BioEssays 21:1052–1058

Eldredge N, Gould SJ (1972) Punctuated equilibria: an alternative to phyletic gradualism. In: Schopf TJM (ed) Models in paleobiology. Freeman, San Francisco, CA, pp 82–115

Farlow A, Long H, Arnoux S, Sung W, Doak TG, Nordborg M, Lynch M (2015) The spontaneous mutation rate in the fission yeast *Schizosaccharomyces pombe*. Genetics 201:737–744

Felsenstein J (1981) Evolutionary trees from DNA sequences: a maximum likelihood approach. J Mol Evol 17:368–376

Firth C, Kitchen A, Shapiro B, Suchard MA, Holmes EC, Rambaut A (2010) Using time-structured data to estimate evolutionary rates of double-stranded DNA viruses. Mol Biol Evol 27:2038–2051

Fisher RA (1936) The measurement of selective intensity. Proc R Soc B 121:58–62

Fitch WM (1976) Molecular evolutionary clocks. In: Ayala FJ (ed) Molecular evolution. Sinaeuer Associates, Sunderland, MA, pp 160–178

Fitch WM, Beintema JJ (1990) Correcting parsimonious trees for unseen nucleotide substitutions: the effect of dense branching as exemplified by ribonuclease. Mol Biol Evol 7:438–443

Freese E (1962) On the evolution of base composition at DNA. J Theor Biol 3:82–101

Gaut B, Muse SV, Clark WD, Clegg MT (1992) Relative rates of nucleotide substitution at the *rbcL* locus of monocotyledonous plants. J Mol Evol 35:292–303

Gaut B, Yang L, Takuno S, Eguiarte LE (2011) The patterns and causes of variation in plant nucleotide substitution rates. Annu Rev Ecol Evol Syst 42:245–266

Geoghegan JL, Duchêne S, Holmes EC (2017) Comparative analysis estimates the relative frequencies of co-divergence and cross-species transmission within viral families. PLOS Pathog 13:e1006215

Gibson B, Eyre-Walker A (2019) Investigating evolutionary rate variation in bacteria. J Mol Evol 87:317–326

Gillespie JH (1984) The molecular clock may be an episodic clock. Proc Natl Acad Sci USA 81:8009–8013

Gillespie JH (1989) Lineage effects and the index of dispersion of molecular evolution. Mol Biol Evol 6:636–647

Gillespie JH (1991) The causes of molecular evolution. Oxford University Press, Oxford, UK

Gillespie JH (1993) Substitution processes in molecular evolution. I. Uniform and clustered substitutions in a haploid model. Genetics 134:971–981

Goldstein DB, Ruiz Linares A, Cavalli-Sforza LL, Feldman MW (1995) Genetic absolute dating based

on microsatellites and the origin of modern humans. Proc Natl Acad Sci USA 92:6723–6727
Goodman M (1961) The role of immunologic differences in the phyletic development of human behavior. Hum Biol 33:131–162
Haldane JBS (1957) The cost of natural selection. J Genet 55:511–524
Hanlon VCT, Otto SP, Aitken SN (2019) Somatic mutations substantially increase the per-generation mutation rate in the conifer *Picea sitchensis*. Evol Lett 3:348–358
Hasegawa M, Kishino H, Yano T (1989) Estimation of branching dates among primates by molecular clocks of nuclear DNA which slowed down in Hominoidea. J Hum Evol 18:461–476
Heath TA, Moore BR (2014) Bayesian inference of species divergence times. In: Chen M-H, Kuo L, Lewis PO (eds) Bayesian phylogenetics: methods, algorithms, and applications. CRC Press, Boca Raton, FL, pp 277–318
Heath TA, Huelsenbeck JP, Stadler T (2014) The fossilized birth-death process for coherent calibration of divergence-time estimates. Proc Natl Acad Sci USA 111:E2957–E2966
Hedges SB, Parker PH, Sibley CG, Kumar S (1996) Continental breakup and the ordinal diversification of birds and mammals. Nature 381:226–229
Ho SYW (2014) The changing face of the molecular evolutionary clock. Trends Ecol Evol 29:496–503
Ho SYW, Duchêne S (2014) Molecular-clock methods for estimating evolutionary rates and timescales. Mol Ecol 23:5947–5965
Ho SYW, Lanfear R, Bromham L, Phillips MJ, Soubrier J, Rodrigo AG, Cooper A (2011) Time-dependent rates of molecular evolution. Mol Ecol 20:3087–3101
Ho SYW, Chen AZY, Lins LSF, Duchêne DA, Lo N (2016) The genome as an evolutionary timepiece. Genome Biol Evol 8:3006–3010
Hodgkinson A, Eyre-Walker A (2011) Variation in the mutation rate across mammalian genomes. Nat Rev Genet 12:756–766
Howell N, Smejkal CB, Mackey DA, Chinnery PF, Turnbull DM, Herrnstadt C (2003) The pedigree rate of sequence divergence in the human mitochondrial genome: there is a difference between phylogenetic and pedigree rates. Am J Hum Genet 72:659–670
Jukes TH, Kimura M (1984) Evolutionary constraints and the neutral theory. J Mol Evol 21:90–92
Kay KM, Whittall JB, Hodges SA (2006) A survey of nuclear ribosomal internal transcribed spacer substitution rates across angiosperms: an approximate molecular clock with life history effects. BMC Evol Biol 6:36
Kern AD, Hahn MW (2018) The neutral theory in light of natural selection. Mol Biol Evol 35:1366–1371
Kikuno R, Hayashida H, Miyata T (1985) Rapid rate of rodent evolution. Proc Japan Acad 61:153–156
Kim S-H, Elango N, Warden C, Vigoda E, Yi SV (2006) Heterogeneous genomic molecular clocks in primates. PLOS Genet 2:e163
Kimura M (1967) On the evolutionary adjustment of spontaneous mutation rates. Genet Res 9:23–34
Kimura M (1968) Evolutionary rate at the molecular level. Nature 217:624–626
Kimura M (1969) The rate of molecular evolution considered from the standpoint of population genetics. Proc Natl Acad Sci USA 63:1181–1188
Kimura M (1983) The neutral theory of molecular evolution. Cambridge University Press, Cambridge
Kimura M (1987) Molecular evolutionary clock and the neutral theory. J Mol Evol 26:24–33
King JL, Jukes TH (1969) Non-Darwinian evolution. Science 164:788–798
Kishino H, Thorne JL, Bruno WJ (2001) Performance of a divergence time estimation method under a probabilistic model of rate evolution. Mol Biol Evol 18:352–361
Kohne DE (1970) Evolution of higher-organism DNA. Q Rev Biophys 3:327–375
Kreitman M, Akashi H (1995) Molecular evidence for natural selection. Annu Rev Ecol Syst 26:403–422
Kumar S (2005) Molecular clocks: four decades of evolution. Nat Rev Genet 6:654–662
Kumar S, Hedges SB (2016) Advances in time estimation methods for molecular data. Mol Biol Evol 33:863–869
Kumar S, Subramanian S (2002) Mutation rates in mammalian genomes. Proc Natl Acad Sci USA 99:803–808
Laird CD, McConaughy BL, McCarthy BJ (1969) Rate of fixation of nucleotide substitutions. Nature 224:149–154
Lanfear R, Welch JJ, Bromham L (2010) Watching the clock: studying variation in rates of molecular evolution between species. Trends Ecol Evol 25:495–503
Lanfear R, Ho SYW, Davies TJ, Moles AT, Aarssen L, Swenson NG, Warman L, Zanne AE, Allen AP (2013) Taller plants have lower rates of molecular evolution. Nat Commun 4:1879
Langley CH, Fitch WM (1974) An examination of the constancy of the rate of molecular evolution. J Mol Evol 3:161–177
Lee MSY, Ho SYW (2016) Molecular clocks. Curr Biol 26:R387–R407
Lee MSY, Soubrier J, Edgecombe GD (2013) Rates of phenotypic and genomic evolution during the Cambrian explosion. Curr Biol 23:1–7
Lee MSY, Cau A, Naish D, Dyke GJ (2014) Morphological clocks in paleontology, and a mid-Cretaceous origin of crown Aves. Syst Biol 63:442–449
Lynch M (2010) Evolution of the mutation rate. Trends Genet 26:345–352
Lynch M, Ackerman MS, Gout J-F, Long H, Sung W, Thomas WK, Foster PL (2016) Genetic drift, selection and the evolution of the mutation rate. Nat Rev Genet 17:704–714
Magallón S, Gómez-Acevedo S, Sánchez-Reyes LL, Hernández-Hernández T (2015) A metacalibrated time-tree documents the early rise of flowering plant phylogenetic diversity. New Phytol 207:437–453

Manceau M, Marin J, Morlon H, Lambert A (2020) Model-based inference of punctuated molecular evolution. Mol Biol Evol 37:3308–3323

Margoliash E (1963) Primary structure and evolution of cytochrome *c*. Proc Natl Acad Sci USA 50:672–679

Martin W, Gierl A, Saedler H (1989) Molecular evidence for pre-Cretaceous angiosperm origins. Nature 339:46–48

Mayr E (1963) Animal species and evolution. Harvard University Press, Cambridge, MA

Meyer M, Fu Q, Aximu-Petri A, Glocke I, Nickel B, Arsuaga J-L, Martínez I, Gracia A, Bermúdez de Castro JM, Carbonell E, Pääbo S (2014) A mitochondrial genome sequence of a hominin from Sima de los Huesos. Nature 505:403–406

Miyata T, Hayashida H, Kikuno R, Hasegawa M, Kobayashi M, Koike K (1982) Molecular clock of silent substitution: at least six-fold preponderance of silent changes in mitochondrial genes over those in nuclear genes. J Mol Evol 19:28–35

Molak M, Ho SYW (2015) Prolonged decay of molecular rate estimates for metazoan mitochondrial DNA. PeerJ 3:e821

Mooers AØ, Harvey PH (1994) Metabolic rate, generation time, and the rate of molecular evolution in birds. Mol Phylogenet Evol 3:344–350

Moorjani P, Sankararaman S, Fu Q, Przeworski M, Patterson N, Reich D (2016) A genetic method for dating ancient genomes provides a direct estimate of human generation interval in the last 45,000 years. Proc Natl Acad Sci USA 113:5652–5657

Morgan GJ (1998) Emile Zuckerkandl, Linus Pauling, and the molecular evolutionary clock, 1959–1965. J Hist Biol 31:155–178

Muse SV, Gaut BS (1997) Comparing patterns of nucleotide substitution rates among chloroplast loci using the relative ratio test. Genetics 146:393–399

Nabholz B, Glémin S, Galtier N (2008) Strong variations of mitochondrial mutation rate across mammals—the longevity hypothesis. Mol Biol Evol 25:120–130

Nabholz B, Glémin S, Galtier N (2009) The erratic mitochondrial clock: variations of mutation rate, not population size, affect mtDNA diversity across birds and mammals. BMC Evol Biol 9:54

Nei M, Suzuki Y, Nozawa M (2010) The neutral theory of molecular evolution in the genomic era. Annu Rev Genomics Hum Genet 11:265–289

Nguyen JMT, Ho SYW (2016) Mitochondrial rate variation among lineages of passerine birds. J Avian Biol 47:690–696

Ohta T (1972) Evolutionary rate of cistrons and DNA divergence. J Mol Evol 1:150–157

Ohta T (1973) Slightly deleterious mutant substitutions in evolution. Nature 246:96–98

Ohta T, Gillespie JH (1996) Development of neutral and nearly neutral theories. Theor Pop Biol 49:128–142

Ohta T, Kimura M (1971) On the constancy of the evolutionary rate of cistrons. J Mol Evol 1:18–25

Orr AJ, Padovan A, Kainer D, Külheim C, Bromham L, Bustos-Segura C, Foley W, Haff T, Hsieh J-F, Morales-Suarez A, Cartwright RA, Lanfear R (2020) A phylogenomic approach reveals a low somatic mutation rate in a long-lived plant. Proc R Soc B 287:20192364

Ossowski S, Schneeberger K, Lucas-Lledó JI, Warthmann N, Clark RM, Shaw RG, Weigel D, Lynch M (2010) The rate and molecular spectrum of spontaneous mutations in *Arabidopsis thaliana*. Science 327:92–94

Pagel M, Venditti C, Meade A (2006) Large punctuational contribution of speciation to evolutionary divergence at the molecular level. Science 314:119–121

Papadopoulou A, Anastasiou I, Vogler AP (2010) Revisiting the insect mitochondrial molecular clock: the mid-Aegean trench calibration. Mol Biol Evol 27:1659–1672

Paraskevis D, Magiorkinis G, Magiorkinis E, Ho SYW, Belshaw R, Allain J-P, Hatzakis A (2013) Dating the origin and dispersal of hepatitis B virus infection in humans and primates. Hepatology 57:908–916

Park C, Qian W, Zhang J (2012) Genomic evidence for elevated mutation rates in highly expressed genes. EMBO Rep 13:1123–1129

Peck KM, Lauring AM (2018) Complexities of viral mutation rates. J Virol 92:e01031–e01017

Pereira SL, Baker AJ (2006) A mitogenomic timescale for birds detects variable phylogenetic rates of molecular evolution and refutes the standard molecular clock. Mol Biol Evol 23:1731–1740

Polly PD (2001) On morphological clocks and paleophylogeography: towards a timescale for *Sorex* hybrid zones. Genetica 112–113:339–357

Puttick MN, Thomas GH, Benton MJ (2016) Dating placentalia: morphological clocks fail to close the molecular fossil gap. Evolution 70:873–886

Pybus OG, Rambaut A (2009) Evolutionary analysis of the dynamics of viral infectious disease. Nat Rev Genet 10:540–550

Rambaut A (2000) Estimating the rate of molecular evolution: incorporating non-contemporaneous sequences into maximum likelihood phylogenies. Bioinformatics 16:395–399

Rannala B, Yang Z (2007) Inferring speciation times under an episodic molecular clock. Syst Biol 56:453–466

Rocha EPC, Danchin A (2004) An analysis of determinants of amino acids substitution rates in bacterial proteins. Mol Biol Evol 21:108–116

Rocha EPC, Maynard Smith J, Hurst LD, Holden MTG, Cooper JE, Smith NH, Feil EJ (2006) Comparisons of d*N*/d*S* are time dependent for closely related bacterial genomes. J Theor Biol 239:226–235

Ronquist F, Klopfstein S, Vilhelmsen L, Schulmeister S, Murray DL, Rasnitsyn AP (2012) A total-evidence approach to dating with fossils, applied to the early radiation of the Hymenoptera. Syst Biol 61:973–999

Sanderson MJ (1997) A nonparametric approach to estimating divergence times in the absence of rate constancy. Mol Biol Evol 14:1218–1231
Sanderson MJ (2002) Estimating absolute rates of molecular evolution and divergence times: a penalized likelihood approach. Mol Biol Evol 19:101–109
Sanjuán R, Nebot MR, Chirico N, Mansky LM, Belshaw R (2010) Viral mutation rates. J Virol 84:9733–9748
Sarich VM, Wilson AC (1967a) Immunological time scale for hominid evolution. Science 158:1200–1203
Sarich VM, Wilson AC (1967b) Rates of albumin evolution in primates. Proc Natl Acad Sci USA 58:142–148
Sarich VM, Wilson AC (1973) Generation time and genomic evolution in primates. Science 179:1144–1147
Scally A (2016) The mutation rate in human evolution and demographic inference. Curr Opin Genet Dev 41:36–43
Schmid-Siegert E, Sarkar N, Iseli C, Calderon S, Gouhier-Darimont C, Chrast J, Cattaneo P, Schütz F, Farinelli L, Pagni M, Schneider M, Voumard J, Jaboyedoff M, Fankhauser C, Hardtke CS, Keller L, Pannell JR, Reymond A, Robinson-Rechavi M, Xenarios I, Reymond P (2017) Low number of fixed somatic mutations in a long-lived oak tree. Nat Plants 3:926–929
Seligmann H (2010) Positive correlations between molecular and morphological rates of evolution. J Theor Biol 264:799–807
Shapiro B, Rambaut A, Drummond AJ (2006) Choosing appropriate substitution models for the phylogenetic analysis of protein-coding sequences. Mol Biol Evol 23:7–9
Shapiro B, Ho SYW, Drummond AJ, Suchard MA, Pybus OG, Rambaut A (2011) A Bayesian phylogenetic method to estimate unknown sequence ages. Mol Biol Evol 28:879–887
Smeds L, Qvarnström A, Ellegren H (2016) Direct estimate of the rate of germline mutation in a bird. Genome Res 26:1211–1218
Smith DR (2015) Mutation rates in plastid genomes: they are lower than you might think. Genome Biol Evol 7:1227–1234
Smith SA, Donoghue MJ (2008) Rates of molecular evolution are linked to life history in flowering plants. Science 322:86–89
Smith NGC, Eyre-Walker A (2003) Partitioning the variation in mammalian substitution rates. Mol Biol Evol 20:10–17
Smith AB, Peterson KJ (2002) Dating the time of origin of major clades: molecular clocks and the fossil record. Annu Rev Earth Planet Sci 30:65–88
Snir S, Wolf YI, Koonin EV (2012) Universal pacemaker of genome evolution. PLOS Comput Biol 8:e1002785
Soubrier J, Steel M, Lee MSY, Der Sarkissian C, Guindon S, Ho SYW, Cooper A (2012) The influence of rate heterogeneity among sites on the time dependence of molecular rates. Mol Biol Evol 29:3345–3358
Springer MS, Murphy WJ, Eizirik E, O'Brien SJ (2003) Placental mammal diversification and the Cretaceous-Tertiary boundary. Proc Natl Acad Sci USA 100:1056–1061
Stebbins GL, Lewontin RC (1972) Comparative evolution at the levels of molecules, organisms and populations. In: Le Cam LM, Neyman J, Scott EL (eds) Proceedings of the sixth Berkeley symposium on mathematical statistics and probability. Volume V: Darwinian, neo-Darwinian, and non-Darwinian evolution. University of California Press, Berkeley, CA
Sturtevant AH (1937) Essays on evolution. I. On the effects of selection on mutation rate. Q Rev Biol 12:467–477
Sueoka N (1962) On the genetic basis of variation and heterogeneity of DNA base composition. Proc Natl Acad Sci USA 48:582–592
Sung W, Tucker AE, Doak TG, Choi E, Thomas WK, Lynch M (2012) Extraordinary genome stability in the ciliate *Paramecium tetraurelia*. Proc Natl Acad Sci USA 109:19339–19344
Takahata N (1987) On the overdispersed molecular clock. Genetics 116:169–179
Takahata N (2007) Molecular clock: an *anti*-neo-Darwinian legacy. Genetics 176:1–6
Thomas JA, Welch JJ, Lanfear R, Bromham L (2010) A generation time effect on the rate of molecular evolution in invertebrates. Mol Biol Evol 27:1173–1180
Thorne JL, Kishino H, Painter IS (1998) Estimating the rate of evolution of the rate of molecular evolution. Mol Biol Evol 15:1647–1657
Tian D, Wang Q, Zhang P, Araki H, Yang S, Kreitman M, Nagylaki T, Hudson R, Bergelson J, Chen J-Q (2008) Single-nucleotide mutation rate increases close to insertions/deletions in eukaryotes. Nature 455:105–108
Tong KJ, Lo N, Ho SYW (2016) Reconstructing evolutionary timescales using phylogenomics. Zool Syst 41:343–351
Wang Z, Zhang J (2009) Why is the correlation between gene importance and gene evolutionary rate so weak? PLOS Genet 5:e1000329
Wang M, Jiang Y-Y, Kim KM, Qu G, Jo H-F, Mittenthal JE, Zhang H-Y, Caetano-Anollés G (2011) A universal molecular clock of protein folds and its power in tracing the early history of aerobic metabolism and planet oxygenation. Mol Biol Evol 28:567–582
Wang L, Ji Y, Hu Y, Hu H, Jia X, Jiang M, Zhang X, Zhao L, Zhang Y, Jia Y, Qin C, Yu L, Huang J, Yang S, Hurst LD, Tian D (2019) The architecture of intra-organism mutation rate variation in plants. PLOS Biol 17:e3000191
Webster AJ, Payne RJH, Pagel M (2003) Molecular phylogenies link rates of evolution and speciation. Science 301:478
Weir JT, Schluter D (2008) Calibrating the avian molecular clock. Mol Ecol 17:2321–2328
Weller C, Wu M (2015) A generation-time effect on the rate of molecular evolution in bacteria. Evolution 69:643–652

Wilson AC, Sarich VM (1969) A molecular time scale for human evolution. Proc Natl Acad Sci USA 63:1088–1093
Wilson AC, Carlson SS, White TJ (1977) Biochemical evolution. Annu Rev Biochem 46:573–639
Wolfe KH, Li W-H, Sharp PM (1987) Rates of nucleotide substitution vary greatly among plant mitochondrial, chloroplast, and nuclear DNAs. Proc Natl Acad Sci USA 84:9054–9058
Wray GA, Levinton JS, Shapiro LH (1996) Molecular evidence for deep Precambrian divergences among metazoan phyla. Science 274:568–573
Wu C-I, Li W-H (1985) Evidence for higher rates of nucleotide substitutions in rodents than in man. Proc Natl Acad Sci USA 82:1741–1745
Yang Z (1996) Among-site rate variation and its impact on phylogenetic analyses. Trends Ecol Evol 11:367–372
Yang Z (2014) Molecular evolution: a statistical approach. Oxford University Press, Oxford, UK
Yue J-X, Li J, Wang D, Araki H, Tian D, Yang S (2010) Genome-wide investigation reveals high evolutionary rates in annual model plants. BMC Plant Biol 10:242
Zhang J, Yang J-R (2015) Determinants of the rate of protein sequence evolution. Nat Rev Genet 16:409–420
Zhu YO, Siegal ML, Hall DW, Petrov DA (2014) Precise estimates of mutation rate and spectrum in yeast. Proc Natl Acad Sci USA 111:E2310–E2318
Zuckerkandl E (1978) Multilocus enzymes, gene regulation, and genetic sufficiency. J Mol Evol 12:57–89
Zuckerkandl E, Pauling L (1962) Molecular disease, evolution, and genic heterogeneity. In: Kasha M, Pullman B (eds) Horizons in biochemistry. Academic, New York, pp 189–225
Zuckerkandl E, Pauling L (1965) Evolutionary divergence and convergence in proteins. In: Bryson V, Vogel HJ (eds) Evolving genes and proteins. Academic, New York, pp 97–166

Molecular Evolution: A Brief Introduction

2

Soojin V. Yi

Abstract

Molecular evolution is an expansive and highly interdisciplinary field of research that investigates the evolution of biological molecules and molecular phenomena over time. A notable feature of the field is its ability to embrace and adapt to novel molecular methods, technology, and data while developing and applying a rigorous theoretical framework of population genetics to data interpretation. In the early days of molecular biology, as protein and DNA sequences began to accumulate, molecular evolutionary analyses contributed to the development of several fundamental concepts that remain impactful even after several decades. From preliminary comparisons of protein sequences from distantly related species emerged the idea of a constant molecular clock. In turn, this idea became one of the main inspirations for the neutral theory of molecular evolution, which provides the basis for widely used statistical approaches to test selection using molecular data, including genome sequences. The nearly neutral theory emphasizes that the evolutionary dynamics of many mutations are governed by genetic drift because their effects on fitness are borderline neutral, and this theory can explain many broad patterns of molecular evolution. As the field of molecular evolution embraces the so-called 'omics' era, these foundational ideas continue to provide guiding principles.

Keywords

Mutation · Molecular clock · Neutral theory · Nearly neutral theory · Effective population size

2.1 Introduction

The field of molecular evolution can be broadly defined as the study of how biological molecules change over time in response to evolutionary forces. The biological molecules of interest extend from individual genes, RNAs, and proteins to whole chromosomes, genomes, and other genomic information that can be collected from organisms (such as transcriptomes and proteomes). Molecular evolutionary research also encompasses processes such as transposition, duplication, and interaction between different biological molecules. Ultimately these studies aim to provide insights into how organisms evolve. As such, the two roots of molecular evolution are evolutionary biology and molecular biology.

Biological molecules were first used in evolutionary analysis long before the emergence and

S. V. Yi (✉)
School of Biological Sciences, Institute for Bioengineering and Bioscience, Georgia Institute of Technology, Atlanta, GA, USA

S. Y. W. Ho (ed.), *The Molecular Evolutionary Clock*, https://doi.org/10.1007/978-3-030-60181-2_2

growth of modern molecular biology. In the early 1900s, George Nuttall used blood serums to infer relatedness between species (Nuttall 1904). Immunological methods continued to be used in the 1960s by visionaries such as Morris Goodman and Allan Wilson to infer primate phylogenies and divergence times (Goodman 1961, 1962, 1963; Sarich and Wilson 1967; Wilson and Sarich 1969).

Although these early studies were revolutionary in their time, it was the advent of protein sequencing technologies that finally allowed scientists to generate data and infer the relationships between protein molecules within and between species (Zuckerkandl et al. 1960; Zuckerkandl and Pauling 1962; Margoliash 1963; Doolittle and Blombäck 1964). It was from these analyses that the molecular clock hypothesis emerged, which remains one of the most famous and influential concepts in molecular evolution. In turn, the foundational concepts around the molecular clock hypothesis inspired the neutral theory of molecular evolution and the nearly neutral theory of molecular evolution. Together, these core concepts form the basis of many modern evolutionary genetic studies.

In this chapter, I will provide a brief overview of these foundational concepts. I will also discuss how these concepts influenced each other's development, as a way of introducing some of the main questions and parameters in the field, including mutation rates, selection coefficients, effective population size, and substitution rates. Finally, at the close of the chapter, some examples of how molecular evolutionary research applies these concepts to genomics-era data will be discussed.

2.2 The Molecular Evolutionary Clock

In the early 1960s, a group of biologists began to investigate similarities and differences between the same proteins in different species (Zuckerkandl and Pauling 1962; Margoliash 1963; Doolittle and Blombäck 1964). Emile Zuckerkandl and Linus Pauling studied haemoglobins between species using a protein 'fingerprinting' technique, involving a combination of chromatography and electrophoresis. The initial goal of their study was to infer the phylogeny of primates, but they soon expanded their analyses to include haemoglobins of other animals, including cow, pig, shark, bony fish, worm, and lungfish (Zuckerkandl et al. 1960). Using these data, they quantified the time of divergence between different species and between different chains of haemoglobins (Zuckerkandl and Pauling 1962). At the centre of their inference was the fundamental idea that the number of differences between amino acids of different species could provide information on the time of divergence between the species. They discovered that the rates of amino acid replacements in haemoglobin from different species were roughly constant over time (Zuckerkandl and Pauling 1962). Analyses of other proteins sequenced around the same time, such as those of cytochrome *c* (Margoliash 1963) and fibrinopeptide (Doolittle and Blombäck 1964), revealed similar patterns: amino acid differences of proteins between species were proportional to the divergence times between species inferred from the fossil record. These findings led Zuckerkandl and Pauling (1965) to famously state, 'there may exist a molecular evolutionary clock' (Fig. 2.1). Thus, in its initial proposal, the term 'molecular clock' referred to constant rates of amino acid changes in proteins over time.

These early studies also illuminated several characteristics of protein evolution that laid foundations for the neutral theory and the nearly neutral theory (Fig. 2.1). Margoliash (1963) compared amino acid sequences of cytochrome *c* from horses, humans, pigs, rabbits, chicken, tuna, and baker's yeast, and noted that cytochrome *c* appears to evolve 'slowly'. He concluded that cytochrome *c* might be more appropriately used to infer relationships between distantly related species. Doolittle and Blombäck (1964) analysed amino acid sequences of fibrinopeptides and proposed that some portions of the proteins were more suitable for evolutionary analyses, because they might serve 'little function'. Thus, differences in the 'speed' of the

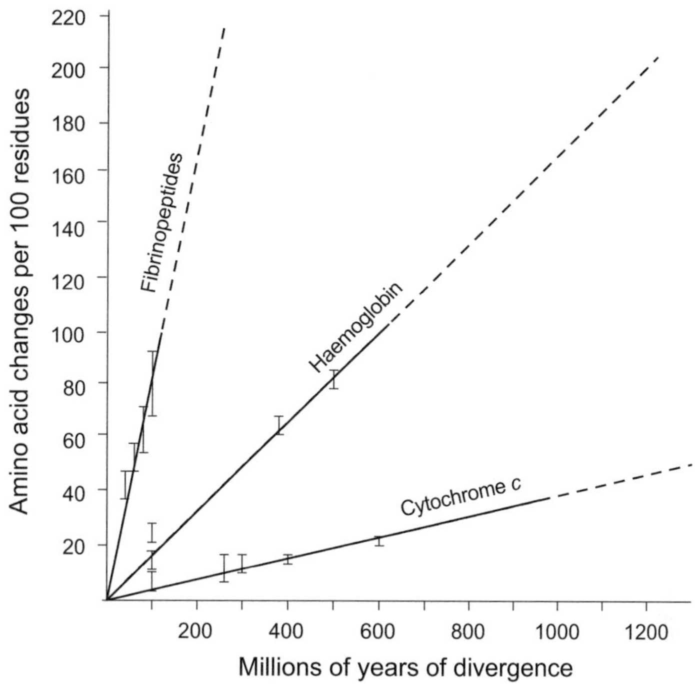

Fig. 2.1 Rates of amino acid changes in fibrinopeptides, haemoglobin, and cytochrome *c* (Zuckerkandl and Pauling 1962; Margoliash 1963; Doolittle and Blombäck 1964). The three proteins show different rates of change per unit time. For each protein, however, the rate of change per unit time appears to be approximately constant. This phenomenon was referred to as the 'molecular evolutionary clock' by Zuckerkandl and Pauling (1965). Figure modified from Dickerson (1971)

molecular clock between proteins, and the potential role of functional importance in determining this speed, were already recognized in these early studies. Moreover, Zuckerkandl and Pauling (1965) offered a mathematical derivation for the molecular clock, positing that 'rates of mutations that have not been eliminated by natural selection' might be roughly constant over time. These ideas foreshadowed the neutral theory of molecular evolution first proposed by Motoo Kimura.

2.3 The Molecular Clock and the Neutral Theory of Molecular Evolution

The idea of a molecular evolutionary clock was controversial from its first appearance (see Dietrich 1998; Morgan 1998). It was proposed during a time when biologists considered natural selection as the predominant evolutionary force. Consequently, most of the evolutionary changes observed between proteins of different species, or 'substitutions', were viewed as the result of natural selection, especially of directional selection. In parallel, molecular variation within populations (polymorphism) was thought to be maintained by balancing selection.

The proposal of the molecular evolutionary clock implied that amino acid substitutions at proteins should occur at constant rates over time. This was diametrically opposed to the notion that natural selection was the main cause of amino acid substitutions. According to population genetic theory, the rate of amino acid substitution caused by natural selection was determined by several parameters, namely the strength of selection (measured by the 'selection coefficient'), effective population size, and adaptive mutation rates (Kimura 1983). These parameters should be specific to different mutations as well as to the different populations and species in which they occur. Although there could be special scenarios where the combination of all of these parameters produces similar substitution rates between species, such scenarios could not be easily generalized across different species and over time. The scientific community and the incipient field of molecular evolution were in dire need of an alternative explanation to understand the pervasively constant rate of protein evolution.

The neutral theory of molecular evolution (Kimura 1968, 1983; King and Jukes 1969) provided just such an explanation. Kimura (1968) proposed that the majority of substitutions at the protein or DNA level were caused by random genetic drift of selectively 'neutral' mutations, and that only a minor fraction of changes at the molecular level were due to adaptive evolution. The neutral theory provided an elegant and simple explanation for the molecular evolutionary clock. Namely, if we assume that most of the mutations that lead to amino acid substitutions between species are neutral, then the rate of substitution is identical to the rate of mutation (Kimura 1968, 1983). Briefly, let us consider a population of N diploid individuals, so that the total number of alleles in the population is $2N$. If each allele can mutate to a new allele at a rate of u, the number of new mutations in this population is $2Nu$. A substitution occurs when any of these new mutations becomes fixed in the population. If the mutations are neutral, each of the $2Nu$ mutations will have the same probability of reaching fixation, namely $1/2N$. The total number of substitutions is the total number of mutations multiplied by the probability of fixation of each mutation. Thus, the substitution rate equals the mutation rate. King and Jukes (1969) also stated that 'in the absence of selection constraints, the substitution rate reaches the maximum value set by the mutation rate'. Independent of other parameters such as effective population size, a constant molecular evolutionary clock could be achieved if mutation rates were equal between species (Kimura and Ohta 1971a).

Another important inspiration for Kimura was the high rate of amino acid substitution in the genome. Using the data available at that time, Kimura (1968) estimated that there has been 1 nucleotide substitution every 2 years in a mammalian genome, which would far exceed the 'limit' of evolution according to the cost of natural selection (Haldane 1957; Kimura 1960). Although Kimura's original estimate was inflated due to limited and incorrect genomic data at that time (such as the number of total nucleotides in the genome, and the lack of understanding of noncoding regions; see Takahata 2007), it was apparent that there were many more substitutions between species than could be sustained by natural selection. Kimura (1968) posited that the high rate of amino acid substitutions in the genome was another indicator that most mutations are neutral and, by definition, free from the constraint or limit of natural selection. In addition, protein electrophoresis data from fruit flies (Hubby and Lewontin 1966; Lewontin and Hubby 1966) demonstrated that the amount of protein polymorphism in natural populations was in fact very high. Kimura (1968) and Kimura and Ohta (1971b) interpreted this high rate of polymorphism as a snapshot of an equilibrium between high rates of neutral mutations and random genetic drift.

The neutral theory of molecular evolution has had a tremendous impact both on molecular evolution and on evolutionary biology as a whole (Nei 2005). Under the neutral theory, it was straightforward to explain differences in evolutionary rates among proteins, by assuming that numbers of selectively unconstrained sites differed between proteins (King and Jukes 1969; Kimura and Ohta 1971a). Higher rates at synonymous sites compared with nonsynonymous sites were also explained by the preponderance of neutral mutations in the former compared with the latter (Kimura 1977). Similarly, the faster evolution of pseudogenes compared with their functional counterparts was attributed to the increase of unconstrained sites following the loss of function (Li et al. 1981). These studies solidified the idea that evolutionary rates of specific sequences reflect their proportion of sites that are free to vary. This concept continues to be extremely useful for identifying functionally important sites in comparative genomic analysis. For example, genomic regions that experience few nucleotide changes during evolution are considered to be candidates for functionally constrained sites (e.g., Margulies et al. 2003; Woolfe et al. 2004).

In addition, Kimura's insightful recognition that the high rate of substitutions (between species) and the high level of polymorphism (within species) reflected two phases of selectively neutral mutations (Kimura and Ohta 1971b) laid foundations for many statistical tests of natural selection using molecular data. Specifically, several statistical approaches explicitly test whether the observed patterns of within-species polymorphism and between-species divergence are consistent with the null model of neutral evolution, as a means to detect underlying selective forces (Hudson et al. 1987; Tajima 1989; McDonald and Kreitman 1991). Broadly, the neutral theory of molecular evolution led to the development of firm scientific frameworks to contrast the impacts of natural selection versus genetic drift in the study of molecular data.

2.4 Nearly Neutral Theory

Ohta (1973, 1974) and Ohta and Kimura (1971) extended the neutral theory by pointing out that mutations with very small selection coefficients, while not strictly neutral, are also highly subject to genetic drift and consequently behave as if they are neutral. Ohta (1972b, 1973, 1974) further developed the nearly neutral theory of molecular evolution, emphasizing the significance of slightly deleterious mutations whose selection coefficients lie near the inverse of the effective population size ($|N_e s| \sim 1$ or $s \sim 1/N_e$). For example, in a population of 10^3 individuals, the fixation probability of a slightly deleterious mutation with selection coefficient of -10^{-3} is 43% of the fixation probability of the neutral mutations (Ohta 1973). According to the nearly neutral theory, patterns of molecular evolution are more consistent with the abundance of nearly neutral mutations than of strictly neutral mutations (Fig. 2.2). In one of the first papers on the nearly neutral theory, Ohta (1973) argued that the high incidences of compensatory substitutions could be explained by assuming slightly deleterious mutations. Since its early days, the nearly neutral theory of molecular evolution has proven powerful and applicable to many other aspects of genome evolution (e.g., see Akashi et al. 2012).

A key feature that separates the neutral theory and the nearly neutral theory is the impact of effective population size on substitution rates. Unlike strictly neutral mutations, the definition of nearly neutral mutations depends critically on the effective population size. Due to the inverse relationship between the effective population size and the range of nearly neutral mutations, the proportion of mutations whose evolutionary fate is largely determined by genetic drift is greater in smaller populations. Consequently, if there are a large number of nearly neutral mutations (namely $|N_e s| \sim 1$), there will be more substitutions per unit time in small populations. Thus, the nearly neutral theory predicts that substitution rates should be negatively correlated with the effective population size (Ohta 1972b, 1974).

It is convenient to test the predictions of the nearly neutral theory using protein-coding DNA sequences. Because mutations at nonsynonymous sites are more likely to be deleterious compared with those at synonymous sites, the ratio of nonsynonymous to synonymous substitutions could be influenced by the fixation of nearly neutral mutations. Using the sequences of 49 genes available at that time, Ohta (1995) showed that the DNA sequences of primates have greater ratios of nonsynonymous to synonymous substitutions than those of artiodactyls and rodents. This was consistent with the prediction of the nearly neutral theory, because primates have smaller effective population sizes than artiodactyls or rodents do. Comparison of protein-coding sequences from the whole genomes of human, rhesus macaque, mouse, and rat demonstrated the same pattern (Rhesus Macaque Genome Sequencing and Analysis Consortium 2007; Kosiol et al. 2008). Humans and rhesus macaques have greater genome-wide ratios of nonsynonymous to synonymous substitutions than the two rodents (Fig. 2.3), providing strong support that nearly neutral mutations contribute to protein evolution in these genomes. The nearly neutral theory continues to be extremely insightful in explaining many observations of molecular evolution (e.g., Ohta 2002, 2011).

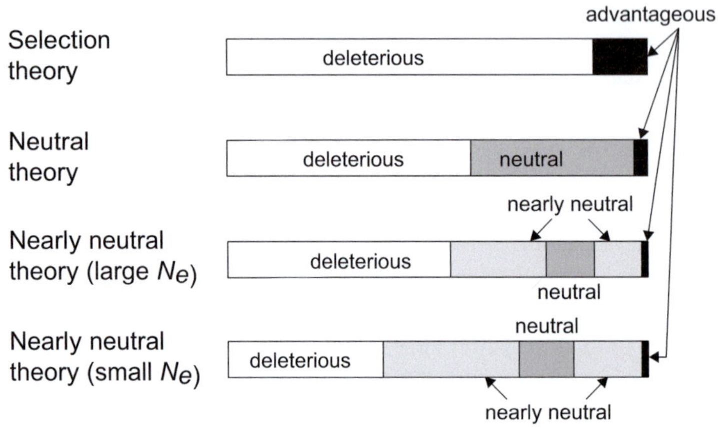

Fig. 2.2 Different proportions of various types of mutations according to the selection theory, neutral theory, and nearly neutral theory. The neutral theory proposes that a large number of mutations at the molecular level are neutral. The nearly neutral theory proposes that many mutations are 'borderline' between neutral and selected (either advantageous or deleterious). Importantly, according to the nearly neutral theory, the proportion of nearly neutral mutations is larger in populations with small effective population sizes. Figure modified from Ohta (2002)

2.5 Origin of Mutations and Substitution Rates

While the neutral theory was inspired by analyses of protein sequences, the nearly neutral theory was in part motivated by the need to reconcile the molecular evolutionary clock with the patterns observed in DNA sequences. As discussed above, according to the neutral theory, neutral sites should evolve at the rate at which mutations occur, and mutation rates should be

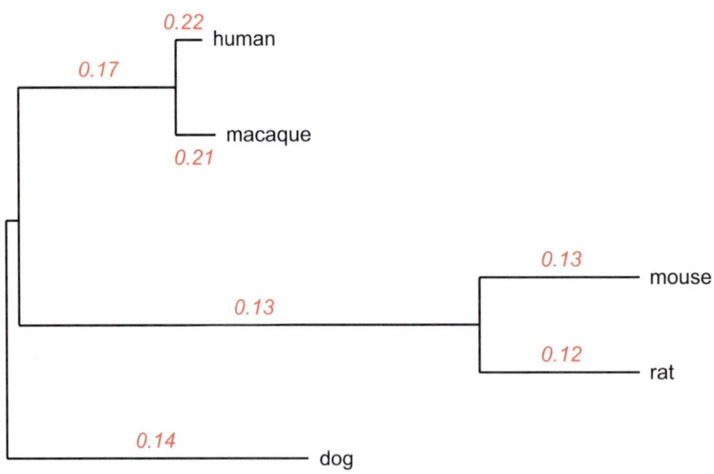

Fig. 2.3 Comparison of orthologous genes among five mammalian species. The branch lengths are drawn according to the numbers of synonymous substitutions. Primate branches are shorter than rodent branches, as explained by the generation-time effect. In comparison, the ratios of nonsynonymous to synonymous substitutions (shown in red above branches) are greater in primates than in rodents. This pattern is consistent with the prediction of the nearly neutral theory. Figure modified from Kosiol et al. (2008)

similar between lineages according to the molecular evolutionary clock. However, analyses of DNA that were performed around the same time as the emergence of the neutral theory found that evolutionary rates of DNA sequences were quite variable between lineages. For example, evolutionary rates inferred from DNA hybridization showed faster evolution of rodent sequences compared with primate sequences (Laird et al. 1969; Kohne 1970). In fact, even earlier immunological studies indicated that evolutionary rates of blood proteins were different between primate lineages, with humans and other apes being particularly slow-evolving (Goodman 1961, 1962, 1963). These studies showed that, in general, lineages with longer generation times had lower evolutionary rates, a pattern referred to as the 'generation-time effect'. This pattern has been supported by numerous subsequent studies (e.g., Wu and Li 1985; Yi et al. 2002; Thomas et al. 2010).

The generation-time effect is consistent with the mechanisms of germline mutations. It has been long considered that most mutations arise due to DNA replication errors in germlines (Haldane 1947; Muller 1954). Since germline DNA replication occurs in each generation, species with shorter generations will have more mutations per given time than organisms with longer generations. Then why were the protein evolutionary rates constant?

The nearly neutral theory answers this question in the following way. Given the same amount of time, fewer mutations should occur in species with longer generations (due to the generation-time effect). However, generation time is often highly correlated with population size. Organisms with longer generations tend to have smaller effective population sizes. Therefore, if mutations were nearly neutral, even though they would occur less frequently in species with longer generations, they are more likely to be fixed due to the small effective population sizes. The generation-time effect would partially be cancelled out due to the increase in substitution rates (Ohta 1972a).

It should be noted that there are many other traits that correlate with generation time. Depending on the data set, other factors might better explain the observed variation in evolutionary rates (e.g., Welch et al. 2008; Tsantes and Steiper 2009). In addition, some mutations are not caused by DNA replication errors. For example, in many species, cytosines are chemically modified by the addition of the methyl group, a process referred to as DNA methylation (Suzuki and Bird 2008; Yi 2012). Primary targets of DNA methylation in many genomes are cytosines followed by guanine, or 'CpGs'. For chemical reasons, methylated CpGs are highly prone to being converted to TpGs (Bird 1980). Because DNA methylation itself is not dependent on cell replication (e.g., Vandiver et al. 2015), mutations caused by DNA methylation might not exhibit a generation-time effect. Kim et al. (2006) used data from primates and showed that mutations at CpGs indeed occur at similar rates between lineages, while those at non-CpG sites show variation consistent with the generation-time effect. Kim et al. (2006) further noted that protein-coding sequences generally have many CpG sites, whereas non-coding sequences have fewer CpG sites. Consequently, we might observe a less pronounced generation-time effect in protein-coding sequences than in non-coding sequences, due to their asymmetric CpG composition (Kim et al. 2006). Subsequent studies have supported this conclusion (e.g., Moorjani et al. 2016).

2.6 Molecular Evolution: Past, Present, and Future

As stated in the opening of this chapter, the two main components of the field of molecular evolution are evolutionary biology and molecular biology. While inspired by discoveries in molecular biology, evolutionary biology, especially population genetics, continues to provide guiding principles and theories that can help us to interpret data from molecular biology. The neutral theory, despite having been proposed without knowledge of the genomic data, has withstood the test of time and continues to offer the most widely used null model for inferring evolutionary forces in molecular data.

The significance of slightly deleterious mutations is even more appreciated in the era of genomics. A notable extension of the nearly neutral theory is the idea that the fixation of slightly deleterious mutations enabled the evolution of complex genomic architecture (Lynch and Conery 2003; Lynch 2007). For example, introns, repetitive sequences, and duplicate genes might have reached fixation in genomes of species with small effective population sizes despite their selective disadvantages (Lynch and Conery 2003), as predicted by the nearly neutral theory. The 'drift-barrier hypothesis' formalizes the idea that the efficiency of natural selection is limited by the extent of genetic drift conferred by the effective population size (Sung et al. 2012; Lynch et al. 2016).

The biological molecules of interest for molecular evolutionists have constantly changed as technical advancements yielded the riches of new types of data. At present, molecular evolutionary studies have completely embraced genomic sequence data (or other genome-scale data such as transcriptomes, lipidomes, and proteomes). The synergy between molecular biology and evolutionary biology is growing even stronger in the era of genomics. Recent technological advances in genomics are leading to explosive growth in genome sequence analyses using evolutionary principles. The idea of using genome sequences to infer relatedness between individuals and demographic history is now no longer confined to the realm of molecular evolution. For example, a new fervour for genealogical DNA tests that utilize the principles of molecular evolution and molecular population genetics is currently expanding on a global scale.

We are in the midst of a massive expansion of access to novel biological data, fuelled by innovative developments in genomic, epigenetic, and molecular technology. In the near future, researchers will have the ability to explore everything from epigenetic modification on a cellular scale to genomic features at a population level from many species. Molecular evolutionists are readily embracing these new data, discovering general principles that can help molecular biologists interpret their findings and generate ideas for future investigations.

References

Akashi H, Osada N, Ohta T (2012) Weak selection and protein evolution. Genetics 192:15–31

Bird A (1980) DNA methylation and the frequency of CpG in animal DNA. Nucleic Acids Res 8:1499–1504

Dickerson RE (1971) The structure of cytochrome *c* and the rates of molecular evolution. J Mol Evol 1:26–45

Dietrich MR (1998) Paradox and persuasion: negotiating the place of molecular evolution within evolutionary biology. J Hist Biol 31:85–111

Doolittle RF, Blombäck B (1964) Amino-acid sequence investigations of fibrinopeptides from various mammals: evolutionary implications. Nature 202:147–152

Goodman M (1961) The role of immunologic differences in the phyletic development of human behavior. Hum Biol 33:131–162

Goodman M (1962) Evolution of the immunologic species specificity of human serum proteins. Hum Biol 34:104–150

Goodman M (1963) Man's place in the phylogeny of the primates as reflected in serum proteins. In: Washburn SL (ed) Classification and human evolution. Aldine Press, Chicago, IL, pp 204–234

Haldane JBS (1947) The mutation rate of the gene for hemophilia, and its segregation ratios in males and females. Ann Eugenics 13:262–272

Haldane JBS (1957) The cost of natural selection. J Genet 55:511–524

Hubby JL, Lewontin RC (1966) A molecular approach to the study of genic heterozygosity in natural populations. I. The number of alleles at different loci in *Drosophila pseudoobscura*. Genetics 54:577–594

Hudson RR, Kreitman M, Aguadé M (1987) A test of neutral molecular evolution based on nucleotide data. Genetics 116:153–159

Kim S-H, Elango N, Warden CW, Vigoda E, Yi S (2006) Heterogeneous genomic molecular clocks in primates. PLOS Genet 2:e163

Kimura M (1960) Optimum mutation rate and degree of dominance as determined by the principle of minimum genetic load. J Genet 57:21–34

Kimura M (1968) Evolutionary rate at the molecular level. Nature 217:624–626

Kimura M (1977) Preponderance of synonymous changes as evidence for the neutral theory of molecular evolution. Nature 267:275–276

Kimura M (1983) The neutral theory of molecular evolution. Cambridge University Press, Cambridge

Kimura M, Ohta T (1971a) On the rate of molecular evolution. J Mol Evol 1:1–17
Kimura M, Ohta T (1971b) Protein polymorphism as a phase of molecular evolution. Nature 229:467–469
King JL, Jukes T (1969) Non-Darwinian evolution. Science 164:788–798
Kohne C (1970) Evolution of higher-organism DNA. Q Rev Biophys 3:327–375
Kosiol C, Vinar T, da Fonseca RR, Hubisz MJ, Bustamante CD, Nielsen R, Siepel A (2008) Patterns of positive selection in six mammalian genomes. PLOS Genet 4:e1000144
Laird CD, McConaughy BL, McCarthy BJ (1969) Rate of fixation of nucleotide substitutions in evolution. Nature 224:149–154
Lewontin RC, Hubby JL (1966) A molecular approach to the study of genetic heterozygosity in natural populations. II. Amount of variation and degree of heterozygosity in natural populations of *Drosophila pseudoobscura*. Genetics 54:595–609
Li W-H, Gojobori T, Nei M (1981) Pseudogenes as a paradigm of neutral evolution. Nature 292:237–239
Lynch M (2007) The origins of genome architecture. Sinauer Associates, Sunderland, MA
Lynch M, Conery JS (2003) The origins of genome complexity. Science 302:1401–1404
Lynch M, Ackerman MS, Gout J-F, Long H, Sung W, Thomas WK, Foster PL (2016) Genetic drift, selection and the evolution of the mutation rate. Nat Rev Genet 17:704–714
Margoliash E (1963) Primary structure and evolution of cytochrome *c*. Proc Natl Acad Sci USA 50:672–679
Margulies EH, Blanchette M, Comparative Sequencing Program NISC, Haussler D, Green ED (2003) Identification and characterization of multi-species conserved sequences. Genome Res 13:2507–2518
McDonald JH, Kreitman M (1991) Adaptive protein evolution at the *Adh* locus in *Drosophila*. Nature 351:652–654
Moorjani P, Amorim CEG, Arndt PF, Przeworski M (2016) Variation in the molecular clock of primates. Proc Natl Acad Sci USA 113:10607–10612
Morgan GJ (1998) Emile Zuckerkandl, Linus Pauling, and the molecular evolutionary clock, 1959–1965. J Hist Biol 31:155–178
Muller HJ (1954) The nature of the genetic effects produced by radiation. In: Hollaender A (ed) Radiation biology. McGraw-Hill, New York, pp 351–473
Nei M (2005) Selectionism and neutralism in molecular evolution. Mol Biol Evol 22:2318–2342
Nuttall GHF (1904) Blood immunity and blood relationship. Cambridge University Press, Cambridge
Ohta T (1972a) Evolutionary rate of cistrons and DNA divergence. J Mol Evol 1:150–157
Ohta T (1972b) Population size and rate of evolution. J Mol Evol 1:305–314
Ohta T (1973) Slightly deleterious mutant substitutions in evolution. Nature 246:96–98
Ohta T (1974) Mutational pressure as the main cause of molecular evolution and polymorphism. Nature 252:351–354
Ohta T (1995) Synonymous and nonsynonymous substitutions in mammalian genes and the nearly neutral theory. J Mol Evol 40:56–63
Ohta T (2002) Near-neutrality in evolution of genes and gene regulation. Proc Natl Acad Sci USA 99:16134–16137
Ohta T (2011) Near-neutrality, robustness, and epigenetics. Genome Biol Evol 3:1034–1038
Ohta T, Kimura M (1971) On the constancy of the evolutionary rate of cistrons. J Mol Evol 1:18–25
Rhesus Macaque Genome Sequencing and Analysis Consortium (2007) Evolutionary and biomedical insights from the rhesus macaque genome. Science 316:222–234
Sarich VM, Wilson AC (1967) Immunological time scale for hominid evolution. Science 158:1200–1203
Sung W, Ackerman MS, Miller SF, Doak TG, Lynch M (2012) Drift-barrier hypothesis and mutation-rate evolution. Proc Natl Acad Sci USA 109:18488–18492
Suzuki MM, Bird A (2008) DNA methylation landscapes: provocative insights from epigenomics. Nat Rev Genet 9:465–476
Tajima F (1989) Statistical method for testing the neutral mutation hypothesis by DNA polymorphism. Genetics 123:585–595
Takahata N (2007) Molecular clock: an anti-neo-Darwinian legacy. Genetics 176:1–6
Thomas JA, Welch JJ, Lanfear R, Bromham L (2010) A generation time effect on the rate of molecular evolution in invertebrates. Mol Biol Evol 27:1173–1180
Tsantes C, Steiper ME (2009) Age at first reproduction explains rate variation in the strepsirrhine molecular clock. Proc Natl Acad Sci USA 106:18165–18170
Vandiver AR, Idrizi A, Rizzardi L, Feinberg AP, Hansen KD (2015) DNA methylation is stable during replication and cell cycle arrest. Sci Rep 5:17911
Welch JJ, Bininda-emonds ORP, Bromham L (2008) Correlates of substitution rate variation in mammalian protein-coding sequences. BMC Evol Biol 8:53
Wilson AC, Sarich VM (1969) A molecular time scale for human evolution. Proc Natl Acad Sci USA 63:1088–1093
Woolfe A, Goodson M, Goode DK, Snell P, McEwen GK, Vavouri T, Smith SF, North P, Callaway H, Kelly K, Walter K, Abnizova I, Gilks W, Edwards YJK, Cooke JE, Elgar G (2004) Highly conserved non-coding sequences are associated with vertebrate development. PLOS Biol 3:e7
Wu C-I, Li W-H (1985) Evidence for higher rates of nucleotide substitution in rodents than in man. Proc Natl Acad Sci USA 82:1741–1745
Yi S (2012) Birds do it, bees do it, worms and ciliates do it too: DNA methylation from unexpected corners of the tree of life. Genome Biol 13:174

Yi S, Ellsworth DL, Li WH (2002) Slow molecular clocks in Old World monkeys, apes, and humans. Mol Biol Evol 19:2191–2198

Zuckerkandl E, Pauling L (1962) Molecular disease, evolution, and genic heterogeneity. In: Kasha M, Pullman B (eds) Horizons in biochemistry. Academic, New York, pp 189–225

Zuckerkandl E, Pauling L (1965) Evolutionary divergence and convergence in proteins. In: Bryson V, Vogel HJ (eds) Evolving genes and proteins. Academic, New York, pp 97–166

Zuckerkandl E, Jones RT, Pauling L (1960) A comparison of animal hemoglobin by tryptic peptide pattern analysis. Proc Natl Acad Sci USA 46:1349–1360

3 Spontaneous Mutation Rates

Susanne P. Pfeifer

Abstract

There is a long-standing interest in the study of mutations—from the quest to enhance evolutionary inference related to the genetic underpinnings of disease, to the improvement of our understanding of the chronology of human evolution, to characterizing relationships between species. There is substantial uncertainty in historical estimates obtained from indirect methods: classical genetic approaches, going back to Haldane's work in 1935 that utilized information from incidence of genetic disorders; and phylogenetic approaches, based on Kimura's observation that under neutrality the mutation rate is equal to the rate of divergence. However, recent advances in high-throughput sequencing have made it possible to estimate mutation rates directly from parent-offspring trios and multigenerational pedigrees. Moreover, the combination of mutation accumulation studies with high-throughput sequencing has led to nearly complete, largely unbiased insights into the genome-wide spontaneous mutation rate in several experimentally tractable organisms. This chapter will focus on the basic concepts underlying the different methods used to estimate spontaneous mutation rates and will summarize current knowledge regarding the evolution of mutation rates across taxa.

Keywords

Spontaneous mutation rate · Mutation–selection balance · Disease incidence-based approach · Neutral Theory of Molecular Evolution · Phylogenetic analysis · Pedigree studies · Mutation accumulation study

3.1 Introduction

Mutation is the ultimate source of genetic variation and is thus a critical process in evolution. The rate at which new (i.e., de novo) mutations arise is of fundamental interest not only for our understanding of evolutionary outcomes, but also for determining the genetic underpinnings of health and disease specifically. As such, research on the evolution of mutation rate itself has garnered considerable scientific interest for decades (e.g., see reviews by Lynch 2010a; Lynch et al. 2016).

The spontaneous mutation rate represents the probability of a de novo mutation occurring in a genomic region (scaled to a single site, a particular gene, or an entire genome) per unit time,

S. P. Pfeifer (✉)
Center for Evolution and Medicine, School of Life Sciences, Arizona State University, Tempe, AZ, USA
e-mail: susanne.pfeifer@asu.edu

S. Y. W. Ho (ed.), *The Molecular Evolutionary Clock*, https://doi.org/10.1007/978-3-030-60181-2_3

measured either as an absolute quantity (per days or years depending on the species of interest) or taxon-specific quantity (per generation). It has proven notoriously difficult to measure with high accuracy. This owes to several reasons: first and foremost, mutation rates are extremely low in most species (particularly in eukaryotes), making it challenging to practically observe them (Drake et al. 1998; Lynch 2006; Lynch et al. 2006). Second, the majority of spontaneous mutations are either neutral (i.e., no change in fitness of an organism) or deleterious (i.e., decreasing fitness)—with the extreme case being lethal, and thus, unobservable in a population (see review by Bank et al. 2014). Third, even for mutations that can be observed, the rate of occurrence strongly depends on the genomic context, such as the nucleotides immediately neighbouring the site of interest (Hwang and Green 2004) or the methylation status (Nachman and Crowell 2000). This causes mutation rates to vary across genomes. As such, there is no single mutation rate for any given organism. Lastly, mutation rate is a quantitative trait, with exogenous agents (such as chemical mutagens or radiation), physiological factors (such as age and sex), and selective pressures causing fluctuations across both fine and large scales (Haldane 1947; Lynch 2008).

With these circumstances and limitations in mind, there are four general types of approaches that can be taken to estimate the spontaneous mutation rate: methods based on disease incidence; phylogenetic analysis; whole-genome sequence data obtained from pedigrees; and, for amenable organisms, mutation accumulation studies of laboratory populations. This chapter will focus on the basic concepts underlying these different methods to estimate spontaneous mutation rate, highlight their limitations, and finally summarize our current knowledge pertaining to spontaneous mutation rate in a variety of organisms. This overview will focus upon single-nucleotide changes, because considerably fewer studies have investigated rates associated with insertions and deletions.

3.2 Methods to Estimate Spontaneous Mutation Rates

3.2.1 Mutation–Selection Balance and the Disease Incidence-Based Approach

Classical genetics approaches rely on screens for highly penetrant, monogenic Mendelian mutants with major phenotypic effects to derive locus-specific rate estimates from crosses and pedigrees. Nearly a century ago, Muller (1928), an experimental geneticist, implemented a phenotypic survey of balanced lethals by breeding a large number of individuals and their offspring and scoring for mutations. In this way, he was able to gain the first insights into the mutation rate of the model organism *Drosophila melanogaster*. However, due to the rarity of mutations, this scoring method required a large number of individuals (hundreds to thousands), thus restricting its usage to species with short generations and large numbers of offspring.

While conceptually simple, the approach used by Muller (1928) has several notable limitations. Most importantly, it cannot provide an estimate of the mutation rate per site because the genomic region producing the mutant phenotype is generally unknown. As a consequence, estimates of mutation rates obtained using this approach can differ by several orders of magnitude even within the same species (Schultz 2001). Nevertheless, the method has provided valuable first insights into the mutation spectrum of several unicellular species (e.g., bacteriophages, *Escherichia coli*, and *Saccharomyces cerevisiae*) as well as multicellular model organisms easily reared in a laboratory (e.g., *Drosophila melanogaster*, *Arabidopsis thaliana*, and maize) (Schultz 2001).

At the same time, Haldane (1927), a statistical geneticist, formulated the mathematical framework describing equilibrium allele frequencies in 'mutation-selection balance'—the idea that deleterious alleles exist in populations because their purging via purifying selection is counterbalanced by a continual influx of new mutations. The observation itself was not novel (it had already been noted by Danforth in 1923), but only with the population genetic theory developed by Haldane did it become possible to indirectly estimate spontaneous mutation rates based on this logic. Specifically, for autosomal dominant mutations (and assuming that the dominance coefficient, h, is equal to 1), the equilibrium allele frequency, q, in a randomly mating population is equal to μ/s, where μ is the mutation rate and s is the selective effect of the deleterious mutation. As a result, estimating the spontaneous mutation rate μ is, at least in principle, straightforward as long as the frequency of the mutant alleles and the strength of selection can be determined.

Using this expectation, Haldane (1932) provided one of the first indirect estimates of the de novo mutation rate in humans from haemophilia, a recessive X-linked disorder. For recessive X-linked mutations, the mutation rate can be calculated as $\mu = qs/3$, whereby q denotes the frequency of haemophilia in males and s the strength of selection. Haldane (1932) assumed s to be (close to) 1, reasoning that most haemophiliacs do not contribute offspring to the next generation. A few years later, Haldane (1935) himself provided an update to the approximate per-locus mutation rate (2×10^{-5}) by utilizing an estimate of the frequency of haemophilic men in London as a proxy for the frequency of the mutant allele.

Following Haldane's pioneering work, these basic genetic principles have been widely employed to estimate spontaneous per-generation mutation rates, especially in humans, by counting the number of individuals affected by a monogenic autosomal dominant or X-linked recessive Mendelian disorder that has emerged from their unaffected parents (Cooper and Krawczak 1993) (Fig. 3.1a). These studies have produced estimates of per-locus rates ranging from 10^{-6} to 10^{-4} (Vogel and Motulsky 1997). However, it is important to note that the approach is, by design, limited to mutations with fitness effects sufficiently large to make the mutant phenotype easily observable in all carriers, such as well-known disease genes. This most certainly biases it towards genomic regions with high mutation rates. On the other hand, the method might underestimate the spontaneous mutation rate if only a subset of mutations produces a visible mutant phenotype, and if subtle phenotypes or those with incomplete penetrance are missed.

Accurately calculating mutation rates is also complicated by the fact that the mutational target size (i.e., the size of the genomic region in which mutations would lead to the mutant phenotype), as well as the strength of selection, need to be known a priori; however, both of these can often be estimated only with a high degree of uncertainty. Controlling for the mutational target size, Kondrashov (2003) estimated a human mutation rate of 2×10^{-8} per site per generation by examining alleles at 20 known disease-causing loci in affected individuals. A few years later, aggregating data across a wider range of loci, Lynch (2010b) estimated a slightly lower human mutation rate of 1.28×10^{-8} per site per generation.

Lastly, although these methods based on disease incidence have provided us with first insights into spontaneous mutation rates, they are naturally limited to specific regions of the genome. Due to the heterogeneity of mutation, estimates from a few selected loci are unlikely to be representative of the process on a genome-wide scale.

3.2.2 The Neutral Theory of Molecular Evolution and the Phylogenetic Approach

In 1968, Kimura noted that for strictly neutral mutations, the mutation rate is equal to the rate of fixation. In other words, he suggested the existence of a molecular clock ticking at a constant speed throughout evolutionary time (see Chap. 2). Specifically, the neutral theory states that the number of substitutions K that accumulate in a lineage over time T is equal to $(\mu/G)T$, where μ is the per-generation mutation rate and G is the generation time (Kimura 1968). As a result, historically averaged estimates of the spontaneous mutation rate can be inferred from phylogenetic data using the extent of sequence divergence at orthologous, putatively neutral genomic regions

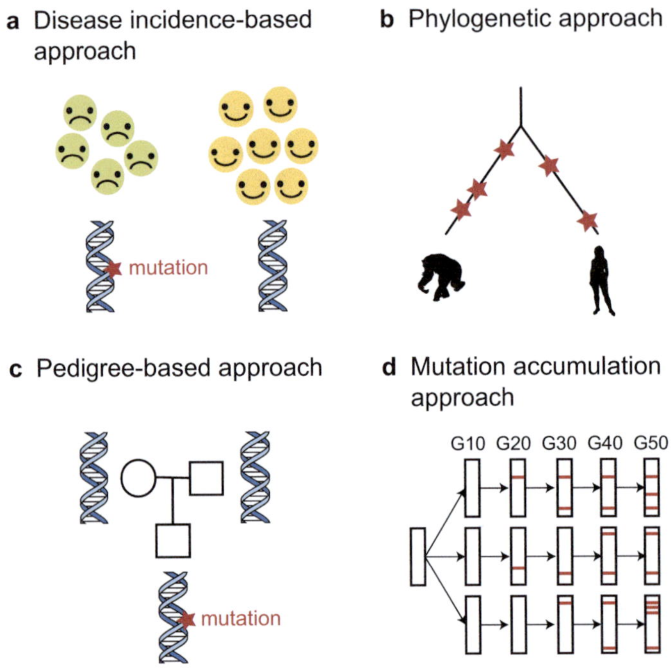

Fig. 3.1 Methods to estimate the spontaneous mutation rate. (**a**) Disease incidence-based methods estimate locus-specific mutation rates by counting the number of individuals affected by a monogenic autosomal dominant or X-linked recessive Mendelian disorder that has emerged from their unaffected parents. (**b**) Phylogenetic methods estimate historically averaged mutation rates using the extent of sequence divergence at orthologous, putatively neutral genomic regions between two species. (**c**) Pedigree-based methods estimate spontaneous mutation rates on a short timescale (usually a few generations) by comparing genome sequences of individuals from present-day parent-offspring pedigrees. (**d**) Mutation accumulation studies directly measure mutation rates in laboratory populations by forcing initially identical ancestral homozygous lines through extreme population bottlenecks and propagating them for many generations (G) to build up spontaneous mutations, before individual lines are sequenced to survey mutations (adapted from Baer et al. 2007). Red stars (**a**–**c**) or lines (**d**) indicate mutations

between two species, as long as both the generation time and the time to the most recent common ancestor (the divergence time) are known a priori (Kondrashov and Crow 1993; Drake et al. 1998; Nachman and Crowell 2000) (Fig. 3.1b).

By estimating the genetic distance between humans and chimpanzees in genomic regions where selection was not believed to be a confounding factor, Nachman and Crowell (2000) estimated a human mutation rate of 2.5×10^{-8} per site per generation. This estimate is in agreement with a per-site estimate of 2×10^{-8} by Kondrashov (2003) and with a per-locus estimate of 2×10^{-5} by Haldane (1935), assuming 10^3 nonsynonymous sites per gene (Vogel and Motulsky 1997) as an approximation (Nachman 2004).

Although the phylogenetic approach is easily applicable to a wide range of organisms, uncertainties in both the underlying assumptions and the resulting parameter estimates place limits on the accuracy that can realistically be achieved. Pseudogenes (non-functional copies of genes) or fourfold degenerate sites (positions of codons at which all four different nucleotides specify the same amino acid) are frequently taken as proxies for regions or sites in the genome that evolve neutrally (i.e., governed exclusively by genetic drift). However, the pervasive effects of purifying selection can cause the unintentional inclusion of non-neutral sites, particularly in species with large effective population sizes (Ohta 1973). Specifically, the inclusion of deleterious alleles will cause an underestimation of mutation rate because natural selection will reduce their probability of fixation relative to the neutral expectation (hence, they are less likely to contribute to divergence). Other effects exacerbate this issue, such as biased gene conversion (Duret and Galtier 2009) and mutation saturation (i.e., several mutations at the same site), which can lead to over- and underestimates of genetic distance, respectively (see review by Ségurel et al. 2014).

Perhaps even more clouded by uncertainty are estimates of generation times, because present-day generation times might not accurately reflect ancestral ones; and divergence times, because securely dated fossils are unavailable for many species. Indeed, rather than remaining constant among lineages, the speed of the molecular clock might have changed substantially throughout evolutionary history as a result of lineage-specific changes in life-history traits, making it difficult to infer absolute mutation rates from taxon-specific estimates (Yi et al. 2002; Elango et al. 2006, 2009; Kim et al. 2006; Ségurel et al. 2014; Amster and Sella 2016; Gao et al. 2016; Moorjani et al. 2016; Scally 2016b). Taken together, uncertainties in these crucial parameters often cause the phylogenetic approach to underestimate the spontaneous mutation rate.

3.2.3 De Novo Mutations in Pedigrees and the Genome Sequencing Approach

The methods based on disease incidence and phylogenetic analysis provide extrapolated and historical mutation rates, respectively. In contrast, comparisons of high-throughput genome sequences of present-day parent-offspring trios or larger multigenerational pedigrees (Fig. 3.1c) offer direct, comprehensive, and largely unbiased insights into the number and genome-wide distribution of de novo mutations (i.e., mutations that are present in the offspring but not in their parents). By their nature, however, they are limited to surveying spontaneous mutations on a short timescale (usually a few generations), which are extremely rare. Thus, the chief difficulty confronting researchers are errors introduced during sequencing, which outnumber genuine spontaneous mutations in most species, and which can lead to the false inference of de novo mutations (i.e., false positives) (Pfeifer 2017b).

The effect of false positives on mutation rate estimates can be attenuated either by experimentally validating identified de novo mutations through an independent technology with a low error rate (such as Sanger sequencing) or by computationally weeding them out using a set of highly stringent statistical filters. Due to the costs associated with additional sequencing, the latter strategy has become widely adopted in the field. However, it is important to note that, while mitigating false positives, this filtering might also result in the loss of genuine de novo mutations (i.e., false negatives) (Ségurel et al. 2014). To estimate the spontaneous mutation rate accurately from high-throughput sequencing data, it is thus necessary to infer both false-positive as well as false-negative rates, which remains a challenging endeavour (Pfeifer 2017a). Moreover, the application of computational filters, combined with frequent uneven sequencing coverage in complex genomic regions, will narrow the number of loci at which genuine de novo mutations can be detected. This makes it necessary to obtain an unbiased estimate of the length of the genomic regions available to the study when estimating a per-site mutation rate. Note that this is not as simple as knowing the genome size, because it is necessary to know the exact number of bases for which reliable sequence data were obtained for any given sequencing experiment.

While easily applicable to a variety of organisms, direct estimation of de novo mutations by high-throughput sequencing requires the availability of high-quality genomic resources in the species of interest which, thus far, has limited its usage. In humans, large-scale sequencing studies of pedigrees have yielded mutation rate estimates of ~10^{-8} per site per generation (reviewed by Ségurel et al. 2014), a twofold decrease compared with earlier indirect estimates. Another potential issue arises from the samples themselves: because germline cells are often either difficult to obtain or unavailable for the species of interest, somatic tissue (e.g., blood or saliva) is frequently used as a surrogate, which can obscure mutational signatures. Specifically, post-zygotic mutations that took place before germline specification might be present in the soma of both parent and offspring (Scally 2016b). As a result, these mutations will incorrectly be inferred as standing variation, contributing to the underestimation of the spontaneous mutation rate (Campbell et al. 2012; Acuna-Hidalgo et al. 2015; Scally 2016a). This problem can be ameliorated by sequencing additional generations to validate stable Mendelian transmission, though naturally only half of the variants will be segregating in the next generation (Ségurel et al. 2014; Scally 2016b; Pfeifer 2017a).

Taken together, mutation rate estimates from genomic sequencing data from pedigrees are likely to represent underestimates of the true per-site mutation rate. Given the considerable variation in mutation rates that has been observed among individuals, families, and populations (Ségurel et al. 2014; Scally 2016b), it is also important to note that this approach only offers insights into the mutation rates of the studied individuals, who are mere representatives of the entire population. Owing to sampling error, the number of individuals included will directly influence the degree of uncertainty in the mutation rate estimates that are obtained (Tran and Pfeifer 2018).

3.2.4 Experimentally Tractable Organisms and the Mutation Accumulation Approach

Previous research has illustrated that the majority of spontaneous mutations are neutral, nearly neutral, or deleterious (see review by Bank et al. 2014). The last of these categories is particularly problematic for the study of spontaneous mutation rates in natural populations because mutations of large deleterious effect are unlikely to be observed. As such, the approaches discussed above should not be viewed as estimators of the total de novo mutation rate because, at a minimum, the contribution of lethal mutations is not being counted. Importantly, the fitness of a mutation depends not only on the selection coefficient (s) but also on the effective population size (N_e), so the number of mutations that will behave neutrally will vary with population size (see Chap. 2). In fact, in her extension of the neutral theory, Ohta (1973) demonstrated that deleterious mutations will behave as effectively neutral in diploid populations as long as s is smaller than $1/(2N_e)$. Thus, in experimentally tractable organisms, the confounding impacts of nearly neutral and deleterious mutations can be overcome by artificially keeping the effective population size sufficiently small to observe a broader spectrum of mutations. Specifically, to allow the accumulation of nearly neutral mutations, a studied population will repeatedly be forced through a demographic bottleneck. This collapses the effective population size to such an extent that genetic drift overpowers weak and moderate selection (Lynch et al. 2016) (Fig. 3.1d).

Mutation accumulation experiments generally start from a series of identical ancestral homozygous lines that are subjected to these extreme bottlenecks: single individuals for self-fertilizing or clonally reproducing species, and single full-sibling pair (brother and sister) for sexually reproducing species. These are then propagated for many generations to build up spontaneous mutations, before individual lines are sequenced and mutations surveyed to allow mutation rates to be estimated directly (Keightley et al. 2014). The approach offers a more detailed view of the mutational spectrum, but it is time consuming and is practically limited to species for which inbred lines can be produced in the laboratory. These organisms include microbial taxa (*Saccharomyces cerevisiae*, Lynch et al. 2008), some plants (e.g., Ossowski et al. 2010), and invertebrate model organisms such as *Caenorhabditis elegans* (Denver et al. 2004, 2009) and *Drosophila melanogaster* (Haag-Liautard et al. 2007; Keightley et al. 2009).

It remains questionable whether (or at least to what extent) laboratory inbred lines represent the mutation patterns expected in natural populations. Moreover, mutation accumulation experiments often yield underestimates of mutation rates

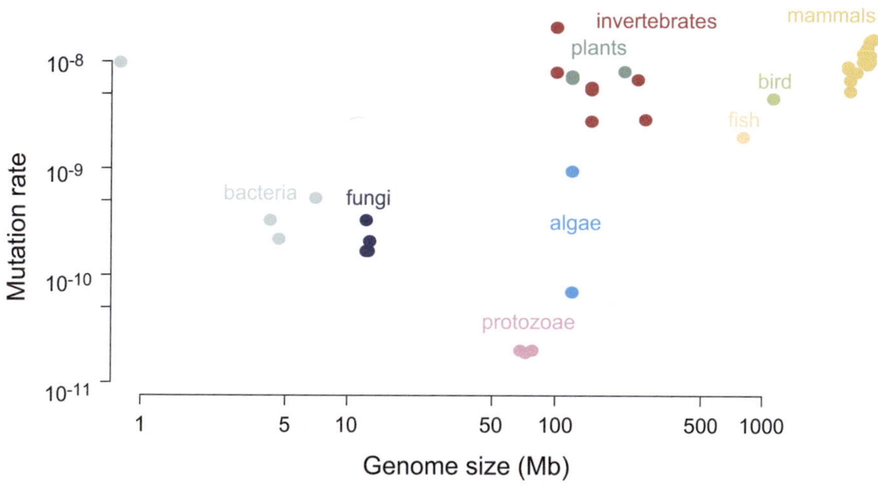

Fig. 3.2 Spontaneous mutation rates in a variety of organisms. The log–log scaling of mutation rate (measured in base pairs per generation) with genome size (measured in Mb). Data sources: Haag-Liautard et al. (2007), Lynch et al. (2008), Awadalla et al. (2010), Ossowski et al. (2010), Roach et al. (2010), Conrad et al. (2011), Denver et al. (2012), Kong et al. (2012), Lee et al. (2012), Michaelson et al. (2012), Sung et al. (2012a, b, 2015), Schrider et al. (2013), Keightley et al. (2014, 2015), Venn et al. (2014), Zhu et al. (2014), Behringer and Hall (2015), Besenbacher et al. (2015, 2019), Farlow et al. (2015), Ness et al. (2015), Rahbari et al. (2016), Uchimura et al. (2015), Yang et al. (2015), Yuen et al. (2015), Kucukyildirim et al. (2016), Smeds et al. (2016), Wong et al. (2016), Xie et al. (2016), Feng et al. (2017), Jónsson et al. (2017), Milholland et al. (2017), Pfeifer (2017a), Tatsumoto et al. (2017), Long et al. (2018), Thomas et al. (2018)

because sterile mutations cannot propagate and lethals still cannot be observed. It is also important to note that during the relaxed selection regime, mutations can arise that themselves might modify the mutation rate (Sniegowski et al. 1997).

3.3 Spontaneous Mutation Rates Across Taxa

Rates of spontaneous mutation, though remarkably similar within taxa, vary by several orders of magnitude across species (ranging from 10^{-11} in single-celled eukaryotes to 10^{-8} in primates), roughly scaling with genome size (Fig. 3.2). This pattern is suggestive of an evolutionary optimum for an organism's genome-wide mutation rate (Drake et al. 1998). Two hypotheses have been put forward to explain this observation. First, the reduction of an organism's spontaneous mutation rate might be intrinsically limited by the biochemical and physiological costs associated with improving replication fidelity (e.g., Drake 1991; Drake et al. 1998). Second, in order to limit the influx of deleterious mutations, purifying selection might act to reduce rates of spontaneous mutations by improving replication fidelity. However, purifying selection will be overpowered by genetic drift when the fitness advantage of further reducing the rate of spontaneous mutations in a diploid organism is smaller than $1/2N_e$, where N_e is the effective population size. Hence, by setting the efficiency of selection, genetic drift determines a lower boundary for an organism's spontaneous mutation rate—this idea is known as the 'drift-barrier hypothesis' (Lynch 2010b). This is an elegant explanation for the observation that species with larger effective population sizes (i.e., more efficient selection) generally exhibit lower per-generation rates of spontaneous mutations. However, further work, extending previous studies by including a wider set of species, is required to evaluate this pattern carefully across phylogenetic lineages.

References

Acuna-Hidalgo R, Bo T, Kwint MP, van de Vorst M, Pinelli M, Veltman JA, Hoischen A, Vissers LELM, Gilissen C (2015) Post-zygotic point mutations are an underrecognized source of *de novo* genomic variation. Am J Hum Genet 97:67–74

Amster G, Sella G (2016) Life history effects on the molecular clock of autosomes and sex chromosomes. Proc Natl Acad Sci USA 113:1588–1593

Awadalla P, Gauthier J, Myers RA, Casals F, Hamdan FF, Griffing AR, Côté M, Henrion E, Spiegelman D, Tarabeux J, Piton A, Yang Y, Boyko A, Bustamante C, Xiong L, Rapoport JL, Addington AM, DeLisi JLE, Krebs M-O, Joober R, Millet B, Fombonne E, Mottron L, Zilversmit M, Keebler J, Daoud H, Marineau C, Roy-Gagnon M-H, Dubé M-P, Eyre-Walker A, Drapeau P, Stone EA, Lafrenière RG, Rouleau GA (2010) Direct measure of the *de novo* mutation rate in autism and schizophrenia cohorts. Am J Hum Genet 87:316–324

Baer CF, Miyamoto MM, Denver DR (2007) Mutation rate variation in multicellular eukaryotes: causes and consequences. Nat Rev Genet 8:619–631

Bank C, Ewing GB, Ferrer-Admettla A, Foll M, Jensen JD (2014) Thinking too positive? Revisiting current methods of population genetic selection inference. Trends Genet 30:540–546

Behringer MG, Hall DW (2015) Genome-wide estimates of mutation rates and spectrum in *Schizosaccharomyces pombe* indicate CpG sites are highly mutagenic despite the absence of DNA methylation. G3-Genes Genom Genet 6:149–160

Besenbacher S, Liu S, Izarzugaza JM, Grove J, Belling K, Bork-Jensen J, Huang S, Als TD, Li S, Yadav R, Rubio-García A, Lescai F, Demontis D, Rao J, Ye W, Mailund T, Friborg RM, Pedersen CNS, Xu R, Sun J, Liu H, Wang O, Cheng X, Flores D, Rydza E, Rapacki K, Damm Sørensen J, Chmura P, Westergaard D, Dworzynski P, Sørensen TIA, Lund O, Hansen T, Xu X, Li N, Bolund L, Pedersen O, Eiberg H, Krogh A, Børglum AD, Brunak S, Kristiansen K, Schierup MH, Wang J, Gupta R, Villesen P, Rasmussen S (2015) Novel variation and *de novo* mutation rates in population-wide *de novo* assembled Danish trios. Nat Commun 6:5969

Besenbacher S, Hvilsom C, Marques-Bonet T, Mailund T, Schierup MH (2019) Direct estimation of mutations in great apes reconciles phylogenetic dating. Nat Ecol Evol 3:286–292

Campbell CD, Chong JX, Malig M, Dumont BL, Vives L, O'Roak BJ, Sudmant PH, Shendure J, Abney M, Ober C, Eichler EE (2012) Estimating the human mutation rate using autozygosity in a founder population. Nat Genet 44:1277–1281

Conrad DF, Keebler JE, DePristo MA, Lindsay SJ, Zhang Y, Casals F, Idaghdour Y, Hartl CL, Torroja C, Garimella KV, Zilversmit M, Cartwright R, Rouleau GA, Daly M, Stone EA, Hurles ME, Awadalla P, 1000 Genomes Project (2011) Variation in genome-wide mutation rates within and between human families. Nat Genet 43:712–714

Cooper DN, Krawczak M (1993) Human gene mutation. BIOS Scientific, Oxford, UK

Danforth CH (1923) The frequency of mutation and the incidence of hereditary traits in man. In: Eugenics, genetics and the family, Scientific papers of the 2nd international congress of eugenics, NY, 1921. Williams & Williams, Baltimore, MD, pp 120–128

Denver DR, Morris K, Lynch M, Thomas WK (2004) High mutation rate and predominance of insertions in the *Caenorhabditis elegans* nuclear genome. Nature 430:679–682

Denver DR, Dolan PC, Wilhelm LJ, Sung W, Lucas-Lledó JI, Howe DK, Lewis SC, Okamoto K, Thomas WK, Lynch M, Baer CF (2009) A genome-wide view of *Caenorhabditis elegans* base-substitution mutation processes. Proc Natl Acad Sci USA 106:16310–16314

Denver DR, Wilhelm LJ, Howe DK, Gafner K, Dolan PC, Baer CF (2012) Variation in base-substitution mutation in experimental and natural lineages of *Caenorhabditis* nematodes. Genome Biol Evol 4:513–522

Drake JW (1991) A constant rate of spontaneous mutation in DNA-based microbes. Proc Natl Acad Sci USA 88:7160–7164

Drake JW, Charlesworth B, Charlesworth D, Crow JF (1998) Rates of spontaneous mutation. Genetics 148:1667–1686

Duret L, Galtier N (2009) Biased gene conversion and the evolution of mammalian genomic landscapes. Annu Rev Genomics Hum Genet 10:285–311

Elango N, Thomas JW, Yi SV (2006) Variable molecular clocks in hominoids. Proc Natl Acad Sci USA 103:1370–1375

Elango N, Lee J, Peng Z, Loh Y-HE, Yi SV (2009) Evolutionary rate variation in Old World monkeys. Biol Lett 5:405–408

Farlow A, Long H, Arnoux S, Sung W, Doak TG, Nordborg M, Lynch M (2015) The spontaneous mutation rate in the fission yeast *Schizosaccharomyces pombe*. Genetics 201:737–744

Feng C, Pettersson M, Lamichhaney S, Rubin C-J, Rafati N, Casini M, Folkvord A, Andersson L (2017) Moderate nucleotide diversity in the Atlantic herring is associated with a low mutation rate. elife 6:e23907

Gao Z, Wyman MJ, Sella G, Przeworski M (2016) Interpreting the dependence of mutation rates on age and time. PLOS Biol 14:e1002355

Haag-Liautard C, Dorris M, Maside X, Macaskill S, Halligan DL, Charlesworth B, Keightley PD (2007) Direct estimation of per nucleotide and genomic deleterious mutation rates in *Drosophila*. Nature 445:82–85

Haldane JBS (1927) A mathematical theory of natural and artificial selection. Part V. Selection and mutation. Math Proc Camb Philos Soc 23:838–844

Haldane JBS (1932) The causes of evolution. Longmans, Green, & Co, London

Haldane JBS (1935) The rate of spontaneous mutation of a human gene. J Genet 31:317–326

Haldane JBS (1947) The mutation rate of the gene for haemophilia, and its segregation ratios in males and females. Ann Eugenics 13:262–271
Hwang DG, Green P (2004) Bayesian Markov chain Monte Carlo sequence analysis reveals varying neutral substitution patterns in mammalian evolution. Proc Natl Acad Sci USA 101:13994–14001
Jónsson H, Sulem P, Kehr B, Kristmundsdottir S, Zink F, Hjartarson E, Hardarson MT, Hjorleifsson KE, Eggertsson HP, Gudjonsson SA, Ward LD, Arnadottir GA, Helgason EA, Helgason H, Gylfason A, Jonasdottir A, Jonasdottir A, Rafnar T, Frigge M, Stacey SN, Magnusson OT, Thorsteinsdottir U, Masson G, Kong A, Halldorsson BV, Helgason A, Gudbjartsson DF, Stefansson K (2017) Parental influence on human germline *de novo* mutations in 1,548 trios from Iceland. Nature 549:519–522
Keightley PD, Trivedi U, Thomson M, Oliver F, Kumar S, Blaxter ML (2009) Analysis of the genome sequences of three *Drosophila melanogaster* spontaneous mutation accumulation lines. Genome Res 19:1195–1201
Keightley PD, Ness RW, Halligan DL, Haddrill PR (2014) Estimation of the spontaneous mutation rate per nucleotide site in a *Drosophila melanogaster* full-sib family. Genetics 196:313–320
Keightley PD, Pinharanda A, Ness RW, Simpson F, Dasmahapatra KK, Mallet J, Davey JW, Jiggins CD (2015) Estimation of the spontaneous mutation rate in *Heliconius melpomene*. Mol Biol Evol 32:239–243
Kim S-H, Elango N, Warden C, Vigoda E, Yi SV (2006) Heterogeneous genomic molecular clocks in primates. PLOS Genet 2:e163
Kimura M (1968) Evolutionary rate at the molecular level. Nature 217:624–626
Kondrashov AS (2003) Direct estimates of human per nucleotide mutation rates at 20 loci causing Mendelian diseases. Hum Mutat 21:12–27
Kondrashov AS, Crow JF (1993) A molecular approach to estimating the human deleterious mutation rate. Hum Mutat 2:229–234
Kong A, Frigge ML, Masson G, Besenbacher S, Sulem P, Magnusson G, Gudjonsson SA, Sigurdsson A, Jonasdottir A, Jonasdottir A, Wong WSW, Sigurdsson G, Walters GB, Steinberg S, Helgason H, Thorleifsson G, Gudbjartsson DF, Helgason A, Magnusson OT, Thorsteinsdottir U, Stefansson K (2012) Rate of *de novo* mutations and the importance of father's age to disease risk. Nature 488:471–475
Kucukyildirim S, Long H, Sung W, Miller SF, Doak TG, Lynch M (2016) The rate and spectrum of spontaneous mutations in *Mycobacterium smegmatis*, a bacterium naturally devoid of the postreplicative mismatch repair pathway. G3-Genes Genom Genet 6:2157–2163
Lee H, Popodi E, Tang H, Foster PL (2012) Rate and molecular spectrum of spontaneous mutations in the bacterium *Escherichia coli* as determined by whole-genome sequencing. Proc Natl Acad Sci USA 109: E2774–E2783
Long H, Doak TG, Lynch M (2018) Limited mutation-rate variation within the *Paramecium aurelia* species complex. G3-Genes Genom Genet 8:2523–2526
Lynch M (2006) The origins of eukaryotic gene structure. Mol Biol Evol 23:450–468
Lynch M (2008) The cellular, developmental and population-genetic determinants of mutation-rate evolution. Genetics 180:933–943
Lynch M (2010a) Evolution of the mutation rate. Trends Genet 26:345–352
Lynch M (2010b) Rate, molecular spectrum, and consequences of human mutation. Proc Natl Acad Sci USA 107:961–968
Lynch M, Koskella B, Schaack S (2006) Mutation pressure and the evolution of organelle genomic architecture. Science 311:1727–1730
Lynch M, Sung W, Morris K, Coffey N, Landry CR, Dopman EB, Dickinson WJ, Okamoto K, Kulkarni S, Hartl DL, Thomas WK (2008) A genome-wide view of the spectrum of spontaneous mutations in yeast. Proc Natl Acad Sci USA 105:9272–9277
Lynch M, Ackerman MS, Gout JF, Long H, Sung W, Thomas WK, Foster PL (2016) Genetic drift, selection and the evolution of the mutation rate. Nat Rev Genet 17:704–714
Michaelson JJ, Shi Y, Gujral M, Zheng H, Malhotra D, Jin X, Jian M, Liu G, Greer D, Bhandari A, Wu W, Corominas R, Peoples A, Koren A, Gore A, Kang S, Lin GN, Estabillo J, Gadomski T, Singh B, Zhang K, Akshoomoff N, Corsello C, McCarroll S, Iakoucheva LM, Li Y, Wang J, Sebat J (2012) Whole-genome sequencing in autism identifies hot spots for *de novo* germline mutation. Cell 151:1431–1442
Milholland B, Dong X, Zhang L, Hao X, Suh Y, Vijg J (2017) Differences between germline and somatic mutation rates in humans and mice. Nat Commun 8:15183
Moorjani P, Amorim CE, Arndt PF, Przeworski M (2016) Variation in the molecular clock of primates. Proc Natl Acad Sci USA 113:10607–10612
Muller HJ (1928) The measurement of gene mutation rate in Drosophila, its high variability, and its dependence upon temperature. Genetics 13:279–357
Nachman MW (2004) Haldane and the first estimates of the human mutation rate. J Genet 83:231–233
Nachman MW, Crowell SL (2000) Estimate of the mutation rate per nucleotide in humans. Genetics 156:297–304
Ness RW, Morgan AD, Vasanthakrishnan RB, Colegrave N, Keightley PD (2015) Extensive *de novo* mutation rate variation between individuals and across the genome of *Chlamydomonas reinhardtii*. Genome Res 25:1739–1749
Ohta T (1973) Slightly deleterious mutant substitutions in evolution. Nature 246:96–98
Ossowski S, Schneeberger K, Lucas-Lledó JI, Warthmann N, Clark RM, Shaw RG, Weigel D, Lynch M (2010) The rate and molecular spectrum of

spontaneous mutations in *Arabidopsis thaliana*. Science 327:92–94
Pfeifer SP (2017a) Direct estimate of the spontaneous germ line mutation rate in African green monkeys. Evolution 71:2858–2870
Pfeifer SP (2017b) From next-generation resequencing reads to a high-quality variant data set. Heredity 118:111–124
Rahbari R, Wuster A, Lindsay SJ, Hardwick RJ, Alexandrov LB, Turki SA, Dominiczak A, Morris A, Porteous D, Smith B, Stratton MR, UK10K Consortium, Hurles ME (2016) Timing, rates and spectra of human germline mutation. Nat Genet 48:126–133
Roach JC, Glusman G, Smit AFA, Huff CD, Hubley R, Shannon PT, Rowen L, Pant KP, Goodman N, Bamshad M, Shendure J, Drmanac R, Jorde LB, Hood L, Galas DJ (2010) Analysis of genetic inheritance in a family quartet by whole-genome sequencing. Science 328:636–639
Scally A (2016a) Mutation rates and the evolution of germline structure. Philos Trans R Soc Lond B 371:20150137
Scally A (2016b) The mutation rate in human evolution and demographic inference. Curr Opin Genet Dev 41:36–43
Schrider DR, Houle D, Lynch M, Hahn MW (2013) Rates and genomic consequences of spontaneous mutational events in *Drosophila melanogaster*. Genetics 194:937–954
Schultz ST (2001) Mutation rates: data. In: eLS. Wiley, Chichester
Ségurel L, Wyman MJ, Przeworski M (2014) Determinants of mutation rate variation in the human germline. Annu Rev Genomics Hum Genet 15:47–70
Smeds L, Qvarnström A, Ellegren H (2016) Direct estimate of the rate of germline mutation in a bird. Genome Res 26:1211–1218
Sniegowski PD, Gerrish PJ, Lenski RE (1997) Evolution of high mutation rates in experimental populations of *E. coli*. Nature 387:703–705
Sung W, Ackerman MS, Miller SF, Doak TG, Lynch M (2012a) Drift-barrier hypothesis and mutation-rate evolution. Proc Natl Acad Sci USA 109:18488–18492
Sung W, Tucker AE, Doak TG, Choi E, Thomas WK, Lynch M (2012b) Extraordinary genome stability in the ciliate *Paramecium tetraurelia*. Proc Natl Acad Sci USA 109:19339–19344
Sung W, Ackerman MS, Gout JF, Miller SF, Williams E, Foster PL, Lynch M (2015) Asymmetric context-dependent mutation patterns revealed through mutation-accumulation experiments. Mol Biol Evol 32:1672–1683
Tatsumoto S, Go Y, Fukuta K, Noguchi H, Hayakawa T, Tomonaga M, Hirai H, Matsuzawa T, Agata K, Fujiyama A (2017) Direct estimation of *de novo* mutation rates in a chimpanzee parent-offspring trio by ultra-deep whole genome sequencing. Sci Rep 7:13561
Thomas GWC, Wang RJ, Puri A, Harris RA, Raveendran M, Hughes DST, Murali SC, Williams LE, Doddapaneni H, Muzny DM, Gibbs RA, Abee CR, Galinski MR, Worley KC, Rogers J, Radivojac P, Hahn MW (2018) Reproductive longevity predicts mutation rates in primates. Curr Biol 28:3193–3197
Tran LAP, Pfeifer SP (2018) Germline mutation rates in Old World monkeys. In: eLS. Wiley, Chichester
Uchimura A, Higuchi M, Minakuchi Y, Ohno M, Toyoda A, Fujiyama A, Miura I, Wakana S, Nishino J, Yagi T (2015) Germline mutation rates and the long-term phenotypic effects of mutation accumulation in wild-type laboratory mice and mutator mice. Genome Res 25:1125–1134
Venn O, Turner I, Mathieson I, de Groot N, Bontrop R, McVean G (2014) Strong male bias drives germline mutation in chimpanzees. Science 344:1272–1275
Vogel F, Motulsky AG (1997) Human genetics: problems and approaches. Springer, Berlin
Wong WS, Solomon BD, Bodian DL, Kothiyal P, Eley G, Huddleston KC, Baker R, Thach DC, Iyer RK, Vockley JG, Niederhuber JE (2016) New observations on maternal age effect on germline *de novo* mutations. Nat Commun 7:10486
Xie Z, Wang L, Wang L, Wang Z, Lu Z, Tian D, Yang S, Hurst LD (2016) Mutation rate analysis via parent-progeny sequencing of the perennial peach. I. A low rate in woody perennials and a higher mutagenicity in hybrids. Proc R Soc B 283:20161016
Yang S, Wang L, Huang J, Zhang X, Yuan Y, Chen J-Q, Hurst LD, Tian D (2015) Parent-progeny sequencing indicates higher mutation rates in heterozygotes. Nature 523:463–467
Yi S, Ellsworth DL, Li WH (2002) Slow molecular clocks in Old World monkeys, apes, and humans. Mol Biol Evol 19:2191–2198
Yuen RKC, Thiruvahindrapuram B, Merico D, Walker S, Tammimies K, Hoang N, Chrysler C, Nalpathamkalam T, Pellecchia G, Liu Y, Gazzellone MJ, D'Abate L, Deneault E, Howe JL, Liu RSC, Thompson A, Zarrei M, Uddin M, Marshall CR, Ring RH, Zwaigenbaum L, Ray PN, Weksberg R, Carter MT, Fernandez BA, Roberts W, Szatmari P, Scherer SW (2015) Whole-genome sequencing of quartet families with autism spectrum disorder. Nat Med 21:185–191
Zhu YO, Siegal ML, Hall DW, Petrov DA (2014) Precise estimates of mutation rate and spectrum in yeast. Proc Natl Acad Sci USA 111:E2310–E2318

Causes of Variation in the Rate of Molecular Evolution

4

Lindell Bromham

Abstract

The genome is an elaborate tapestry of conservation and change. The complex patterns of molecular evolution frustrate simple analyses and can obscure the historical narrative recorded in the genome. In addition to differences in patterns and rates of change between different regions of the genome, there are consistent differences in the tempo and mode of molecular evolution among species. In particular, the fact that species differ in their average rate of molecular evolution makes inference of evolutionary time from genetic divergence very challenging. But variation in the patterns of change between lineages also gives us a rich source of evidence with which to test ideas about the causes of molecular evolutionary change. The genome is not simply a passive recorder of history, but a dynamic engine of change, both creating and responding to a species' changing situation over evolutionary time. We need to develop an understanding of the influences on the rates and patterns of molecular evolution if we are to use genomic data to study evolutionary history.

L. Bromham (✉)
Macroevolution and Macroecology, Division of Ecology and Evolution, Research School of Biology, Australian National University, Canberra, ACT, Australia
e-mail: lindell.bromham@anu.edu.au

Keywords

Mutation rate · Substitution rate · Diversification · Body size · Population size

4.1 Introduction

At the heart of evolutionary biology is the principle that simple processes, operating continuously over long periods, can generate complex outcomes. This is as evident at the molecular level as at the level of phenotype or ecosystem. In this chapter, I will look at how the basic processes of mutation and substitution combine to generate complex patterns of variation in the rate of molecular evolution. My focus will be on the causes of consistent differences in the average rate of molecular evolution between different species, rather than on variation in rate of change across the genome. Consistent variation in rate among species makes estimation of divergence times from molecular data challenging (Welch and Bromham 2005). But if we can understand and predict those patterns of change, it may help us to develop more reliable molecular dating methods, or at least to be able to identify cases where our dates are likely to be systematically misleading (e.g., Phillips 2015). More importantly, studying the way that rates of genome evolution vary among species illuminates the causes of variation in the tempo and mode of molecular evolution. To understand how

S. Y. W. Ho (ed.), *The Molecular Evolutionary Clock*, https://doi.org/10.1007/978-3-030-60181-2_4

species-specific differences in the rate of molecular evolution are generated, the processes that we have to consider are mutation (creation of heritable variation), substitution (fate of heritable variation), and diversification (accumulation of heritable variation).

A mutation is a physicochemical change that occurs in a DNA molecule, which alters the genetic information that is copied and passed on. A mutation happens in one individual, altering the heritable information that the individual is able to pass on to its offspring, should it be fortunate enough to reproduce. We can detect mutations by comparing offspring with their parents. Many mutations will be lost when their carriers fail to leave descendants. For most molecular dating studies, we only detect differences in DNA sequences between populations, species, or lineages. So we need to understand how a mutation that has occurred in a single individual can rise in frequency until it is carried by all members of a population. This is the process of substitution, which can happen through selection (biased reproduction or survival of particular mutations) or through drift (chance variation in mutation frequencies due to random sampling effects). Since mutations occur continually, and some of them rise in frequency and become substitutions, over time each population accumulates more and more genetic changes. Some of these genetic changes will be associated with adaptation to particular niches, some will affect the likelihood or success of reproduction with members of other populations, and some will have little or no effect on fitness. Eventually, a population might acquire sufficient substitutions that it becomes a genetically isolated species. In this way, mutation drives substitution which drives diversification of lineages: this is the engine room of biodiversity.

4.2 Mutation

4.2.1 An Evolutionary Balancing Act

For the purposes of this chapter, I am going to consider a mutation to be a permanent change in the nucleotide base sequence of the genome, such that the original base can no longer be recovered, so the new base sequence will be present in copies made of that genome. Many and varied disasters can befall the genome—double-strand breaks that sever the DNA helix, lesions that prevent the replication machinery from copying a sequence, or structural damage to the bases that prevents proper pairing. Such mutations happen, but if they prevent proper function or replication, they will not tend to be passed on, so will not provide fuel for evolutionary change.

For molecular dating, the kind of mutations we are interested in are those where an accident changes the DNA sequence, but that are not so disastrous that the cell can no longer make a copy of its genome. Furthermore, molecular dating is, by and large, concerned only with mutations in nucleotide sequences where one base is exchanged for another (not insertions, deletions, rearrangements, changes to methylation, etc.). Mutations arise by accident, as damage that is imperfectly repaired or copy errors that are not corrected. However, the rate of mutation is subject to evolutionary modification, through investment of time and resources into DNA repair and copy fidelity, and through the use of energy and material to build, maintain, and operate specialized DNA repair equipment. Varying the level of investment in repair and fidelity will change the mutation rate.

The simplest illustration of the balancing act involved in replication and repair is to consider the action of DNA polymerase molecules. Polymerase copies a strand of DNA by adding nucleotides to the end of a new polynucleotide chain, adding complementary bases to match the opposite strand. Some polymerase molecules have proofreading activity: if the newly added base is not complementary to the base on the template strand, it might be removed and replaced. If the polymerase fails to detect and correct an incorrect base, there are additional mismatch repair pathways that can detect incorrect base pairs, excise the DNA strand, and replace it (for an introduction to DNA replication and repair, see Bromham 2016). The relative balance between the 'forward' activity of the polymerase (adding bases) and the 'backward' activity (proofreading) will alter the fidelity of DNA replication, but it will also alter replication speed. Investing less in proofreading or post-replicative repair increases the cost due to harmful somatic mutations and poor genome copies. Investing more in proofreading and repair increases the costs of replication in terms of time and resources taken to copy the DNA, with more bases excised and replaced (for a useful review, see Kunkel 2009).

Mutations can alter DNA replication accuracy. Changes to the polymerase enzyme can increase fidelity by slowing replication, favouring more exonuclease (proofreading) activity (Herr et al. 2011b; Maslowska et al. 2018). Mutations can also reduce the effectiveness of proofreading and repair, leading to greater replication speed and higher rates of generation of beneficial mutations, but at the cost of reduced genome fidelity and lowered survival rates (Herr et al. 2011a). The same considerations are likely to apply to other aspects of replication that affect copy fidelity, such as the speed of movement of the replication fork (Mertz et al. 2017). The availability of heritable variation that alters DNA replication fidelity, and the clear impacts on fitness of such variation, suggest that the mutation rate itself is subject to evolutionary change (Baer et al. 2007). A similar argument can be made for the repair of incidental DNA damage, for example by mismatch repair (Denamur and Matic 2006).

Most of the research on the evolution of mutation rates has, not surprisingly, been conducted on viruses, bacteria, and yeast. Do the findings of laboratory-based studies on microbes tell us much about the causes of variation in rate of molecular evolution for other species, such as plants, animals, and fungi? DNA repair rates vary among species, among individuals, and among polymerase enzymes within an individual. Can we conclude that these differences are also shaped by selection for a balance between the costs and benefits of mistakes and fidelity, damage and repair? It seems reasonable to suppose that multicellular species carry the costs of low fidelity or inefficient damage repair, though it is harder to establish the metabolic costs of increased repair. Nonetheless, copy fidelity and damage repair do not come for free. All species must ultimately balance the costs of mutation against the costs of repair, and many different factors might weigh into the balance (Bromham 2009). For example, increasing the rate of cell division might not only generate more copy errors per unit time, but might also leave less time for efficient damage repair, increasing the mutation rate (Gao et al. 2016). We expect that different species will find different balance points, depending on the costs of mutation (in terms of both frequency and impact) and the costs of fidelity and repair (in terms of resources used and time taken).

4.2.2 Mutation Risk

Species may face different risks of mutation, in terms of the probability of mutations occurring. Given that the two sources of mutation are replication error and damage, risk of mutation should increase with the number of copies made per unit time (more opportunity for copy errors), and with the relative influence of mutagens (more opportunities for damage). Note that there are also differences in mutation risk across the genome, for example due to transcription-associated mutagenesis (Jinks-Robertson and Bhagwat 2014) or variation in base composition and methylation patterns (Mugal et al. 2015), but here we are focussing only on the overall differences in mutation risk between different species.

4.2.2.1 Environmental Mutagens

We expect species with increased exposure to mutagens to suffer more DNA damage. For example, bacteria living in high-altitude lakes in the Andes must be able to survive conditions of high UV exposure, high salinity, and concentrations of heavy metals that would be lethal to other organisms. They have increased DNA repair activity, for example through the efficient use of photolyase enzymes to reverse UV-related damage (Albarracín et al. 2012). Similarly, some species have evolved highly efficient double-strand-break repair in order to survive the genome-wrecking effects of desiccation, including bdelloid rotifers and *Deinococcus radiodurans* (otherwise known as 'Conan the Bacterium'). Their increased DNA repair ability makes them incidentally resistant to a range of other mutagens such as irradiation (Slade and Radman 2011; Hespeels et al. 2014). These superhero-like abilities to repair DNA illustrate the balancing act of mutation rate evolution: high rates of damage demand greater investment in repair if a species is to survive in a mutagenic environment. Therefore, increased DNA damage does not necessarily increase the mutation rate (the number of DNA base sequence changes passed on to offspring) if most of the damage is repaired.

This balancing act is well illustrated by considering one of the most pervasive mutagens, UV light, which has a number of mutagenic effects. It causes pyrimidine dimers to form: adjacent thymines (Ts) on the same DNA strand pair with each other and so lose the ability to pair with the opposite strand in the helix. This is disastrous not only due to the loss of information from complementary base pairing, but also because the dimers create lesions that block the action of polymerases, preventing replication. A cell that cannot replicate its DNA is a cell that cannot pass on mutations, and therefore does not contribute to the evolutionary future of the lineage. UV light can also cause other forms of DNA damage, for example generating free radicals that trigger the formation of 8-oxoguanine, which can lead to mismatched base pairs in the DNA helix (Ikehata and Ono 2011; Sage et al. 2012).

Cells have a number of systems for ameliorating UV-induced damage, employing both specific responses to UV-signature damage (such as photoreactivation, which reverses pyrimidine dimerization), and more general pathways that remove damaged bases or fix broken strands (such as nucleotide excision repair; Karentz 2015). Mutations that affect the proteins in these pathways can lead to higher rates of light-induced mutation, or to increased ability to repair UV-induced damage (e.g., Friedberg et al. 2002; Tanaka et al. 2002; Tang and Chu 2002). Therefore, efficiency of repair of UV damage is evolvable and can be shaped by selection in response to typical levels of exposure. For example, populations of pond fleas (*Daphnia*) living in transparent ponds have higher rates of DNA repair of UV-induced damage than those living in ponds with lower UV transparency (Miner et al. 2015). Bacteria from high-altitude lakes show a range of abilities to respond to UV damage, consistent with their different levels of exposure in their natural environments (Fernández Zenoff et al. 2006).

UV-damage repair systems are evolvable in response to environmental levels of exposure, so greater mutation risk from environmental mutagens might result in greater investment in repair, effectively cancelling out the influence of environment on mutation rate. This higher rate of repair investment might be inducible on an individual basis, or might be selected for over evolutionary time (Jansen et al. 1998). For example, *Daphnia* with high UV exposure could moderate DNA damage through behavioural responses (light avoidance), through investment in protective mechanisms such as melanin and carotenoids, or through enhanced repair (Miner et al. 2015). The evolutionary adjustment of UV-damage repair might explain why there is little consistent evidence of higher mutation rates in lineages with greater average UV exposure. While some studies have found that UV exposure is correlated with substitution rates in flowering plants (Davies et al. 2004), others have found that higher average UV exposure is associated with lower mutation rates in the chloroplast and mitochondrial genomes (Bromham et al. 2015). Furthermore, while UV exposure increases with altitude, rates of molecular evolution do not (Dowle et al. 2013).

It has been suggested that environmental temperature could influence the mutation rate (e.g., Wright et al. 2011). Higher rates of molecular evolution have been reported in species living in warmer areas (e.g., Wright et al. 2011), but the pattern is not consistent (e.g., Qiu et al. 2014). Most studies have not separately tested for an association between temperature and synonymous substitution rates, which are changes to the DNA sequence that do not affect protein sequences and so are expected to reflect differences in the underlying mutation rates (Davies et al. 2004; Wright et al. 2011; Gillman and Wright 2014; Barrera-Redondo et al. 2018). So it is difficult to evaluate whether the reported links between temperature and rates of molecular evolution are due to an increase in mutation rate or a general increase in pace of evolution, for example due to differences in life history (Bromham and Cardillo 2003), accelerated niche evolution (e.g., Kozak and Wiens 2010), or greater tempo of diversification (e.g., Cardillo 1999). As for other mutagens, it might be that lineages subject to high temperatures invest in mutation mitigation, just as they do for high UV, in which case we might not expect to see raised mutation rates over evolutionary time.

A latitudinal gradient in rates of molecular evolution has been reported (e.g., Lourenço et al. 2012), but a study over a large number of species pairs found only a slight influence of latitude on rates of molecular evolution (Orton et al. 2018). There is some evidence that efficiency of DNA repair can vary with latitude (Svetec et al. 2016), which might contribute to a lack of clear relationships between mutation rates, environmental temperatures, and latitude. Environmental temperature and latitude both correlate with many other traits that are associated with rates of molecular evolution, so confounding factors should be taken into account in any analysis of the association between environment and rates of molecular evolution. For example, temperature scales with species richness and species richness is associated with rates of molecular evolution (Dowle et al. 2013; Bromham et al. 2015). Similarly, body size and life history can vary along geographic temperature gradients (e.g., Blackburn et al. 1999; Angilletta Jr et al. 2004), which could contribute to environmental variation in the rate of molecular evolution.

Currently, the empirical support for a direct effect of environmental temperature or UV on lineage-specific mutation rates is not very strong. This is because most studies of molecular evolutionary rates along environmental gradients have not specifically looked for signals of mutation rate variation, nor controlled for confounding factors that also scale with environment and mutation rates. Alternatively, environmental conditions could increase the mutation rate indirectly, through increasing stress (such as extreme temperatures or inadequate nutrition), which can result in higher mutation rates. For example, enzymes that ameliorate UV-induced mutation in *Daphnia* are less efficient in cases of nutritional stress (Balseiro et al. 2008). DNA repair efficiency can be reduced at suboptimal temperatures (MacFadyen et al. 2004; Berger et al. 2017). For example, *Caenorhabditis* raised at non-optimal temperatures have higher microsatellite mutation rates (Matsuba et al. 2013), and seed beetles (*Callosobruchus*) raised at high temperatures were less able to repair radiation-induced mutations (Berger et al. 2017). This might be a result of resources being directed to stress mitigation, reducing the availability of resources for repair.

Environmental stresses can also interact with other factors that influence mutation rates. For example, higher exposure to mutagens can result in lower growth, so reduced biomass production, which might have knock-on effects on the number of replication errors per unit time (see Karentz 2015). Experiments on the bacterium *Escherichia coli* have demonstrated that opposing effects of nutrient availability and population density can result in a minimum mutation rate at intermediate values of both (Krašovec et al. 2018).

While environmental stress could increase the mutation rate of a whole population, individuals might vary in their ability to mitigate the effects of stress. Some studies of *Drosophila* have revealed that individuals that are low quality, due to suboptimal genotype or poor condition, pass on more mutations to their offspring (Agrawal and Wang 2008; Sharp and Agrawal 2012). Owing to all of these factors and more, lineages under environmental stress might accumulate more mutations (Ram and Hadany 2012; Jiang et al. 2014). However, the contribution of stress-induced elevation of mutation rate to lineage-specific differences in molecular evolution is unclear. For example, stress-induced mutation might cause only transient pulses in mutation generation, because species that experience long-term changes in environmental conditions might either adapt their DNA repair levels or suffer extinction due to mutational load. There might be some interaction between stress-induced mutation and the influence of stress on the selective coefficients of mutations in populations under stress, which might also potentially be confounded by changes in population size due to suboptimal conditions.

4.2.2.2 Copy Errors

DNA replication represents an opportunity for mutation. Every time a base is copied there is a small chance that it will be changed. The error rate is amazingly small, typically ranging from one-in-ten-million to one-in-a-billion bases (Kunkel 2004). But because most genomes are very big—ranging from ten million to over ten billion bases for non-bacterial organisms—even this high copy fidelity will result in new mutations being introduced into every genome copy. The more times the genome is copied, the more opportunities there are for mutations to occur.

The copy-error effect has generally been considered to explain why animals with longer generations have lower mutation rates, on the assumption that they copy their genomes less often per unit time (Ohta 1993; see Chap. 2). The correlation between generation length and rate of molecular evolution is pervasive in animals, including invertebrates (Thomas et al. 2010), mammals (Bromham et al. 1996; Nabholz et al. 2008), birds (Mooers and Harvey 1994), and reptiles (Bromham 2002). A generation-time effect has also been reported for plants, although this is typically based on a proxy for generation time such as herbaceous versus woody habit (Smith and Donoghue 2008), and has not been supported by all studies (Whittle and Johnston 2003). In bacteria, spore formation can be interpreted as increasing the generation length, reducing the number of cell divisions per unit time, and *Firmicutes* inferred to be spore-forming (on the basis of genome composition) have been found to have lower mutation rates than their presumed non-spore-forming relatives (Weller and Wu 2015).

But, while there is broad empirical evidence for a correlation between generation time and rate of molecular evolution, particularly in animals, there are a number of complications in interpreting the cause of the generation-time effect. Variation in rate of molecular evolution does not scale simply with the number of generations per unit time. For example, mice can go through 50 generations for every human generation, yet the mutation rate in mice is only several times higher than that of humans (Bromham 2011). The mismatch may be partly explained by lack of a simple relationship between the number of cell divisions and age at first reproduction (the most common measure of generation time). This can be illustrated by considering the number of cell generations in the germline in the individual, which reflects the number of times the genome is copied from one generation to the next, from the formation of the embryo to the growth and maturation of the individual to the production of gametes. For example, compared with mice, a human has only 6.5 times more cell generations in the male germline (average of 401 cell generations in humans to 62 in mice) and only 1.2 times more cell divisions in the female germline (31 to 25, respectively; Bromham 2011). Similarly, it has been estimated that the number of cell divisions between reproductive events in flowering plants is only doubled in tall forest trees compared with short annuals (Burian et al. 2016). As an added complication, it is possible that some cell divisions in the germline are more mutation-prone than others (Gao et al. 2014).

The relationship between the number of cell generations, generation time, and mutation rate is particularly complicated for plants. Generation time (as reflected in perenniality) scales negatively with stature in plants (Duminil et al. 2009), so we might expect taller plants to have lower rates of molecular evolution (Lanfear et al. 2013; Bromham et al. 2015). But, for a plant that produces reproductive structures at the tips of growth, gametes from taller plants could have been the product of a longer chain of cell divisions than gametes from a shorter plant, so might have had the opportunity to collect more replication errors per generation (Scofield and Schultz 2006).

One explanation for the observed negative relationship between plant height and rate of molecular evolution is that absolute growth rates slow as plants increase in height (Lanfear et al. 2013). Furthermore, the cell line that gives rise to gametes might have lower rates of cell division and growth than the somatic tissue, offsetting the effect of height on number of cell divisions, and therefore dampening the DNA copy-error effect (Burian et al. 2016; Watson et al. 2016). So the number of cell divisions per generation does not necessarily increase with age, longevity, or amount of biomass (Watson et al. 2016). In addition, for long-lived species, age at first breeding might be a poor indicator of average age at reproduction (Petit and Hampe 2006). For species with a long reproductive span, the number of cell divisions per generation might increase with longevity (Lehtonen and Lanfear 2014). So while there is much empirical evidence for an association between generation time and rate of molecular evolution, this might not reflect a simple copy-frequency effect (Gao et al. 2016).

Replication frequency has been implicated in other patterns of variation in rate of molecular evolution that are not closely tied to generation length. For example, highly eusocial bees and wasps have higher rates of molecular evolution than their non-social relatives, which might be a result of increased number of cell divisions in species where queens produce vast numbers of offspring (Bromham and Leys 2005). If the eggs that will become reproductive offspring are produced only after many eggs have developed into non-reproductive workers, then the genome copies passed to the next generation will have gone through a larger number of cell generations than in species that produce few or no workers. Consistent with this, colony size is related to rate of molecular evolution across a wide range of social insects (Rubin et al. 2019). Additional support for this hypothesis comes from comparisons of social parasites with their eusocial relatives. Social parasites typically do not produce non-reproductive workers, so will presumably have fewer genome replications in their average generation length. Consistent with predictions, social parasites have lower rates of molecular evolution than their eusocial relatives (Bromham and Leys 2005). However, these results rest on relatively few phylogenetically independent comparisons; further investigation would be needed to establish whether the effect is general, and to determine whether the effect is due to copy frequency, effective population size, or some other factor. Similarly, raised rates of molecular evolution in species of the fungus *Neurospora* that produce asexual spores have been attributed to increased mitotic divisions in spore production, though this result is based on a very small number of comparisons (Nygren et al. 2011).

One of the strongest cases for a replication-frequency pattern on mutation rate is the phenomenon often referred to as 'male-driven evolution' or 'male-biased mutation'. Male gametes are typically produced in greater abundance than female gametes, and male gametes are commonly the product of more cell generations than female gametes of the same species. So in many taxa, DNA sequences that spend more time in males will go through more replications per unit time than those that spend more time in females, and so will accumulate more copy errors. This has been noted in mammals (higher mutation rates on the Y chromosome than the X) and in birds (higher mutation rates on the Z chromosome than the W) (Wilson Sayres and Makova 2011). Plants provide an interesting test of the hypothesis because of the diversity of patterns of inheritance of organelle DNA. For example, in some conifer lineages both the chloroplast and mitochondrial genomes are inherited from the female parent, but in others they are both inherited from the male parent, and in other lineages the chloroplast is paternally inherited but the mitochondrion is maternally inherited. Contrasts between lineages with different patterns of organelle inheritance have revealed that the rate of evolution in organellar DNA is greater when it is passed through the paternal line than through the maternal line (Whittle and Johnston 2002). So the average mutation rate for a species will be strongly influenced by mutations occurring in males during the production of gametes (Gao et al. 2016).

Sperm are produced continuously throughout a male's reproductive lifespan, so the average number of DNA replications in the germline should also increase with age of males at reproduction, and it should also increase with the amount of sperm produced. Testis size in primates correlates with the predicted degree of sperm competition due to multiple matings (Harcourt et al. 1995), and also correlates positively with rate of molecular evolution (Wong 2014). Because of the strong effect of sperm production on the average number of cell replications in the germline, it has been suggested that the length of reproductive lifespan for males is a better indicator of the effect of DNA copy errors on mutation rate than is the age of first breeding for females (Thomas et al. 2018). Note that we expect the same copy-error effect in females in species where oogenesis is continuous, so that the longer the reproductive lifespan, and the higher the fecundity, the more copy errors should accumulate. In this case, we might expect the mutation rate per year to scale more closely with the average age at reproduction than with the age at maturity (Lehtonen and Lanfear 2014). This effect has been noted in highly eusocial bees and wasps, where one or few females each produces a very large number of offspring (Bromham and Leys 2005). But in rockfish, where oogenesis is continuous, mutation rate does not appear to increase with fecundity, counter to the expectation of a copy-number effect (Hua et al. 2015).

4.2.3 Mutation Cost

We have seen that the opportunity for mutations to occur is influenced by the relative impact of mutagenic agents (increasing DNA damage) and the relative number of DNA replications (increasing copy errors). But we have also seen that the response to mutation risk might be to increase investment in mitigation, for example increased efficiency of DNA repair in response to higher environmental mutagens. Given that there is ample evidence that natural populations contain heritable variation in repair efficiency and replication fidelity, and that greater mutation risk can be countered by greater investment in repair and fidelity, why do species have different mutation rates (see Chap. 3)? If most mutations are harmful, should we expect selection to reduce the mutation rate to a minimum in all species by favouring increased repair and copy fidelity to reduce the mutational burden?

One possible explanation for persistent lineage-specific differences in mutation rate is that the relative costs of mutation vary among lineages. If reduction in mutation rate is costly in terms of resources invested and time taken, then it must be balanced by a reduction in the negative impact of mutation (Sniegowski et al. 2000). If costs of mutation vary among species, then we might expect levels of investment in fidelity and repair also to vary, and therefore modulate the average mutation rate for that lineage. Mutation is costly because most mutations are deleterious, but the distribution of selection coefficients is expected to vary with the environment and population dynamics, which will fluctuate over time (Lanfear et al. 2014). Here we will only consider the general ways in which species' characteristics can influence the average costs of mutation.

In the previous section, we considered that the generation-time effect on rates of molecular evolution might not be a straightforward result of replication frequency. It is possible that the observed generation-time effect is, at least in part, due to generation time scaling with other aspects of life history that also have an influence on mutation rate, by altering the balance between the costs of mutation and the costs of repair and fidelity. Several correlates of generation time have been shown to have a significant association with rates of molecular evolution: body size, fecundity, and longevity. In addition to influencing the relative risk of mutation (for example, through a greater number of cell generations), each of these might have an impact on the relative cost of mutations (Bromham 2009).

4.2.3.1 Body Size

Bigger animals and taller plants tend to have lower rates of molecular evolution (e.g., Martin and Palumbi 1993; Bromham 2002; Gillooly et al. 2005; Lanfear et al. 2013; Barrera-Redondo et al. 2018). It has commonly been assumed that the size scaling of substitution rates is an indirect effect of the correlation between size and other life-history traits, such as generation time, population size, or metabolic rate (Martin and Palumbi 1993; Gillooly et al. 2005). Body size might have an indirect effect on the generation of mutations. For example, larger animals have lower mass-specific metabolic rates, so might generate fewer free radicals per unit time and thereby might suffer lower rates of DNA damage. So far, there is little evidence that metabolic rate provides a convincing explanation of variation in rate of molecular evolution, once its covariation with other life-history traits is accounted for (Bromham et al. 1996; Lanfear et al. 2007; Galtier et al. 2009b). It might be that DNA repair rate is adjusted to the expected level of insult from mutagens, whether generated internally or externally.

Body size might have a direct influence on the costs of mutation, by modulating the relative impact of mutation on fitness. A larger animal has more cells, and so has more copies of the genome, each of which is subject to ongoing mutation. Mutation in somatic cells can have a significant effect on fitness, for example by knocking out a critical function in a cell line of the developing embryo or generating a cancerous cell line. The risk of a life-threatening mutation occurring per lifetime should increase with the number of cell generations required to make a body, and the number of cells that must be maintained over the reproductive lifespan. Even within a species, larger individuals can have a greater cancer risk (Nunney 2018). So a species with larger average body size might require greater investment in DNA damage control or replication fidelity in order to maintain the same lifetime risk of death from somatic mutation as a smaller animal (Nunney 1999). Yet there is no evidence that larger animals have a greater lifetime cancer risk than smaller animals (an observation known as Peto's Paradox). This paradox can be explained if lineages of larger animals invest in decreasing their mutation rates in order to reduce rates of spontaneous cancer formation (Caulin and Maley 2011). This risk reduction might occur through selection on specific genes involved in cancer, but increases in DNA repair and copy fidelity could also play a role (Rozhok and DeGregori 2019).

Perhaps we can see the influence of selection reducing the somatic mutation rate if we compare a large, sexually reproducing individual plant with a multi-part clonal plant that grows and reproduces by ramets. Even if it contains as many cells as the larger individual, the cost of mutation might be lower for the colony as a whole, if a mutation in one part reduces the success of that ramet but does not compromise the longevity or reproductive potential of the clonal set as a whole. For example, a study of buttercups found that the older the meadow, the more buttercups with extra petals, and the lower the viability of pollen (Warren 2009). Since the buttercups predominantly reproduce clonally, this was interpreted as evidence of the gradual accumulation of mutations in long-lived clonal sets over time. The collection of somatic mutation in clonal growth can have measurable effects on the genetic diversity of a population of asexual ramets, resulting in a level of genetic diversity greater than in sexual populations of the same species (Gross et al. 2012). So the costs of body size in increasing lifetime mutation risk might be higher for an organism with a single reproductive cell line and an indivisible phenotype (e.g., many animals) than for an organism with multiple reproductive cell lines and a flexible phenotype where parts can be lost without compromising the success of the whole (e.g., many plants).

4.2.3.2 Fecundity

Some studies have suggested that fecundity correlates with rate of molecular evolution, above and beyond its covariation with generation time. For example, a study of rates in mammals found that fecundity scaled with rates of both synonymous (silent) and nonsynonymous (amino-acid replacement) substitutions (Welch et al. 2008). In some species, a link between fecundity and rates of molecular evolution might represent a copy-number effect. For example, where gamete production is continuous throughout life, the more offspring produced, the greater the average number of germline copies. This is a possible explanation for the higher rate of molecular evolution in highly eusocial bees, wasps, and ants (Schmitz and Moritz 1998; Bromham and Leys 2005; Rubin et al. 2019). Similarly, if fecundity scales with the opportunity for multiple matings, or with the length of male reproductive lifespan, then it will also scale with the number of male germline replications. But the copy-number effect clearly does not represent a general explanation for the fecundity effect: in rockfish, rates of molecular evolution are greatest in small-bodied, short-lived species, even though they have lower fecundity than their larger relatives (Hua et al. 2015).

An alternative explanation is that higher fecundity changes the relative costs of mutation. If each parent produces a large number of offspring, then, unless the population is growing rapidly, a greater number of offspring will die (or otherwise fail to reproduce). For a high-fecundity species, the loss of any one offspring due to deleterious mutation represents a lower proportional reduction in reproductive fitness. As long as some offspring survive and reproduce, the production of additional defective copies is a cost that can be borne (as long as investment in poor-quality offspring is low). But for low-fecundity species with relatively higher investment in each offspring, a deleterious mutation can cause a proportionally greater reduction in reproductive success.

The balancing act between fecundity and fidelity is evident in RNA viruses, which have a stupendously high mutation rate. Even with a small genome, the error rate per genome copy is such that between 40% and 100% of all offspring will carry a mutation (Belshaw et al. 2008). These high rates of error per nucleotide base copied push RNA viruses perilously close to the 'error threshold', the mutation rate beyond which there are too few viable offspring to maintain the lineage. RNA virus mutation rates might represent a balance point, modulated by genome length, below which the mutation rate per genome copy is too high to guarantee the production of sufficient functional offspring to allow the lineage to persist (Holmes 2003). In fact, it has been suggested that an association between genome size and mutation rate is a general phenomenon for microbes, an observation known as Drake's rule (Drake et al. 1998).

There is not yet enough evidence or analysis to clarify whether there is a general effect of fecundity on rate of molecular evolution. If such a general effect does exist, it is unclear whether it is due to higher mutation risk (more replications per unit time) or lower mutation cost (less reduction in fitness per mutation), or both, or neither. While increased fecundity might raise the mutation risk through more genome copies per unit time, it could also decrease the cost of each mutation; the level of investment in mutation avoidance would be reduced, in line with reduced parental investment per offspring (Welch et al. 2008; Bromham 2011). Increasing multiple mating without increasing overall fecundity could have the opposite effect, because higher sperm competition selects for increased quality of offspring, which could potentially favour lower mutation rates (Firman and Simmons 2012).

4.2.3.3 Lifespan

Maximum recorded lifespan scales with mitochondrial substitution rates in mammals, birds, and fish, though it is less of a strong predictor of nuclear rates (Nabholz et al. 2008; Welch et al. 2008; Galtier et al. 2009a; Hua et al. 2015). Some researchers have explained this observation in terms of metabolically induced DNA damage. Aerobic metabolism in the mitochondria produces oxygen free radicals, which can damage cellular biomolecules including DNA. The mitochondrial theory of ageing suggests that metabolically induced mitochondrial damage contributes to age-related decline and therefore limits lifespan.

Some studies have found a link between mitochondrial mutation and variation in lifespan between individuals, but others have failed to support this connection (see discussion in Hua et al. 2015). However, there is little evidence that mass-specific metabolic rate *per se* is a good predictor of variation in rate of molecular evolution (Bromham et al. 1996; Lanfear et al.

2007; Galtier et al. 2009a). Furthermore, the longevity effect on molecular evolution has been noted in taxa that show no signs of senescence with advancing age (Hua et al. 2015). Instead, it might be that, just as for externally produced mutagens, the level of DNA repair is adjusted to the level of risk arising from cellular metabolites, balanced against the average cost of mutation.

Longevity could increase mutation risk, if it results in increased reproductive lifespan and therefore more cell generations in the germline. This might explain why the male mutation bias is stronger in longer-lived organisms (Goetting-Minesky and Makova 2005). Longevity might also increase the cost of mutation, if fitness-harming somatic mutations accumulate over time. Therefore, selection could drive lower rates of mutation in longer-lived organisms to counter the increase in lifetime risk of fitness-harming mutation.

It is important to recognize that the somatic mutation rate (which generates the cost of body size and longevity) need not be the same as the germline mutation rate (which can be protected from risk and receive enhanced repair). Germline cells might receive heightened levels of repair, through greater investment in damage detection and maintenance activities, potentially at the expense of investment in damage control in somatic cells (Maklakov and Immler 2016). Germline DNA might be further protected from damage by being kept in a relatively quiescent state, through reduced metabolic activity or lower growth rates (Allen 1995; Burian et al. 2016). Protection of the germline can further dissemble a simple linear relationship between size, generation time, longevity, and mutation rate. This might explain the lack of a proportional relationship between plant age and the number of mutations transmitted to the next generation (Watson et al. 2016). The problem is that longevity rarely varies without being correlated with changes in other life-history traits that can also influence rate of molecular evolution. It would be interesting to compare the relative effects of longer lifespan against more germline cell generations by comparing rates of molecular evolution in semelparous species with high sperm competition with their non-semelparous relatives, or related species with similar longevity but different mating strategies.

4.3 Substitution

So far we have considered the different evolutionary forces that shape the mutation rate, which affects the number of permanent changes introduced to an individual's genome. But the rate of change of nucleotide sequences in the genome is not wholly governed by the mutation rate. Many mutations that occur will fail to be passed on to offspring and, of those that are, not all will persist in the population over the long term. If a mutation rises in frequency over generations until all individuals in the population carry a copy of that mutation, it becomes a substitution (one base has been substituted for another). We will be able to detect the substitution as a change in the DNA sequence that is characteristic of a particular population or lineage. Mutation is the individual-level generation of heritable variants: substitution is the population-level replacement of all other heritable variants at that sequence position. While variation in mutation rate should be correlated with variation in the substitution rate, there are additional factors that can shape variation in substitution rates even when the mutation rate is static.

There are two broad ways in which a mutation can rise in frequency until it replaces all other variants at that position in the DNA sequence. The first is by chance: in a population where not all individuals reproduce, or where different versions of DNA sequences have differing numbers of descendants, the frequency of variants present in one generation might not be an exact representation of the frequencies in the previous generation, simply due to random sampling error. This sampling error will cause allele frequencies to fluctuate. By chance, one variant might undergo, on average, more increases in frequency than decreases, and might even 'wander' all the way to a frequency of 100%.

We can appreciate the effects of random sampling with any simple sampling experiment, such as flipping a coin or pulling coloured beans from a bag. But there are two important features of genetic sampling that allow substitution to occur by chance. First, the sampling occurs in a series, so that accidentally sampling an excess of one variant over another in one generation results in an excess of that variant in the next generation. If in the following generation the frequency increases again by chance, it will creep closer to 100%. Once any variant reaches 100%, we say it has become fixed in the population, because the alternative variants have disappeared. Serial sampling can generate a random walk in allele frequencies that might wander all the way to fixation, with the concomitant loss of all other variants at that locus. If the subsampling is severe (few variants in the parent generation make it into the offspring generation) then there is a high potential for skewing frequencies.

A second feature of genetic sampling that allows substitution to occur by chance is that frequencies are determined entirely by replication: barring repeat mutation, all copies of a variant are descendants from an original copy. This is important because if the frequency of any one variant wanders all the way to 100% then substitution occurs; it is not possible for the frequencies of that variant to change again in the next generation. All individuals in the population carry identical copies of the same variant so there can be no offspring born with any other variant. Similarly, if any variant wanders all the way to a frequency of 0%, there are no more copies of the mutation that can be copied to the next generation, so no offspring can inherit that variant from its parents. These two things together—replication and finite population size—make substitution due to chance (genetic drift) a possibility, particularly for mutations that have little or no effect on fitness.

The second process that can drive substitution is selection: if a mutation, by virtue of its heritable effects, has a higher chance of ending up in successful offspring than other variants present in the population, then it is expected to increase in frequency until it replaces all other variants in the population (if its advantage is maintained). The relative influence of biased transmission (selection) and random sampling (drift) is dependent on circumstance. For example, the state of the environment will affect the relative fitness of variants, and aspects of population size and structure will influence the number of individuals that pass on their genes to the next generation. Because the likelihood of a mutation going to fixation is influenced by its selective coefficient, and because the selective coefficient is dependent on the interaction between genome, population, and environment, substitution rates can vary across the genome (Wong and Seguin 2015). But in this chapter we are only concerned with factors that influence the genome-wide, lineage-specific rate of molecular evolution, so we will consider only the ways in which substitution rates can differ consistently between species over long time periods.

Effective population size is one of the major determinants of patterns and rates of substitution, because it mediates the relative influences of drift and selection (Ohta 1987; Charlesworth 2009; Lanfear et al. 2014). The substitution rate of completely neutral mutations, those that have no effect on fitness, is determined primarily by the mutation rate (Kimura 1983; Lehmann 2014) and is generally assumed to be unaffected by population size (though some patterns of demography can result in fluctuations in the neutral substitution rate; Balloux and Lehmann 2012). Advantageous mutations are fixed more efficiently in large populations where the influence of random sampling on allele frequencies is less severe, but, on the whole, advantageous mutations are rare (Eyre-Walker and Keightley 2007). Strongly deleterious mutations are, by definition, unlikely to be passed on to the next generation, so do not contribute to the substitution rate.

But there are many mutations that are neither advantageous nor strongly deleterious (see Chap. 2). For example, a mutation might slightly reduce the efficiency or structural stability of an enzyme, yet not prevent it from functioning. These slightly deleterious mutations are not severe enough to prevent survival and reproduction under most circumstances, so they can be passed on to the next generation. In a large population, selection is efficient enough to eventually eliminate even slightly disadvantageous mutations. But in small populations, where the effects of random sampling are the most marked, these slightly deleterious mutations can occasionally go to fixation by chance. Therefore, in a small population, some mutations with a slightly negative effect on fitness will behave as if neutral, and their fate will be governed by drift. Because these 'nearly neutral' mutations are defined by both the selective coefficient and the relative strength of drift, the proportion of mutations that fall into this category is determined not only by the distribution of selective effects but also by the effective population size. Because the proportion of nearly neutral mutations increases with decreasing effective population size (Castellano et al. 2018), and because nearly neutral mutations are more numerous than advantageous changes (Eyre-Walker and Keightley 2007), we expect smaller populations to have a higher overall substitution rate (driven by drift on nearly neutral changes) than larger populations (driven by positive selection).

All other things being equal, anything that causes a long-term reduction in population size should increase the substitution rate, relative to similar lineages with larger population sizes. It is surprisingly hard to test this prediction empirically, because effective population size is not independently known for many populations (although it is commonly estimated from the genetic data themselves using assumptions about the substitution process). Furthermore, the relationship between mutation rate, substitution rate, and population size can change over time with changing selection pressures, and with population expansion and decline (Lanfear et al. 2014; Hua and Bromham 2017). Given that substitution rates are typically estimated from phylogenetic branch lengths, which reflect changes that have accumulated over millions of years, a 'snapshot' of population size from the tips of the phylogeny might have relatively little explanatory power unless it adequately reflects long-term average population sizes for those lineages.

Most empirical studies of the effect of population size on substitution rates rely on comparing distantly related species with distinct differences in population size. However, these species would usually also differ in many aspects of their biology that might also influence rates, confounding the search for evidence of a relationship between population size and substitution rates. But there is some evidence from studies of closely related pairs of species that those with smaller population sizes tend to have higher substitution rates. For example, a comparison of island endemic species with their mainland relatives found a significant trend towards higher ratios of nonsynonymous to synonymous changes, as predicted (Woolfit and Bromham 2005), though this result is not always strong or consistent (James et al. 2016). Similarly, comparisons between endosymbiotic microbes and their free-living relatives found higher substitution rates in the species confined to small populations in insect guts (Woolfit and Bromham 2003); this result is also consistent with a population-size effect, but additional driving factors cannot be ruled out.

The relevant measure here is not necessarily census population size (number of individuals): effective population size reflects the number of individuals that contribute alleles to the next generation. For example, higher rates of molecular evolution in eusocial bees, wasps, and ants have also been interpreted in terms of reduced effective population size, because the number of reproductive individuals in a colony is very small despite the large overall number of individuals (Schmitz and Moritz 1998; Bromham and Leys 2005; Rubin et al. 2019). Similarly, if strong sexual selection results in a reduction in the number of contributing parents (e.g., a small number of males father a disproportionate share of offspring), this should reduce the effective population size and increase the substitution rate, although this effect could be conflated with higher male-biased mutation due to increased sperm production (Bartosch-Härlid et al. 2003). A wider range of empirical tests of the effect of population size on substitution rates would be welcome.

4.4 Diversification

Rate of molecular evolution, as measured from phylogenetic branch lengths, is correlated with diversification rate, as measured by phylogenetic clade size (Barraclough and Savolainen 2001; Pagel et al. 2006; Eo and DeWoody 2010; Lanfear et al. 2010; Duchêne and Bromham 2013; Ezard et al. 2013; Bromham et al. 2015, 2018). At first, this might look suspiciously like an artefact of measurement: more lineages, more nodes in the phylogeny, more inferred substitutions (Hugall and Lee 2007). Yet the same pattern has been found for a wide range of sequences and lineages using a number of different methods, each of which addresses some measurement biases. This growing body of evidence suggests that we must look for reasons why lineages with higher rates of diversification tend to have higher rates of molecular evolution than related lineages with lower diversification rates (Hua and Bromham 2017).

What could be the cause of a link between rate of molecular evolution and rate of lineage diversification? One explanation is that speciation generates genetic change, as species adapt to new environments or undergo selection for reproductive isolation (Venditti and Pagel 2009). Selection on specific traits is unlikely to be reflected in genome-wide rates of molecular evolution. Advantageous mutations are rare: increased strength of selection will apply to only a small number of mutations, and those directly adjacent to them (hitchhiking). Such sites seem unlikely to be included in the kinds of sequences classically included in phylogenetic analyses ('housekeeping genes'), which are exactly the kinds of sequences that have furnished evidence for the association between diversification rate and substitution rate. However, other processes associated with speciation might influence genome-wide substitution rates, which would be reflected in all kinds of sequences compared between lineages.

Repeated reductions in population size, caused by founder effects associated with speciation occurring in small population isolates, could generate bursts of nearly neutral substitutions (Venditti and Pagel 2008). In the words of Ernst Mayr (1970): 'small populations … are great opportunities for a genetic revolution'. Changes in population size will influence the substitution rate of nearly neutral mutations, and selection will affect the fixation rate of advantageous mutations. Both of these classes of mutations, advantageous and nearly neutral, must have some effect on phenotype (in the broad sense, including patterns of gene expression) in order to have nonzero effects on fitness. Therefore, both of the proposed mechanisms linking speciation to rate of molecular evolution—population size influencing nearly neutral substitutions and selection affecting advantageous substitutions—should be detectable only in the nonsynonymous substitution rate (changes that might influence phenotype), not in the rate of synonymous substitutions (which do not affect the encoded proteins and so have negligible selection coefficients).

Yet the association between diversification rate and rate of molecular evolution has been demonstrated for synonymous substitution rates (Lanfear et al. 2010; Duchêne and Bromham 2013; Bromham et al. 2015). The observed link between variation in synonymous substitution rate and diversification rate is not easy to explain in terms of selection associated with speciation events, or changes in population size influencing the proportion of nearly neutral mutations (because synonymous mutations should be essentially neutral and therefore unaffected by population size). Synonymous substitution rates are expected primarily to reflect the underlying mutation rate. So what could cause a link between diversification rate and mutation rate? It is difficult to think of a convincing explanation for speciation rate affecting the mutation rate. One possible explanation is that repeated reductions in population size might result in a degradation of repair efficiency through the substitution of slightly deleterious mutations in replication and repair pathways (Lynch et al. 2016). But a link between effective population size and mutation rate is not evident in all species (e.g., Lanfear et al. 2014; Castellano et al. 2018).

Could an increased mutation rate, for example due to changes in life-history traits (see Sect. 4.2), directly lead to an increased rate of diversification? To do so, increased mutation rate would need to have some impact on increasing speciation rates, decreasing extinction rates, or both. If adaptation is mutation-limited, then it is possible that increased mutation could increase speciation or lower extinction, if it facilitates rapid response to changing environments. But such a strategy would presumably come at the expense of a greater rate of deleterious mutations. Alternatively, increased mutation rate could influence the speciation rate by providing fuel for the development of hybrid incompatibility between separated populations. This might be through the selection of changes that increase reproductive isolation, such as changes in flowering time, particularly in traits that both increase ecological separation and promote nonrandom mating (Servedio et al. 2011). Or it might simply be that the accumulation of slightly deleterious changes and associated compensatory changes drive the erosion of genome compatibility through Bateson–Dobzhansky–Muller incompatibilities (e.g., Wang et al. 2013). In this way, life history can be linked to changes in mutation rate, which contribute to variation in the substitution rate, resulting in variation in diversification rate (Bromham et al. 2015).

In addition, we must consider the possibility that the rate of molecular evolution and diversification rate are indirectly linked through other traits. For example, if body size influences both the mutation rate and the diversification rate, then it could generate a correlation between mutation rate and diversification rate even if the two are not directly functionally linked. For a more detailed discussion of the links in the explanatory chain connecting mutation rates, substitution rates, and diversification rates, see Hua and Bromham (2017).

4.5 Implications

Most contemporary molecular dating analyses use methods that allow rates of molecular evolution to vary among lineages (see Chap. 5 and others in this book). However, these methods typically draw rates from a convenient distribution and rely on a stochastic model of rate change (Bromham et al. 2018). In other words, most molecular dating methods aim to account for rate variation without explicitly modelling mechanisms or causes of rate changes. There are few methods that model the way that rates of molecular evolution evolve in concert with species biology (Lartillot and Delsuc 2012; Nabholz et al. 2013). This is not, in and of itself, a problem. After all, most base substitution models allow different rates of transitions and transversions without incorporating any information on how that bias towards transitions results from a particular process of damage (e.g., UV-induced thymine dimers), differential rates of repair (e.g., bias in mismatch repair), or selection (e.g., more transitions are synonymous). Just as the approach to base transition frequencies is to model the rate difference without reference to the underlying causes of the difference, so the approach in molecular dating has been to model variation in rate of molecular evolution without incorporating any a priori expectation of how those rates will vary among lineages.

Can we be confident that these stochastic models are adequately capturing rate variation? We should be worried that different molecular dating analyses can give dramatically different answers, even when applied to similar—or even identical—data sets (e.g., dos Reis et al. 2014; Foster et al. 2016). The lack of agreement between date estimates from different studies shows that inference is highly sensitive to assumptions made in the analyses. This does not tell us whether the answer is right or wrong, but it does tell us that the assumptions are driving the estimates, over and above the signal from the data.

How can we use our understanding of rate variation to improve molecular date estimates? In some cases, we can check whether our inferred rate estimates across the phylogeny are consistent with our predictions on the basis of our understanding of rate variation. For example, parasitic plants have higher rates of molecular evolution than their non-parasitic relatives (Bromham et al. 2013), and encouragingly these higher rates can be detected by a range of models of rate variation (Bellot and Renner 2014). But the different models distribute these higher rates in different ways, some at the tips of the parasitic clade, some on the deeper lineages, and some at the root (Bromham 2019). This results in a doubling of the inferred age of the clade under some models. It is not clear, in the absence of additional information on the timing of origin of this clade, which molecular dates are correct.

We could use an understanding of the causes and patterns of rate variation to highlight cases in which the date estimates might be subject to error. For example, it has been suggested that surprisingly old molecular dates for the radiation of placental mammals (e.g., Bininda-Emonds et al. 2007) might be due to the well-established link between body size and rates of molecular evolution in mammals. If early mammals were, on average, smaller than those species usually included in phylogenies, then the rate estimates might be too low for the basal lineages in the phylogeny, making the molecular date estimates too old (Bromham 2003; Phillips 2015). Similarly, reduction in body size in birds over the Cretaceous–Palaeogene boundary could have accelerated the rate of molecular evolution, causing molecular dating based on rates in younger lineages to overestimate the age of the radiation (Berv and Field 2017). This is a topic of much debate, but serves as a useful case study for examining the challenges of molecular dating when rates evolve with species biology.

Ideally, though, we would use our understanding of the causes of variation in rates of molecular evolution to inform the rates models themselves, using information on traits likely to correlate with rate variation to derive empirically informed distributions of plausible substitution rates. Such models are not easy to develop and apply, but are hopefully on their way to becoming a reality.

Acknowledgements I thank Simon Ho and Minh Bui for helpful suggestions and encouragement.

References

Agrawal AF, Wang AD (2008) Increased transmission of mutations by low-condition females: evidence for condition-dependent DNA repair. PLOS Biol 6:e30

Albarracín VH, Pathak GP, Douki T, Cadet J, Borsarelli CD, Gärtner W, Farias ME (2012) Extremophilic *Acinetobacter* strains from high-altitude lakes in Argentinean Puna: remarkable UV-B resistance and efficient DNA damage repair. Orig Life Evol Biosph 42:201–221

Allen J (1995) Separate sexes and the mitochondrial theory of ageing. J Theor Biol 180:135–140

Angilletta MJ Jr, Steury TD, Sears MW (2004) Temperature, growth rate, and body size in ectotherms: fitting pieces of a life-history puzzle. Integr Comp Biol 44:498–509

Baer C, Miyamoto MM, Denver DR (2007) Mutation rate variation in multicellular eukaryotes: causes and consequences. Nat Rev Genet 8:619–631

Balloux F, Lehmann L (2012) Substitution rates at neutral genes depend on population size under fluctuating demography and overlapping generations. Evolution 66:605–611

Balseiro E, Souza MS, Modenutti B, Reissig M (2008) Living in transparent lakes: low food P:C ratio decreases antioxidant response to ultraviolet radiation in *Daphnia*. Limnol Oceanogr 53:2383–2390

Barraclough TG, Savolainen V (2001) Evolutionary rates and species diversity in flowering plants. Evolution 55:677–683

Barrera-Redondo J, Ramírez-Barahona S, Eguiarte LE (2018) Rates of molecular evolution in tree ferns are associated with body size, environmental temperature, and biological productivity. Evolution 72:1050–1062

Bartosch-Härlid A, Berlin S, Smith NGC, Møller AP, Ellegren H (2003) Life history and the male mutation bias. Evolution 57:2398–2406

Bellot S, Renner SS (2014) Exploring new dating approaches for parasites: the worldwide Apodanthaceae (Cucurbitales) as an example. Mol Phylogenet Evol 80:1–10

Belshaw R, Gardner A, Rambaut A, Pybus OG (2008) Pacing a small cage: mutation and RNA viruses. Trends Ecol Evol 23:188–193

Berger D, Stångberg J, Grieshop K, Martinossi-Allibert I, Arnqvist G (2017) Temperature effects on life-history trade-offs, germline maintenance and mutation rate under simulated climate warming. Proc R Soc B 284:20171721

Berv JS, Field DJ (2017) Genomic signature of an avian Lilliput effect across the K-Pg extinction. Syst Biol 67:1–13

Bininda-Emonds ORP, Cardillo M, Jones KE, MacPhee RDE, Beck RMD, Grenyer R, Price SA, Vos R, Gittleman JL, Purvis A (2007) The delayed rise of present-day mammals. Nature 446:507–512

Blackburn TM, Gaston KJ, Loder N (1999) Geographic gradients in body size: a clarification of Bergmann's rule. Divers Distrib 5:165–174
Bromham L (2002) Molecular clocks in reptiles: life history influences rate of molecular evolution. Mol Biol Evol 19:302–309
Bromham L (2003) Molecular clocks and explosive radiations. J Mol Evol 57:S13–S20
Bromham L (2009) Why do species vary in their rate of molecular evolution? Biol Lett 5:401–404
Bromham L (2011) The genome as a life-history character: why rate of molecular evolution varies between mammal species. Philos Trans R Soc B 366:2503–2513
Bromham L (2016) An introduction to molecular evolution and phylogenetics. Oxford University Press, Oxford, UK
Bromham L (2019) Six impossible things before breakfast: assumptions, models, and belief in molecular dating. Trends Ecol Evol 34:474–486
Bromham L, Cardillo M (2003) Testing the link between the latitudinal gradient in species richness and rates of molecular evolution. J Evol Biol 16:200–207
Bromham L, Leys R (2005) Sociality and rate of molecular evolution. Mol Biol Evol 22:1393–1402
Bromham L, Rambaut A, Harvey PH (1996) Determinants of rate variation in mammalian DNA sequence evolution. J Mol Evol 43:610–621
Bromham L, Cowman PF, Lanfear R (2013) Parasitic plants have increased rates of molecular evolution across all three genomes. BMC Evol Biol 13:126
Bromham L, Hua X, Lanfear R, Cowman P (2015) Exploring the relationships between mutation rates, life history, genome size, environment and species richness in flowering plants. Am Nat 185:507–524
Bromham L, Duchêne S, Hua X, Ritchie A, Duchêne D, Ho SYW (2018) Bayesian molecular dating: opening up the black box. Biol Rev 93:1165–1191
Burian A, Barbier de Reuille P, Kuhlemeier C (2016) Patterns of stem cell divisions contribute to plant longevity. Curr Biol 26:1385–1394
Cardillo M (1999) Latitude and rates of diversification in birds and butterflies. Proc R Soc B 266:1221–1225
Castellano D, James J, Eyre-Walker A (2018) Nearly neutral evolution across the *Drosophila melanogaster* genome. Mol Biol Evol 35:2685–2694
Caulin AF, Maley CC (2011) Peto's Paradox: evolution's prescription for cancer prevention. Trends Ecol Evol 26:175–182
Charlesworth B (2009) Effective population size and patterns of molecular evolution and variation. Nat Rev Genet 10:195–205
Davies TJ, Savolainen V, Chase MW, Moat J, Barraclough TG (2004) Environmental energy and evolutionary rates in flowering plants. Proc R Soc B 271:2195–2200
Denamur E, Matic I (2006) Evolution of mutation rates in bacteria. Molec Microbiol 60:820–827
dos Reis M, Donoghue PCJ, Yang Z (2014) Neither phylogenomic nor palaeontological data support a Palaeogene origin of placental mammals. Biol Lett 10:20131003
Dowle E, Morgan-Richards M, Trewick S (2013) Molecular evolution and the latitudinal biodiversity gradient. Heredity 110:501–510
Drake J, Charlesworth B, Charlesworth D, Crow J (1998) Rates of spontaneous mutation. Genetics 148:1667–1686
Duchêne D, Bromham L (2013) Rates of molecular evolution and diversification in plants: chloroplast substitution rates correlate with species richness in the Proteaceae. BMC Evol Biol 13:65
Duminil J, Hardy OJ, Petit RJ (2009) Plant traits correlated with generation time directly affect inbreeding depression and mating system and indirectly genetic structure. BMC Evol Biol 9:177
Eo SH, DeWoody JA (2010) Evolutionary rates of mitochondrial genomes correspond to diversification rates and to contemporary species richness in birds and reptiles. Proc R Soc B 277:3587–3592
Eyre-Walker A, Keightley PD (2007) The distribution of fitness effects of new mutations. Nat Rev Genet 8:610–618
Ezard THG, Thomas GH, Purvis A (2013) Inclusion of a near-complete fossil record reveals speciation-related molecular evolution. Methods Ecol Evol 4:745–753
Fernández Zenoff V, Siñeriz F, Farías ME (2006) Diverse responses to UV-B radiation and repair mechanisms of bacteria isolated from high-altitude aquatic environments. Appl Environ Microbiol 72:7857–7863
Firman RC, Simmons LW (2012) Male house mice evolving with post-copulatory sexual selection sire embryos with increased viability. Ecol Lett 15:42–46
Foster CSP, Sauquet H, van der Merwe M, McPherson H, Rossetto M, Ho SYW (2016) Evaluating the impact of genomic data and priors on Bayesian estimates of the angiosperm evolutionary timescale. Syst Biol 66:338–351
Friedberg EC, Wagner R, Radman M (2002) Specialized DNA polymerases, cellular survival, and the genesis of mutations. Science 296:1627–1630
Galtier N, Blier PU, Nabholz B (2009a) Inverse relationship between longevity and evolutionary rate of mitochondrial proteins in mammals and birds. Mitochondrion 9:51–57
Galtier N, Jobson RW, Nabholz B, Glemin S, Blier PU (2009b) Mitochondrial whims: metabolic rate, longevity and the rate of molecular evolution. Biol Lett 5:413–416
Gao J-J, Pan X-R, Hu J, Ma L, Wu J-M, Shao Y-L, Ai S-M, Liu S-Q, Barton SA, Woodruff RC, Zhang Y-P, Fu Y-X (2014) Pattern of mutation rates in the germline of *Drosophila melanogaster* males from a large-scale mutation screening experiment. G3-Genes Genom Genet 4:1503–1514
Gao Z, Wyman MJ, Sella G, Przeworski M (2016) Interpreting the dependence of mutation rates on age and time. PLOS Biol 14:e1002355

Gillman LN, Wright SD (2014) Species richness and evolutionary speed: the influence of temperature, water and area. J Biogeogr 41:39–51

Gillooly JF, Allen AP, West GB, Brown JH (2005) The rate of DNA evolution: Effects of body size and temperature on the molecular clock. Proc Natl Acad Sci USA 102:140–145

Goetting-Minesky MP, Makova KD (2005) Mammalian male mutation bias: impacts of generation time and regional variation in substitution rates. J Mol Evol 63:537–544

Gross CL, Nelson PA, Haddadchi A, Fatemi M (2012) Somatic mutations contribute to genotypic diversity in sterile and fertile populations of the threatened shrub, *Grevillea rhizomatosa* (Proteaceae). Ann Bot 109:331–342

Harcourt AH, Purvis A, Liles L (1995) Sperm competition: mating system, not breeding season, affects testes size of primates. Funct Ecol 9:468–476

Herr AJ, Ogawa M, Lawrence NA, Williams LN, Eggington JM, Singh M, Smith RA, Preston BD (2011a) Mutator suppression and escape from replication error-induced extinction in yeast. PLOS Genet 7: e1002282

Herr AJ, Williams LN, Preston BD (2011b) Antimutator variants of DNA polymerases. Crit Rev Biochem Mol Biol 46:548–570

Hespeels B, Knapen M, Hanot-Mambres D, Heuskin AC, Pineux F, Lucas S, Koszul R, Doninck K (2014) Gateway to genetic exchange? DNA double-strand breaks in the bdelloid rotifer *Adineta vaga* submitted to desiccation. J Evol Biol 27:1334–1345

Holmes EC (2003) Error thresholds and the constraints to RNA virus evolution. Trends Microbiol 11:543–546

Hua X, Bromham L (2017) Darwinism for the genomic age: connecting mutation to diversification. Front Genet 8:12

Hua X, Cowman P, Warren D, Bromham L (2015) Longevity is linked to mitochondrial mutation rates in rockfish: a test using Poisson regression. Mol Biol Evol 32:2633–2645

Hugall AF, Lee MS (2007) The likelihood node density effect and consequences for evolutionary studies of molecular rates. Evolution 61:2293–2307

Ikehata H, Ono T (2011) The mechanisms of UV mutagenesis. J Radiat Res 52:115–125

James JE, Lanfear R, Eyre-Walker A (2016) Molecular evolutionary consequences of island colonization. Genome Biol Evol 8:1876–1888

Jansen MA, Gaba V, Greenberg BM (1998) Higher plants and UV-B radiation: balancing damage, repair and acclimation. Trends Plant Sci 3:131–135

Jiang C, Mithani A, Belfield EJ, Mott R, Hurst LD, Harberd NP (2014) Environmentally responsive genome-wide accumulation of de novo *Arabidopsis thaliana* mutations and epimutations. Genome Res 24:1821–1829

Jinks-Robertson S, Bhagwat AS (2014) Transcription-associated mutagenesis. Annu Rev Genet 48:341–359

Karentz D (2015) Beyond xeroderma pigmentosum: DNA damage and repair in an ecological context. A Tribute to James E. Cleaver. Photochem Photobiol 91:460–474

Kimura M (1983) The neutral theory of molecular evolution. Cambridge University Press, Cambridge

Kozak KH, Wiens JJ (2010) Accelerated rates of climatic-niche evolution underlie rapid species diversification. Ecol Lett 13:1378–1389

Krašovec R, Richards H, Gifford DR, Belavkin RV, Channon A, Aston E, McBain AJ, Knight CG (2018) Opposing effects of final population density and stress on *Escherichia coli* mutation rate. ISME J 12:2981–2987

Kunkel TA (2004) DNA replication fidelity. J Biol Chem 279:16895–16898

Kunkel TA (2009) Evolving views of DNA replication (in)fidelity. Cold Spring Harb Symp Quant Biol 74:91–101

Lanfear R, Thomas JA, Welch JJ, Bromham L (2007) Metabolic rate does not calibrate the molecular clock. Proc Natl Acad Sci USA 104:15388–15393

Lanfear R, Ho SYW, Love D, Bromham L (2010) Mutation rate influences diversification rate in birds. Proc Natl Acad Sci USA 107:20423–20428

Lanfear R, Ho SYW, Davies TJ, Moles AT, Aarssen L, Swenson NG, Warman L, Zanne AE, Allen AP (2013) Taller plants have lower rates of molecular evolution: the rate of mitosis hypothesis. Nat Commun 4:1879

Lanfear R, Kokko H, Eyre-Walker A (2014) Population size and the rate of evolution. Trends Ecol Evol 29:33–41

Lartillot N, Delsuc F (2012) Joint reconstruction of divergence times and life-history evolution in placental mammals using a phylogenetic covariance model. Evolution 66:1773–1787

Lehmann L (2014) Stochastic demography and the neutral substitution rate in class-structured populations. Genetics 197:351–360

Lehtonen J, Lanfear R (2014) Generation time, life history and the substitution rate of neutral mutations. Biol Lett 10:20140801

Lourenço J, Glémin S, Chiari Y, Galtier N (2012) The determinants of the molecular substitution process in turtles. J Evol Biol 26:38–50

Lynch M, Ackerman MS, Gout J-F, Long H, Sung W, Thomas WK, Foster PL (2016) Genetic drift, selection and the evolution of the mutation rate. Nat Rev Genet 17:704–714

MacFadyen EJ, Williamson CE, Grad G, Lowery M, Jeffrey WH, Mitchell DL (2004) Molecular response to climate change: temperature dependence of UV-induced DNA damage and repair in the freshwater crustacean *Daphnia pulicaria*. Glob Chang Biol 10:408–416

Maklakov AA, Immler S (2016) The expensive germline and the evolution of ageing. Curr Biol 26:R577–R586

Martin AP, Palumbi SR (1993) Body size, metabolic rate, generation time and the molecular clock. Proc Natl Acad Sci USA 90:4087–4091

Maslowska KH, Makiela-Dzbenska K, Mo J-Y, Fijalkowska IJ, Schaaper RM (2018) High-accuracy lagging-strand DNA replication mediated by DNA polymerase dissociation. Proc Natl Acad Sci USA 115:4212–4217

Matsuba C, Ostrow DG, Salomon MP, Tolani A, Baer CF (2013) Temperature, stress and spontaneous mutation in *Caenorhabditis briggsae* and *Caenorhabditis elegans*. Biol Lett 9:20120334

Mayr E (1970) Populations, species and evolution. Belknap Press, Cambridge, MA

Mertz T, Harcy V, Roberts S (2017) Risks at the DNA replication fork: effects upon carcinogenesis and tumor heterogeneity. Genes 8:46

Miner BE, Kulling PM, Beer KD, Kerr B (2015) Divergence in DNA photorepair efficiency among genotypes from contrasting UV radiation environments in nature. Mol Ecol 24:6177–6187

Mooers AØ, Harvey PH (1994) Metabolic rate, generation time and the rate of molecular evolution in birds. Mol Phylogenet Evol 3:344–350

Mugal CF, Arndt PF, Holm L, Ellegren H (2015) Evolutionary consequences of DNA methylation on the GC content in vertebrate genomes. G3-Genes Genom Genet 5:441–447

Nabholz B, Glemin S, Galtier N (2008) Strong variations of mitochondrial mutation rate across mammals–the longevity hypothesis. Mol Biol Evol 25:120–130

Nabholz B, Uwimana N, Lartillot N (2013) Reconstructing the phylogenetic history of long-term effective population size and life-history traits using patterns of amino acid replacement in mitochondrial genomes of mammals and birds. Genome Biol Evol 5:1273–1290

Nunney L (1999) Lineage selection and the evolution of multistage carcinogenesis. Proc R Soc B 266:493–498

Nunney L (2018) Size matters: height, cell number and a person's risk of cancer. Proc R Soc B 285:20181743

Nygren K, Strandberg R, Wallberg A, Nabholz B, Gustafsson T, García D, Cano J, Guarro J, Johannesson H (2011) A comprehensive phylogeny of *Neurospora* reveals a link between reproductive mode and molecular evolution in fungi. Mol Phylogenet Evol 59:649–663

Ohta T (1987) Very slightly deleterious mutations and the molecular clock. J Mol Evol 26:1–6

Ohta T (1993) An examination of the generation time effect on molecular evolution. Proc Natl Acad Sci USA 90:10676–10680

Orton MG, May JA, Ly W, Lee DJ, Adamowicz SJ (2018) Is molecular evolution faster in the tropics? Heredity 122:513–524

Pagel M, Venditti C, Meade A (2006) Large punctuational contribution of speciation to evolutionary divergence at the molecular level. Science 314:119–121

Petit RJ, Hampe A (2006) Some evolutionary consequences of being a tree. Annu Rev Ecol Evol Syst 37:187–214

Phillips MJ (2015) Geomolecular dating and the origin of placental mammals. Syst Biol 65:546–557

Qiu F, Kitchen A, Burleigh JG, Miyamoto MM (2014) Scombroid fishes provide novel insights into the trait/rate associations of molecular evolution. J Mol Evol 78:338–348

Ram Y, Hadany L (2012) The evolution of stress-induced hypermutation in asexual populations. Evolution 66:2315–2328

Rozhok A, DeGregori J (2019) Somatic maintenance impacts the evolution of mutation rate. BMC Evol Biol 19:172

Rubin BER, Jones BM, Hunt BG, Kocher SD (2019) Rate variation in the evolution of non-coding DNA associated with social evolution in bees. Philos Trans R Soc B 374:20180247

Sage E, Girard P-M, Francesconi S (2012) Unravelling UVA-induced mutagenesis. Photochem Photobiol Sci 11:74–80

Schmitz J, Moritz RFA (1998) Sociality and the rate of rDNA sequence evolution in wasps (Vespidae) and honeybees (*Apis*). J Mol Evol 47:606–612

Scofield DG, Schultz ST (2006) Mitosis, stature and evolution of plant mating systems: low-Φ and high-Φ plants. Proc R Soc B 273:275–282

Servedio MR, Doorn G, Kopp M, Frame AM, Nosil P (2011) Magic traits in speciation: 'magic' but not rare? Trends Ecol Evol 26:389–397

Sharp NP, Agrawal AF (2012) Evidence for elevated mutation rates in low-quality genotypes. Proc Natl Acad Sci USA 109:6142–6146

Slade D, Radman M (2011) Oxidative stress resistance in *Deinococcus radiodurans*. Microbiol Mol Biol Rev 75:133–191

Smith SA, Donoghue MJ (2008) Rates of molecular evolution are linked to life history in flowering plants. Science 322:86–89

Sniegowski PD, Gerrish PJ, Johnson T, Shaver A (2000) The evolution of mutation rates: separating causes from consequences. BioEssays 22:1057–1066

Svetec N, Cridland JM, Zhao L, Begun DJ (2016) The adaptive significance of natural genetic variation in the DNA damage response of *Drosophila melanogaster*. PLOS Genet 12:e1005869

Tanaka A, Sakamoto A, Ishigaki Y, Nikaido O, Sun G, Hase Y, Shikazono N, Tano S, Watanabe H (2002) An ultraviolet-B-resistant mutant with enhanced DNA repair in Arabidopsis. Plant Physiol 129:64–71

Tang J, Chu G (2002) Xeroderma pigmentosum complementation group E and UV-damaged DNA-binding protein. DNA Repair 1:601–616

Thomas JA, Welch JJ, Lanfear R, Bromham L (2010) A generation time effect on the rate of molecular evolution in invertebrates. Mol Biol Evol 27:1173–1180

Thomas GW, Wang RJ, Puri A, Harris RA, Raveendran M, Hughes D, Murali S, Williams L, Doddapaneni H, Muzny D, Gibbs RA, Abee CR, Galinski MR, Worley KC, Rogers J, Radivojac P, Hahn MW (2018) Reproductive longevity predicts mutation rates in primates. Curr Biol 28:3193–3197

Venditti C, Pagel M (2008) Speciation and bursts of evolution. Evol Educ Outreach 1:274–280

Venditti C, Pagel M (2009) Speciation as an active force in promoting genetic evolution. Trends Ecol Evol 25:14–20

Wang RJ, Ané C, Payseur BA (2013) The evolution of hybrid incompatibilities along a phylogeny. Evolution 67:2905–2922

Warren J (2009) Extra petals in the buttercup (*Ranunculus repens*) provide a quick method to estimate the age of meadows. Ann Bot 104:785–788

Watson JM, Platzer A, Kazda A, Akimcheva S, Valuchova S, Nizhynska V, Nordborg M, Riha K (2016) Germline replications and somatic mutation accumulation are independent of vegetative life span in *Arabidopsis*. Proc Natl Acad Sci USA 113:12226–12231

Welch JJ, Bromham L (2005) Molecular dating when rates vary. Trends Ecol Evol 20:320–327

Welch JJ, Bininda-Emonds ORP, Bromham L (2008) Correlates of substitution rate variation in mammalian protein-coding sequences. BMC Evol Biol 8:53

Weller C, Wu M (2015) A generation-time effect on the rate of molecular evolution in bacteria. Evolution 69:643–652

Whittle C-A, Johnston MO (2002) Male-driven evolution of mitochondrial and chloroplastidial DNA sequences in plants. Mol Biol Evol 19:938–949

Whittle C-A, Johnston MO (2003) Broad-scale analysis contradicts the theory that generation time affects molecular evolutionary rates in plants. J Mol Evol 56:223–233

Wilson Sayres MA, Makova KD (2011) Genome analyses substantiate male mutation bias in many species. BioEssays 33:938–945

Wong A (2014) Covariance between testes size and substitution rates in primates. Mol Biol Evol 31:1432–1436

Wong A, Seguin K (2015) Effects of genotype on rates of substitution during experimental evolution. Evolution 69:1772–1785

Woolfit M, Bromham L (2003) Increased rates of sequence evolution in endosymbiotic bacteria and fungi with small effective population sizes. Mol Biol Evol 20:1545–1555

Woolfit M, Bromham L (2005) Population size and molecular evolution on islands. Proc Biol Sci 272: 2277–2282

Wright S, Ross H, Jeanette Keeling D, McBride P, Gillman L (2011) Thermal energy and the rate of genetic evolution in marine fishes. Evol Ecol 25:525–530

Part II

Molecular Dating

Principles of Molecular Dating

5

Susana Magallón

Abstract

Time-calibrated trees are a fundamental starting point for investigating organismal evolution. The use of molecular sequence data to infer the time of lineage origin and diversification was initially based on the assumption that molecular rates were homogeneous across lineages, giving rise to the original concept of the molecular clock. Evidence of vast rate heterogeneity, even among closely related species, prompted the development of clock models that account for among-lineage rate heterogeneity. In particular, relaxed clocks allow each branch in the phylogenetic tree to have a unique rate. Relaxed clocks include numerical, semiparametric, and parametric methods, the last of these in fully Bayesian implementations. Absolute temporal information is typically used to separate the time and rate components of the branches in a phylogenetic tree. Temporal information can be obtained from fossils, which provide minimum calibration ages; from extrinsic events linked to cladogenesis, which can provide maximum calibration ages; or from age intervals estimated in independent analyses. Importantly, molecular clocks differ in terms of how temporal information is introduced. In node-dating methods, temporal information is used to calibrate internal phylogenetic nodes, and the resulting time-trees typically include only extant taxa. In tip-dating methods, temporal information is provided by fossils that are included in the data matrix. In the resulting time-tree, fossils appear as tips of extinct branches. In the fossilized birth–death process, temporal information is provided by fossils which, together with molecular data from extant species, influence the diversification process that generates the tree prior. In the resulting time-tree, fossils are extinct tips or sampled ancestors of lineages.

Keywords

Branch lengths · Calibrations · Fossilized birth–death process · Fossils · Molecular clock · Node-dating · Nonparametric rate smoothing · Penalized likelihood · Tip-dating

5.1 Introduction

Comparative methods today encompass a body of models and analytical techniques to investigate organismal evolution. These rely on phylogenetic trees that express evolutionary relationships among species as branching structures, in which branch lengths typically represent the product of the rate of character evolution and elapsed time. While phylogenetic relationships are of critical

S. Magallón (✉)
Instituto de Biología, Universidad Nacional Autónoma de México, Ciudad de México, Mexico
e-mail: s.magallon@ib.unam.mx

S. Y. W. Ho (ed.), *The Molecular Evolutionary Clock*, https://doi.org/10.1007/978-3-030-60181-2_5

importance, explicit information about the absolute times of origin and diversification of lineages allows us to place evolutionary events in the context of a global temporal scale, and relative to each other. Therefore, time-trees, which express the temporal component of phylograms, are a fundamental starting point for exploring organismal evolution (Bromham and Penny 2003), including the rate of character change, morphological evolution, biogeographic history, biome assembly, and co-diversification, among many other topics.

In a very general way, the relationship between genetic distance and elapsed time allows us to evaluate the temporal dimension in a phylogenetic tree. Emile Zuckerkandl and Linus Pauling (1962, 1965) relied on the assumption that rates of molecular evolution are constant among lineages and through time, developing the concept of the molecular clock. This assumption was supported by subsequent observations that the genetic differences among pairs of living species are approximately proportional to the time since their lineages diverged, based on the fossil record (Doolittle and Blombäck 1964; Zuckerkandl and Pauling 1965; Morgan 1998). The molecular clock is a model of substitution rate constancy among lineages. It rests on the assumption that the genetic distance between two lineages reflects, with probabilistic regularity, the passage of time since they diverged from their common ancestor. Accordingly, the genetic distance between a pair of species whose divergence can be calibrated with reference to an absolute date, such as a fossil, provides a scale against which the divergence times of other lineages can be estimated.

There are substantial challenges to the use of the molecular clock, including correctly estimating genetic distances and obtaining accurate calibration dates from the fossil record. However, a fundamental obstacle is the absence of molecular rate constancy among lineages (Sanderson 1998), even among closely related species, as documented by vast empirical evidence (see Chap. 1). Many different causes underlie among-lineage molecular rate differences, including intrinsic biological traits and extrinsic environmental or physical factors, such as life form, generation time, metabolic rate, or exposure to UV light (dos Reis et al. 2016; Bromham et al. 2018; Chap. 4).

The original formulation of the molecular clock model (i.e., the strict clock) assumes that molecular rates are constant among lineages and through time. As such, all branches in the phylogenetic tree can be characterized by a single rate. While this model might be appropriate among intraspecific populations, it is unrealistic when considering different species because heterogeneous molecular rates are pervasive. It has been recognized that rate constancy among lineages is not a general feature of molecular evolution, but exceptional, and that variable among-lineage molecular rates are ubiquitous. This led to the proposal of different molecular clock models to accommodate rate heterogeneity.

Different molecular clock models account for different ways in which molecular rate heterogeneity is distributed among lineages. These models roughly correspond to two distinct modes: those that allow few but usually large changes, and those that allow many different rates that vary relatively little from each other (Welch and Bromham 2005). In the former type, the number of distinct molecular rates is usually much smaller than the number of branches in the tree. Ho and Duchêne (2014) and Bromham et al. (2018) refer to these models as 'multi-rate' clocks. Branches with the same rate can be closely grouped (local clocks) or scattered across the phylogenetic tree (discrete clocks; Ho and Duchêne 2014). In multi-rate clock models, an important concern is identifying the number of distinct rates in a tree, and assigning these rates to the branches of the tree.

The second type corresponds to relaxed-clock models, which assign a unique substitution rate to each branch in the phylogenetic tree. Rates can be modelled following the principle of temporal autocorrelation (Gillespie 1991; see below), which considers that attributes that determine molecular rates are transmitted from ancestral to descendent lineages and, consequently, that rates among closely related lineages are similar. This principle might be appropriate for molecular change among closely related species (Ho 2009). However, it is not clear if it adequately reflects molecular rate differences among lineages that diverged in the distant past, where substantial extinction has taken place, or when taxon sampling is incomplete. In uncorrelated relaxed-clock models, rates are sampled from a statistical distribution and are independent of those on adjacent branches (Drummond et al. 2006; Rannala and Yang 2007).

5.2 Sources of Absolute Temporal Calibrations

The absence of constant rates among interspecific lineages is an empirical reality that precludes the use of a strict molecular clock. However, even when applying variable-rates clock models, a major difficulty is that absolute substitution rates and time, which determine the lengths of branches in phylogenetic trees, are not identifiable (Fig. 5.1; Rannala 2016). The algorithms used in parametric phylogeny estimation (Felsenstein 1981) produce trees in which branch lengths are the product of the absolute substitution rate (i.e., number of substitutions per unit time) and temporal duration: such trees are known as phylograms. Because for any given branch there is an infinite number of combinations of absolute rates and times that can yield the same length, and leading to the same likelihood score, explicit information about the magnitude of one of these components is necessary to estimate the other. In the absence of absolute information about rates or times to anchor the phylogram, estimated divergence times cannot be linked to an absolute timescale. The most common sources of information about absolute time are the fossil record, extrinsic events at local or global scales assumed to be causally linked to speciation (e.g., tectonics, orography, or temperature shifts), and secondary calibrations (Fig. 5.2; Hipsley and Müller 2014).

5.2.1 The Fossil Record

Fossils are the most common source of independent information for temporal calibration of phylogenetic trees (see Chap. 8). A fossil provides a minimum age for the phylogenetic splitting event that gave rise to the lineage to which it belongs. This is because a certain amount of time elapsed before this lineage acquired morphological distinctiveness—which allows us to recognize it—and became preserved in the fossil record (Fig. 5.2a). The magnitude of this temporal difference is unknown; it varies from lineage to lineage and depends on the rate of morphological evolution and on lineage-specific rates of fossil preservation.

To be used as a calibration, the phylogenetic position of a fossil with respect to extant taxa included in the dating analysis needs to be identified. The placement of the fossil can be estimated simultaneously with divergence times in a total-evidence context, which involves combining molecular and morphological data from extant taxa with morphological data from fossil species. The minimum age provided by the fossil can be implemented with a hard or a soft bound (Yang and Rannala 2006), or as a feature of a probability density. In the latter case, selecting the type of distribution and the values of its parameters are important choices. Useful considerations for selecting the type of probability distribution are available (e.g., Ho and Phillips 2009; Bromham et al. 2018), but providing values for its parameters is mostly a grey area in need of further theoretical and empirical research (Matschiner et al. 2017). For a more detailed treatment of this topic, see Chap. 8.

5.2.2 Extrinsic Events

Extrinsic events at local or global scales that are causally linked with phylogenetic divergences can provide a maximal age or a time estimate for an internal node in the phylogeny (Fig. 5.2b; see also Chap. 9). These events include, for example, tectonic and orographic processes at global or

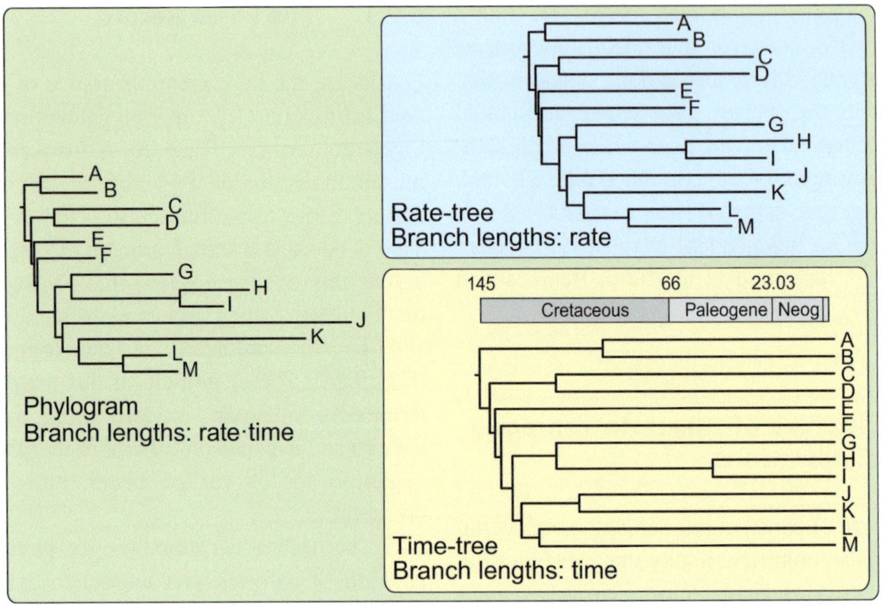

Fig. 5.1 The time and rate components of a phylogram. A phylogram is a phylogenetic tree in which branch lengths represent the product of the molecular substitution rate and time duration. Molecular dating uses numerical, semiparametric, or fully parametric methods, combined with independent information about molecular rates or, more commonly, time calibrations, to separate the two components of branch lengths. Molecular dating produces a time-tree (or chronogram) in which branch lengths represent absolute time units (e.g., million years). The absolute molecular substitution rate (number of substitutions per site per time unit) on each branch is also obtained, which can be used to build a rate-tree. Nevertheless, molecular rates are usually not recorded this way

local scales, as well as events derived from these extrinsic processes, such as climate change, origin of biomes, and major geochemical changes in Earth's history (Falcón et al. 2010; Knoll and Nowak 2017). The critical assumption is that the event was the underlying cause of cladogenesis, which is in itself a hypothesis to be tested (Magallón 2004; Kodandaramaiah 2011). The temporal relationship between the calibrated node and the extrinsic event is imprecise, because geologic, tectonic, orographic, or climatic processes usually take place over millions of years; and lineage splitting became effectively established at an unknown time after the onset of the event (Magallón 2004). Extrinsic global or local events can thus provide temporal information, but rely on a tentative causality hypothesis which, if correct, provides a soft maximal age constraint for a phylogenetic node (Fig. 5.2b). For a more detailed treatment of this topic, see Chap. 9.

5.2.3 Secondary Calibrations

Ages estimated in independent molecular clock analyses can be used as node calibrations when the fossil record of a group is of low quality or is nonexistent, and when other sources of calibration are unavailable (Fig. 5.2c). There are numerous caveats to the use as calibrations of dates derived from independent studies. Each step in a molecular dating exercise is laden with assumptions that might be tenuous and potentially derived from dubious choices, and the resulting inaccuracies and biases will be transferred to the new study. However, secondary calibrations might be the only available source of independent temporal information. In these cases, an important consideration is whether a tree dated in absolute time is strictly necessary; a tree scaled in relative time units (e.g., with internal splits given dates relative to the root) might be sufficient for addressing the question driving the research.

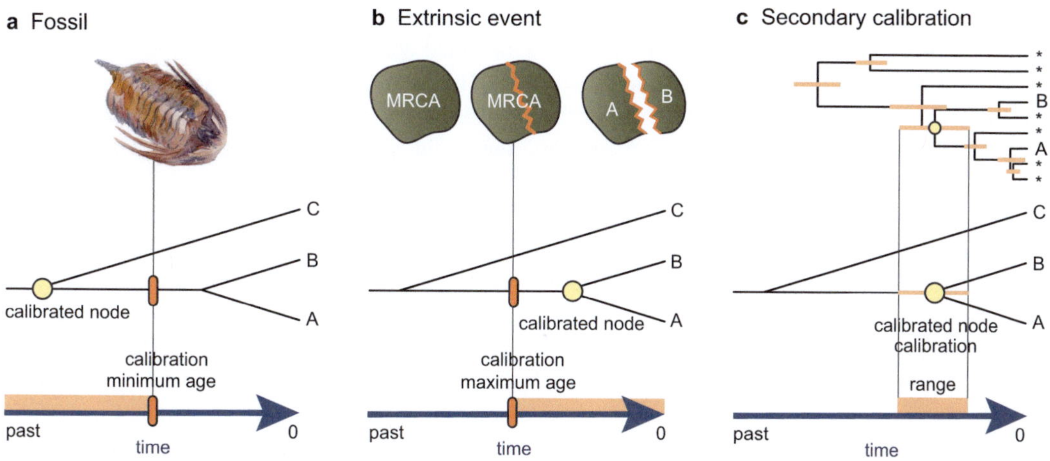

Fig. 5.2 Three main sources of absolute temporal calibrations for molecular dating. (**a**) Fossil calibration. A fossil reliably identified as belonging to an extant clade, such as the clade that includes species A and B, can be used to calibrate (orange bar) the stem node of that clade (yellow circle). Because an unknown amount of time elapsed between the divergence of A + B from C and the acquisition of morphological distinctiveness and fossil preservation, a fossil belonging to clade A + B provides a minimum age for the origin of this clade. The age estimate for the divergence between A + B and C (orange interval) can be any time older than the fossil. (**b**) Calibration with an extrinsic event. Extrinsic global or local events can be used to calibrate an internal node in the phylogeny if causally linked to cladogenesis. The underlying assumption is that an extrinsic event, such as continental breakup due to plate tectonics, separated an ancestral species (MRCA) whose descendants evolved into species A and B on different land masses. The calibration (orange bar) represents the maximum age of the calibrated node (yellow circle), which represents the phylogenetic divergence between A and B. The estimated age for the divergence between A and B can be any time younger than the calibration (orange interval). (**c**) Calibration from an independent molecular clock analysis. A node, for example the divergence between A and B (yellow circle), can be calibrated with the age estimated for the equivalent node in an independent molecular clock analysis. A secondary calibration involves the strong assumption that the independent study obtained an accurate estimate of the age of the node, and should be treated critically and cautiously. The calibration should encompass the full uncertainty interval associated with the estimated age (orange interval)

If the need for a tree scaled to absolute time is inescapable and other sources of temporal information are unavailable, a secondary calibration might be applied tentatively, bearing in mind that any potential error in the primary analysis will be transferred to the new calibration. Implementing a secondary calibration should be done carefully and, at the very least, should involve a rigorous evaluation and consideration of all sources of uncertainty in the primary dating analysis. This should include a critical examination of the choices of molecular data, taxonomic sample, source and reliability of temporal calibrations, models, and how each of these was applied. Secondary calibrations should be applied as an interval that encompasses the associated error estimated in the original study (Fig. 5.2c).

5.3 Types of Relaxed-Clock Methods

Many important decisions are required when conducting a relaxed-clock analysis to estimate divergence times and rates (Sauquet 2013). These include, for example, the choice and density of taxon sampling for the ingroup and the outgroup, the type of molecular data and its phylogenetic informativeness relative to the sampled taxa, whether to include morphological data, and, in the case of parametric relaxed molecular clocks, the models and model parameters for different components of the likelihood. In the case of relaxed clocks, two critical choices need to be made: the type of relaxed clock to be implemented, and the way in which absolute temporal information to separate rates and times will be incorporated into the analysis. These critical choices are closely intertwined, conceptually and methodologically, and should be primarily determined by the research question. Available relaxed-clock methods differ substantially in the input data that they require, in the type and mode of implementation of external calibrations, and in their underlying mathematical machinery. However, these methods share the ability to estimate divergence times while allowing each branch in the tree to have its own molecular rate.

5.3.1 Nonparametric Rate-Smoothing Numeric Relaxed Clock

The earliest implementation of a molecular clock that directly incorporates rate differences among lineages is the nonparametric rate smoothing (NPRS) method developed by Sanderson (1997). NPRS is an entirely numerical method that relies on minimization of substitution rate differences between ancestor and descendent lineages in a phylogenetic tree. Each branch is allowed to have its own substitution rate, under the assumption of temporal rate autocorrelation (Gillespie 1991). The principle of rate autocorrelation is based on population-level processes that assume that molecular rates (or the factors underlying them) are inherited by descendants from their ancestors. The assumption of rate autocorrelation imposes constraints on the amount of rate change between ancestral and descendent branches and, effectively, on the difference in molecular rates among nearby branches in the tree.

Nonparametric rate smoothing uses as data the branch lengths in an independently estimated phylogram, where branch lengths represent expected number of substitutions (absolute rate r multiplied by time t). To obtain divergence-age estimates in terms of absolute time, temporal calibrations on internal nodes are necessary. These calibrations are frequently obtained from fossils but can also be obtained from other sources (Fig. 5.2, and see above). Each node can be constrained to a minimal and/or a maximal age.

The absolute substitution rate on each branch is calculated as $r = b/t$ (where b is the branch length in substitutions per site), which is the maximum-likelihood estimate of the rate in a Poisson process (Sanderson 1997). To account for temporal autocorrelation, the method seeks branch durations (t) that minimize differences in rates (r) between ancestral and descendent branches, contingent on node calibrations. Minimization is solved as a non-linear optimization problem approached through standard numerical techniques (Sanderson 1997). As a result, estimates of divergence times and absolute molecular rates are obtained. Errors associated with estimates of time and absolute rate are obtained by conducting the analysis on phylograms derived from bootstrapped data sets.

The availability of NPRS led to substantial enthusiasm for estimating divergence times on trees for different groups of organisms (e.g., Wikström et al. 2001; Smith et al. 2006). Initial tests showed that age estimation with NPRS outperformed a strict-clock model, especially when rates were nonclocklike and autocorrelated (Sanderson 1997). However, it was later discovered that NPRS tends to introduce excessive rate variation, losing predictive power (Sanderson 2002). Its use has been mostly discontinued in favour of semiparametric or fully parametric options, particularly methods that do not rely on a fixed tree topology.

5.3.2 Penalized-Likelihood Semiparametric Relaxed Clock

A penalized-likelihood relaxed molecular clock was later introduced by Sanderson (2002). It is a semiparametric rate-smoothing method that combines a parametric model that allows each phylogenetic branch to have its own molecular rate. A numerical roughness penalty deters rates from varying excessively among nearby branches. The penalty is based on the level of smoothing applied to the data, which is determined through a cross-validation procedure. As in NPRS, penalized likelihood uses as input data a phylogram in which branch lengths represent expected number of substitutions.

The parametric component of penalized likelihood is a saturated model that assigns a unique molecular rate to each branch in the tree. Instead of using a conventional molecular model to estimate the number of substitutions along a branch, penalized likelihood uses a simpler method in which the number of substitutions is an observation drawn from a Poisson process (Sanderson 2002). The choice of rates is regulated by a roughness penalty (Φ) that avoids drastic changes by penalizing squared rate differences between ancestral and descendent branches. It is equivalent to the rate-smoothing parameter in NPRS, which is based on the principle of temporal rate autocorrelation (Sanderson 2002).

The roughness penalty should adequately reflect rate differences in the data. This is achieved through a smoothing parameter (λ), which can take any value between zero and infinity. When the smoothing parameter is zero, each branch is allowed to have its own rate, and differences between neighbouring branches are not penalized. As the magnitude of the smoothing parameter increases, rate changes between branches are increasingly constrained. When the value of the smoothing parameter is very large, all branches have the same rate, in what constitutes a strict molecular clock.

Each value of the smoothing parameter entails its own solution in terms of absolute rates and estimated divergence times. Therefore, it is critical that the smoothing parameter adequately represents the amount of rate variation in the tree. The magnitude of the smoothing parameter is selected through a cross-validation procedure related to prediction error. The cross-validation procedure starts by removing one terminal branch of the phylogenetic tree. Penalized likelihood and a selected smoothing magnitude are applied to the data of the remaining branches to predict the length of the pruned branch. The squared difference between the predicted and the observed values is calculated, weighted by the inverse of the variance, to reflect the mean of the Poisson process (Sanderson 2002). This step is repeated by pruning each of the terminal branches in turn, and the average of the prediction errors represents the cross-validation score for the smoothing magnitude being tested. The cross-validation is repeated for a range of magnitudes of the smoothing parameter, for example from 0.1 to 100,000. The smoothing magnitude that returns the lowest cross-validation score is chosen as the optimal one.

The roughness penalty increases as rates are more variable across branches. Age constraints can be placed on internal nodes as minimum and/or maximum bounds. The smoothing parameter balances the amount of rate variation in the data identified through cross-validation with goodness-of-fit to the saturated parametric model. Absolute rates and divergence times are estimated by maximizing the penalized likelihood (Ψ), which combines the roughness penalty and the smoothing parameter. The solution consists of the set of branch rates and divergence times that maximize the penalized likelihood for the observed branch lengths in the input phylogram, given the chosen smoothing parameter. Associated errors can be obtained by conducting penalized-likelihood analyses on trees derived from bootstrapped data sets.

The penalized-likelihood method was rewritten in the program treePL to handle very large data sets (Smith and O'Meara 2012). Although its reliance on a fixed phylogram as input is a drawback, this can be compensated for by conducting penalized-likelihood analyses on phylograms derived from bootstrapped data sets. Penalized likelihood has proven to be a robust method for estimating divergence times and rates under a variety of conditions, and remains a useful tool for dealing with very large data sets, both under its r8s and treePL implementations (e.g., O'Meara et al. 2016).

5.3.3 Parametric (Bayesian) Relaxed Clocks

Parametric relaxed molecular clocks, specifically Bayesian relaxed clocks (dos Reis et al. 2016), have revolutionized estimation of absolute branch rates and divergence times (see Chap. 6). They can allow the simultaneous estimation of phylogenetic, substitution, and diversification parameters, while taking advantage of vast amounts of available molecular sequence and genomic data for all types of organisms.

Bayesian relaxed clocks have been implemented in the general Bayesian framework, in which the posterior probability is the product of the prior probability of model parameters and the likelihood of the data, divided by the probability of the data (Nascimento et al. 2017). In the context of relaxed molecular clocks, the data are the aligned molecular sequences (and morphological characters) of the sampled taxa. The likelihood involves two components. One is the model of character evolution, typically a conventional model of nucleotide (or amino acid or codon) substitutions, but might also include a model of morphological character evolution (Lewis 2001). Its parameters include, at least, the rates of transitions among character states and state frequencies. The second component of the likelihood is the tree model, in which the relevant parameters are the topology and the branch lengths. Branch lengths represent the product of the absolute character substitution rate and the branch duration.

The prior probability of each parameter is described by a statistical distribution or a process governed by hyperparameters. Branch durations (and node ages) are potentially influenced by the prior on the tree, which might be based on a birth–death diversification process or one of its variants (e.g., Gernhard 2008), and independent temporal information, which can be introduced in different ways into the analysis (see below; Fig. 5.3). The posterior distribution represents the prior distribution updated with the likelihood calculated from the data. The posterior distribution is usually estimated by Markov chain Monte Carlo sampling, which involves repeatedly calculating the ratio of two posterior probabilities (see Chap. 6). We can extract the marginal posterior probability distributions of any parameters of interest.

5.4 Incorporating Temporal Information

Absolute time calibrations are among the most consequential components of molecular dating analyses. One of the earliest (albeit sometimes implicit) choices in a molecular clock analysis is how to introduce temporal information to separate absolute rates and time in the branches of the phylogeny. Different ways in which temporal information is introduced determine a major categorization of molecular dating methods. Temporal calibrations for internal phylogenetic nodes can be obtained from fossils, extrinsic events, or dates estimated in an independent analysis. Terminal nodes can be calibrated with known sampling dates in data sets from viruses, bacteria, or obtained from ancient DNA (Rambaut 2000; Chap. 10).

More recent methods allow extant species to be combined with fossil taxa in the taxonomic sample for molecular dating (e.g., Pyron 2011; Ronquist et al. 2012a). Diversification models that sample species through time (Stadler 2010; Didier et al. 2012) can combine fossils with extant species in the tree prior (Heath et al. 2014). Given these alternatives, one critical step at the outset of a molecular clock analysis is deciding how absolute temporal information will be introduced. This decision is closely associated with (or depends on) the type of available data, the source of temporal information, and the dating method to be implemented.

5.4.1 Node Dating: Calibrating Internal Nodes

Calibrating internal nodes in a phylogenetic tree (node dating) is the most common way to incorporate absolute chronological information into molecular clock analyses. In node-dating analyses, divergence-time estimation is independent of the type and source of calibrations. The dates provided by calibrations only inform about the age of an internal node (as a minimum value, maximum value, or a probability density), but are

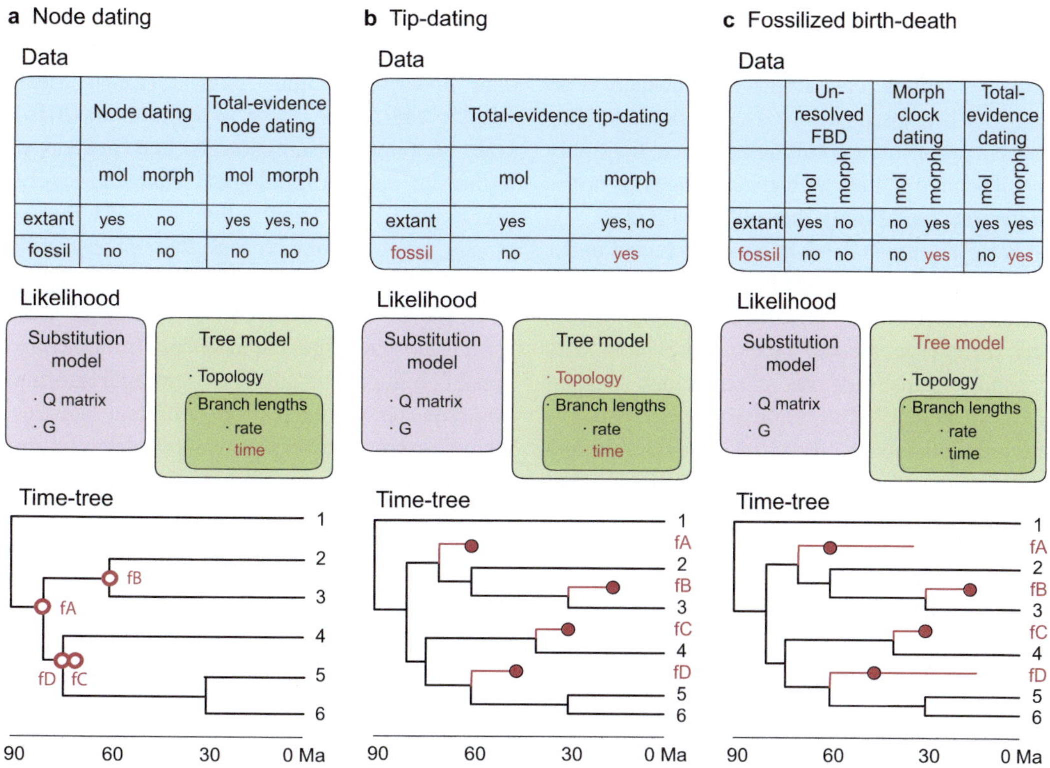

Fig. 5.3 Different ways of incorporating temporal information from fossils in molecular dating analyses. The role of fossils (fA, fB, fC, and fD) is indicated in red. Filled circles indicate the presence of fossils in the tree; empty circles indicate that the age of the fossil informs the age of the node, but the fossil itself is not present in the tree. (**a**) Node dating does not include fossils in the data matrix. Molecular data (node dating) or molecular and morphological data (total-evidence node dating) from extant species can be included in the data matrix. Based on independently inferred relationships, fossils directly inform the age of internal nodes (empty red circles) but do not appear in the resulting time-tree. Fossils may be redundant if two or more calibrate the same node. In addition to fossils, internal nodes can be calibrated with extrinsic events or secondary calibrations. (**b**) Tip-dating includes fossils in the data matrix. Morphological data are scored for fossil and extant species, together with molecular data from extant species (total-evidence tip-dating). Phylogenetic relationships and divergence times are estimated simultaneously, so fossils contribute to inference of topology and to the time component of branch lengths. Fossils (red circles) appear as tips of extinct branches in the estimated time-tree. (**c**) The fossilized birth–death process includes fossils in the data matrix. Input data can be: molecular sequences from extant species only (unresolved FBD); morphological data from both extant and fossil species (morphological clock dating); or molecular data from extant species and morphological data from both extant and fossil species (total-evidence dating). If morphological data are included, phylogenetic relationships and divergence times can be estimated simultaneously. If not, the placement of fossils needs to be specified based on independently inferred relationships. Fossil and extant taxa are part of the diversification process that generates the phylogenetic tree and its branching times. Fossils (red circles) appear as tips of extinct branches (fB and fC) or as sampled ancestors of lineages that became extinct some time after the fossil was sampled (fA and fD)

somewhat detached from the mechanisms of estimating phylogenetic relationships, diversification parameters, and other components of the analysis (Fig. 5.3). The specified calibrations form the basis of age estimation but they do not play a direct role in estimating other model parameters. Except for analyses of time-structured data sets (see Chap. 10), the product of a node-dating analysis is a time-tree that includes only extant terminals. Phylogenetic relationships among extant taxa can be simultaneously estimated in some Bayesian implementations (e.g., BEAST; Bouckaert et al. 2014). When fossils are used as calibrations, they do not appear in the resulting time-tree: their only manifestation is their influence on ages of nodes in the tree, particularly the calibrated nodes.

Node-dating analyses can be conducted under a wide variety of methodological and analytical conditions, including numerical, semiparametric, and parametric relaxed clocks. The input data depend on the type of relaxed-clock method used. In numerical or semiparametric methods, the required input consists of a phylogram and node calibrations, which are indicated as point values or as minimum and/or maximum ages. Because ages are optimized on the basis of phylogram branch lengths (Sanderson 1997, 2002), a character matrix is not required. The phylogram is expected to be derived from molecular data, but it could also be derived from a total-evidence data set. In parametric molecular clocks, a data matrix represents the fundamental input for estimating divergence times and other parameters of the model. In Bayesian node dating, the input data are a character matrix comprising molecular data (for extant taxa) and possibly morphological data for extant and fossil species (total-evidence node dating; Fig. 5.3a). Some implementations require a fixed tree topology, so that the estimates of divergence times are conditioned on a specified set of evolutionary relationships.

The taxonomic sample in the data matrix for a node-dating analysis usually includes only extant taxa, but taxa that recently went extinct (e.g., Tasmanian tiger and woolly mammoth) can also be included, provided that their ages are much younger than the divergence times in the phylogeny. In addition to the species of interest, or a substantial representation of them, the taxonomic selection might need to be expanded to encompass nodes that can be reliably calibrated, for example, by a particularly well phylogenetically placed and confidently dated fossil. This might lead to a denser representation in the ingroup, or expansion of sampling to include more external outgroups until a node that can be reliably calibrated is represented in the phylogenetic tree. The data matrix typically contains only molecular data, but it is possible to include morphological data under appropriate models (Lewis 2001; Wright and Hillis 2014; Wright et al. 2016). One important assumption is that models of molecular and morphological character evolution appropriately capture the rate and variation of the evolutionary process that gave rise to the data.

Because in node-dating analyses the calibrations are largely independent from the process of phylogenetic inference, temporal information can be obtained from different sources, including non-biological ones (e.g., from extrinsic large-scale events or from independent molecular clock analyses; see above and Fig. 5.2). The maximum number of calibrations that can be usefully implemented in a node-dating analysis is equal to the number of internal nodes in the tree. If several sources of calibration are available for a particular node, such as multiple fossils, the oldest will determine the minimum age of the node and the rest will become uninformative (Fig. 5.3a), unless they are used to inform the shape or parameters of a calibration prior distribution. Furthermore, older calibrations will supersede any younger calibrations that are applied to deeper nodes (i.e., closer to the root), rendering them irrelevant because of the tree structure. However, it is possible to combine different calibrating sources for different internal nodes, for example, assigning to the root node a secondary calibration and calibrating one or more internal nodes with fossil-derived information. Importantly, identifying the correct calibration nodes and ages can involve considerable uncertainty.

5.4.2 Tip-Dating: Ages of Phylogenetic Terminals

Extant and fossil taxa can be combined for the joint estimation of phylogenetic relationships and divergence times in a tip-dating analysis. Such an analysis is based on a total-evidence data set that contains molecular sequences from extant species and morphological characters from both fossil and extant species (Fig. 5.3b). It is critically important that scoring of morphological characters is as complete as possible for the sampled species, because these data represent the source of information used to estimate the phylogenetic relationships of the fossil species, and will affect the estimated divergence times. Fossils occupy terminal positions on extinct phylogenetic branches and their ages serve as calibrations for estimating divergence times in the tree. The outcome of a tip-dating analysis is a time-calibrated phylogenetic tree that includes extant and fossil species as terminals.

In tip-dating analyses, calibrations are only provided by fossils (but see below), because they are directly included in the phylogenetic tree. Inference of the phylogeny is an integral part of the analysis, so the placement of fossils with respect to extant species does not need to be specified a priori. The age of each fossil can be introduced as a fixed point, or preferably as a uniform interval (e.g., spanning the stratigraphic interval of the fossil species), a probability density, or a distribution that reflects the uncertainty in radioisotopic dating (Ho and Duchêne 2014). The number of fossils that can be usefully included in a tip-dating analysis can be smaller than, equal to, or larger than the number of internal branches in the tree. The fossils do not need to be limited to the oldest members of each clade; because fossils are tips in the tree, fossils that are younger than the oldest member of a clade can play a useful role in divergence-time estimation. Although it is desirable to include as many fossils as possible, because they will provide greater temporal information for dating the tree, their usefulness is contingent on the availability of phylogenetically informative morphological data.

Tip-dating analyses can be performed in Bayesian frameworks that allow simultaneous inference of phylogenetic relationships and divergence times (e.g., Ronquist et al. 2012b; Bouckaert et al. 2014). In numerical and semiparametric relaxed clocks, tip-dating is possible if the input phylogenetic tree includes extant and fossil species and, importantly, in which branch lengths represent expected number of character state changes. It is technically possible to conduct a tip-dating analysis using numerical or semiparametric relaxed clocks on a phylogram with extant and fossil species in which branch lengths represent evolutionary distances and fossil species are specified as non-extant terminals (Magallón 2010). However, these analyses differ importantly from conventional parametric tip-dating in that the estimation of phylogeny and divergence times are uncoupled from each other.

The most attractive element of tip-dating analyses is the possibility to infer simultaneously phylogenetic relationships and divergence times among extant species and fossils in a total-evidence context. The estimation of phylogenetic relationships among fossils and living taxa frees the need to specify a priori the position of calibrations in the tree, which can be problematic. Furthermore, the requirement of a total-evidence data matrix as input should promote documentation, justification, and discussion of morphological characters (Sauquet and Magallón 2018). One question is the interpretation of the position of fossils as terminals in extinct lineages, namely, whether the fossil represents the time of extinction of the lineage. Most tip-dating methods assume that a lineage does not persist beyond the sampling of a fossil.

Tip-dating analyses, despite their appeal, have often produced unexpected results when applied to empirical data sets. In some studies, fossils have been placed as stem-lineage representatives (i.e., diverged after the separation of the lineage from its extant sister clade, but before the diversification of the crown group of extant species (e.g., Pyron 2011), or in unresolved phylogenetic positions (e.g., Ronquist et al. 2012a). It is not clear if the frequent assignment of fossils to stem lineages correctly reflects their relationships with respect to extant clades, or whether it is a consequence of incompletely documented and/or incompletely scored morphological data. In other studies, estimated ages have been found to be unrealistically old (e.g., Arcila et al. 2015 and references therein).

The exceedingly old ages estimated in tip-dating analyses might be due to the use of models that insufficiently capture the complexity of morphological evolution, including possible correlated evolution and ascertainment bias (dos Reis et al. 2016). It has also been suggested that, because tip-dating does not necessarily place any temporal constraints on internal nodes, age estimation might become extremely sensitive to the tree-generating model used for the prior (dos Reis et al. 2016).

A potentially useful methodological alternative would be to combine a tip-dating analysis including extant species and fossils, with node calibrations derived from fossils that are not included in the data matrix. This presents a means of mitigating the unreasonably old ages estimated in some tip-dating analyses. In a study of the evolution of floral structure at the onset of the diversification of the Pentapetalae clade (eudicot angiosperms; López-Martínez et al. in prep.), the availability of complete and informative morphological characters for fossils and extant species, combined with a strong maximal age constraint on the root node, allowed an improved estimation of the phylogenetic placement of fossils. The analysis also produced age estimates that fell within the general understanding of the evolutionary timeframe of the group being examined.

5.4.3 Fossilized Birth–Death Process: Absolute Time Is a Component of the Tree-Generating Diversification Process

Divergence-time estimation with the fossilized birth-death (FBD) process (Heath et al. 2014) is based on a tree-generating birth–death diversification model, which includes sampling of extant species and fossils (Stadler 2010). In contrast with other molecular dating methods, temporal calibrations in the FBD process do not need to be attached directly to the phylogeny (as in node dating) or incorporated into the data matrix (as in tip-dating), but play a role in the diversification process that generates the tree prior for the topology and divergence times (Fig. 5.3c). The diversification process incorporates extant species for which molecular data are available, and extinct species sampled as fossils of known age, within the same birth–death model (Stadler 2010; Didier et al. 2012; Chap. 11).

The FBD process has been implemented in Bayesian phylogenetic dating methods (e.g., Bouckaert et al. 2014; Höhna et al. 2016). The method requires an input data matrix containing molecular data from extant species and a list of dated fossil occurrences (unresolved FBD). The data matrix can include morphological data from extant and fossil species, either alone (morphological clock dating; Chap. 7) or in addition to the molecular data from extant species (total-evidence dating) (Gavryushkina et al. 2017). The age of each fossil needs to be specified as a point age, or preferably as a temporal interval (Barido-Sottani et al. 2018) that accounts for uncertainty in the age of the fossil. Incomplete sampling of extant clades can be accommodated with an appropriate correction (Zhang et al. 2016).

Absolute temporal information is provided by fossils in the tree-generating birth–death process. The number of fossils that can be meaningfully introduced can be smaller than, equal to, or larger than the number of internal branches in the phylogenetic tree. There can potentially be a very large number of fossils, limited only by the ability to specify their phylogenetic position, or by the informativeness of morphological data (if available) to provide phylogenetic resolution. Fossils can be crown-group members (i.e., being more closely related to a particular set of extant species than to the clade as a whole) or stem-lineage representatives of the extant clade. Studies have suggested that including stem-lineage fossils leads to greater accuracy in estimated divergence times (Gavryushkina et al. 2017).

The outcome of molecular dating with the FBD process is a time-tree containing extant and fossil taxa. Phylogeny estimation is implicitly included in the FBD but, in practice, there is a range of options for specifying phylogenetic relationships. In FBD analyses that include only molecular data, relationships among extant species are estimated, whereas the phylogenetic placements of fossils must be explicitly specified (unresolved FBD). The precision of this specification can range from a broad indication of clade membership to a strict placement of the position of the fossil with respect to other species in the tree. Including morphological data for extant and fossil species, either alone (morphological clock dating) or combined with molecular data (total-evidence dating), allows FBD analyses in which phylogenetic relationships and divergence times are simultaneously estimated.

Estimation of divergence times with the FBD model implies calculation of diversification parameters: the speciation rate λ and extinction rate μ (or net diversification d and turnover ν), the fossil recovery rate ψ (or fossil sampling proportion s), and the sampling proportion of extant species ρ (Heath et al. 2014; Gavryushkina et al. 2017). When every sampled ancestor can come from a different point in time, birth–death models are nonidentifiable. Hence, at least one of these parameters needs to be specified in order to allow the others to be estimated (Gavryushkina et al. 2017).

One distinctive element of the FBD process is its ability to interpret fossils either as terminals of extinct lineages or as ancestors on branches of the tree. These phylogenetic branches might lead to extant species or become extinct some time after the sampling of the fossil (Gavryushkina et al. 2014, 2017). This allows the presence of a fossil to be unlinked from the extinction of the branch to which it belongs.

The most notable difference between the fossilized birth–death process and node-dating and tip-dating is the way in which temporal information from fossils is used. In node-dating or tip-dating methods, this temporal information is used for estimating parameters of the tree model, either only time (in node dating) or both time and topology (in tip dating). In tip-dating, fossils are interpreted only as extinct terminals in the tree. In the FBD process, the temporal information provided by fossils is directly involved in the diversification process that generates the tree prior. The fossils can be resolved as extinct terminals or as direct ancestors.

5.5 Conclusions

Today, molecular clocks are standard components in the bioinformatic toolkit of comparative phylogenetics. Although molecular dating is now mainstream, there are questions and important challenges to be resolved. The availability of genomic data represents an extraordinary asset in phylogenetic inference, broadening our understanding of relationships at all scales of the tree of life. Yet, the use of genomic data in molecular dating is not without problems. The availability of vast amounts of molecular sequence data can improve phylogram branch-length estimates, but does not provide additional information about the temporal component in those branch lengths (e.g., Yang and Rannala 2006; dos Reis and Yang 2013; Chap. 13). More specifically, simulation studies have shown that when the amount of molecular data is increased (Yang and Rannala 2006; dos Reis and Yang 2013; Zhu et al. 2015), posterior estimates of divergence times do not converge on point values, as would be expected in conventional Bayesian estimation. Instead, there are lingering uncertainties that are due to the uncertainty in the calibrations (Inoue et al. 2010; dos Reis et al. 2012).

Other empirical studies have shown that genome-scale data sets do not inherently improve age estimates and, furthermore, that smaller amounts of sequence data can provide comparable results (Foster et al. 2017). Increased computational demand, along with correct choice and assignment of substitution models and clock models, become more critical when applied to genome-scale data sets. Comparable efforts towards increasing the number of informative calibrations might yield more substantial returns.

Currently available relaxed molecular clocks are powerful tools for estimating divergence times and absolute molecular rates in phylogenetic trees. Through the development of molecular clocks, some of the most important advances have involved the use of increasingly realistic models of among-lineage molecular rate heterogeneity. Nevertheless, some prominent molecular dating analyses have estimated ages that seem unexpectedly old, especially in comparison with first appearances in the fossil record. These discrepancies are at least partly due to the use of models that insufficiently account for the extent and drastic pattern of among-lineage molecular rate variation (e.g., Dornburg et al. 2012; Wertheim et al. 2012; Beaulieu et al. 2015). Improvements in modelling rate variation among lineages should be a prime area of research for the further development of molecular clock models and methods.

References

Arcila D, Pyron RA, Tyler JC, Ortí G, Betancur-R R (2015) An evaluation of fossil tip-dating versus node-age calibrations in tetraodontiform fishes (Teleostei: Percomorphaceae). Mol Phylogenet Evol 82:131–145

Barido-Sottani J, Aguirre-Fernández G, Hopkins MJ, Stadler T, Warnock RC (2018) Ignoring stratigraphic age uncertainty leads to erroneous estimates of species divergence times under the fossilized birth-death process. Proc R Soc B 286:20190685

Beaulieu JM, O'Meara BC, Crane P, Donoghue MJ (2015) Heterogeneous rates of molecular evolution and diversification could explain the Triassic age estimate for angiosperms. Syst Biol 64:869–878

Bouckaert R, Heled J, Kühnert D, Vaughan T, Wu C-H, Xie D, Suchard MA, Rambaut A, Drummond AJ (2014) BEAST 2: a software platform for Bayesian evolutionary analysis. PLOS Comput Biol 10: e1003537

Bromham L, Penny D (2003) The modern molecular clock. Nat Rev Genet 4:216–224

Bromham L, Duchêne S, Hua X, Ritchie AM, Duchêne DA, Ho SYW (2018) Bayesian molecular dating: opening up the black box. Biol Rev 93:1165–1191

Didier G, Royer-Carenzi M, Laurin M (2012) The reconstructed evolutionary process with the fossil record. J Theor Biol 315:26–37

Doolittle RF, Blombäck B (1964) Amino-acid sequence investigations of fibrinopeptides from various mammals: evolutionary implications. Nature 202:147–152

Dornburg A, Brandley MC, McGowen MR, Near TJ (2012) Relaxed clocks and inferences of heterogeneous patterns of nucleotide substitution and divergence time estimates across whales and dolphins (Mammalia: Cetacea). Mol Biol Evol 29:721–736

dos Reis M, Yang Z (2013) The unbearable uncertainty of Bayesian divergence time estimation. J Syst Evol 51:30–43

dos Reis M, Inoue J, Hasegawa M, Asher RJ, Donoghue PCJ, Yang Z (2012) Phylogenomic datasets provide both precision and accuracy in estimating the timescale of placental mammal phylogeny. Proc R Soc B 279:3491–3500

dos Reis M, Donoghue PCJ, Yang Z (2016) Bayesian molecular clock dating of species divergences in the genomics era. Nat Rev Genet 17:71–80

Drummond AJ, Ho SYW, Phillips MJ, Rambaut A (2006) Relaxed phylogenetics and dating with confidence. PLOS Biol 4:e88

Falcón LI, Magallón S, Castillo A (2010) Dating the cyanobacterial ancestor of the chloroplast. ISME J 4:777–783

Felsenstein J (1981) Evolutionary trees from DNA sequences: a maximum likelihood approach. J Mol Evol 17:368–376

Foster CSP, Sauquet H, van der Merwe M, McPherson H, Rossetto M, Ho SYW (2017) Evaluating the impact of genomic data and priors on Bayesian estimates of the angiosperm evolutionary timescale. Syst Biol 66:338–351

Gavryushkina A, Welch D, Stadler T, Drummond AJ (2014) Bayesian inference of sampled ancestor trees for epidemiology and fossil calibration. PLOS Comput Biol 10:e1003919

Gavryushkina A, Heath TA, Ksepka DT, Stadler T, Welch D, Drummond AJ (2017) Bayesian total-evidence dating reveals the recent crown radiation of penguins. Syst Biol 66:57–73

Gernhard T (2008) The conditioned reconstructed process. J Theor Biol 253:769–778

Gillespie JH (1991) The causes of molecular evolution. Oxford University Press, Oxford, UK

Heath TA, Huelsenbeck JP, Stadler T (2014) The fossilized birth-death process for coherent calibration of divergence-time estimates. Proc Natl Acad Sci USA 111:E2957–E2966

Hipsley CA, Müller J (2014) Beyond fossil calibrations: realities of molecular clock practices in evolutionary biology. Front Genet 5:138

Ho SYW (2009) An examination of phylogenetic models of substitution rate variation among lineages. Biol Lett 5:421–424

Ho SYW, Duchêne S (2014) Molecular-clock methods for estimating evolutionary rates and timescales. Mol Ecol 23:5947–5965

Ho SYW, Phillips MJ (2009) Accounting for calibration uncertainty in phylogenetic estimation of evolutionary divergence times. Syst Biol 58:367–380

Höhna S, Landis MJ, Heath TA, Boussau B, Lartillot N, Moore BR, Huelsenbeck JP, Ronquist F (2016) RevBayes: Bayesian phylogenetic inference using graphical models and an interactive model-specification language. Syst Biol 65:726–736

Inoue J, Donoghue PCJ, Yang Z (2010) The impact of the representation of fossil calibrations on Bayesian estimation of species divergence times. Syst Biol 59:74–89

Knoll AH, Nowak MA (2017) The timetable of evolution. Sci Adv 3:e1603076

Kodandaramaiah U (2011) Tectonic calibrations in molecular dating. Curr Zool 57:116–124

Lewis PO (2001) A likelihood approach to estimating phylogeny from discrete morphological character data. Syst Biol 50:913–925

Magallón SA (2004) Dating lineages: molecular and paleontological approaches to the temporal framework of clades. Int J Plant Sci 165:S7–S21

Magallón S (2010) Using fossils to break long branches in molecular dating: a comparison of relaxed clocks applied to the origin of angiosperms. Syst Biol 59:384–399

Matschiner M, Musilová Z, Barth JMI, Starostová Z, Salzburger W, Steel M, Bouckaert R (2017) Bayesian phylogenetic estimation of clade ages supports trans-Atlantic dispersal of cichlid fishes. Syst Biol 66:3–22

Morgan GJ (1998) Emile Zuckerkandl, Linus Pauling, and the molecular evolutionary clock, 1959–1965. J Hist Biol 31:155–178

Nascimento FF, dos Reis M, Yang Z (2017) A biologist's guide to Bayesian phylogenetic analysis. Nat Ecol Evol 1:1446–1454

O'Meara BC, Smith SD, Armbruster WS, Harder LD, Hardy CR, Hileman LC, Hufford L, Litt A, Magallón S, Smith SA, Stevens PF, Fenster CB, Diggle PK (2016) Non-equilibrium dynamics and floral trait interactions shape extant angiosperm diversity. Proc R Soc B 283:20152304

Pyron RA (2011) Divergence time estimation using fossils as terminal taxa and the origins of Lissamphibia. Syst Biol 60:466–481

Rambaut A (2000) Estimating the rate of molecular evolution: incorporating non-contemporaneous sequences into maximum likelihood phylogenies. Bioinformatics 16:395–399

Rannala B (2016) Conceptual issues in Bayesian divergence time estimation. Philos Trans R Soc 371:20150134

Rannala B, Yang Z (2007) Inferring speciation tunes under an episodic molecular clock. Syst Biol 56:453–466

Ronquist F, Klopfstein S, Vilhelmsen L, Schulmeister S, Murray DL, Rasnitsyn AP (2012a) A total-evidence approach to dating with fossils, applied to the early radiation of the Hymenoptera. Syst Biol 61:973–999

Ronquist F, Teslenko M, van der Mark P, Ayres DL, Darling A, Höhna S, Larget B, Liu L, Suchard MA, Huelsenbeck JP (2012b) MrBayes 3.2: efficient Bayesian phylogenetic inference and model choice across a large model space. Syst Biol 61:539–542

Sanderson MJ (1997) A nonparametric approach to estimating divergence times in the absence of rate constancy. Mol Biol Evol 14:1218–1231

Sanderson MJ (2002) Estimating absolute rates of molecular evolution and divergence times: a penalized likelihood approach. Mol Biol Evol 19:101–109

Sanderson MJ (1998) Estimating rate and time in molecular phylogenies: beyond the molecular clock? In: Soltis DE, Soltis PS, Doyle JJ (eds) Molecular systematics of plants II. Springer, Boston, MA, USA, pp 242–264

Sauquet H (2013) A practical guide to molecular dating. C R Palevol 12:355–367

Sauquet H, Magallón S (2018) Key questions and challenges in angiosperm macroevolution. New Phytol 219:1170–1187

Smith SA, O'Meara BC (2012) treePL: divergence time estimation using penalized likelihood for large phylogenies. Bioinformatics 28:2689–2690

Smith AB, Pisani D, Mackenzie-Dodds JA, Stockley B, Webster BL, Littlewood DTJ (2006) Testing the molecular clock: molecular and paleontological estimates of divergence times in the Echinoidea (Echinodermata). Mol Biol Evol 23:1832–1851

Stadler T (2010) Sampling-through-time in birth-death trees. J Theor Biol 267:396–404

Welch JJ, Bromham L (2005) Molecular dating when rates vary. Trends Ecol Evol 20:320–327

Wertheim JO, Fourment M, Kosakovsky Pond SL (2012) Inconsistencies in estimating the age of HIV-1 subtypes due to heterotachy. Mol Biol Evol 29:451–456

Wikström N, Savolainen V, Chase MW (2001) Evolution of the angiosperms: calibrating the family tree. Proc R Soc B 268:2211–2220

Wright AM, Hillis DM (2014) Bayesian analysis using a simple likelihood model outperforms parsimony for estimation of phylogeny from discrete morphological data. PLOS ONE 9:e109210

Wright AM, Lloyd GT, Hillis DM (2016) Modeling character change heterogeneity in phylogenetic analyses of morphology through the use of priors. Syst Biol 65:602–611

Yang Z, Rannala B (2006) Bayesian estimation of species divergence times under a molecular clock using multiple fossil calibrations with soft bounds. Mol Biol Evol 23:212–226

Zhang C, Stadler T, Klopfstein S, Heath TA, Ronquist F (2016) Total-evidence dating under the fossilized birth-death process. Syst Biol 65:228–249

Zhu T, dos Reis M, Yang Z (2015) Characterization of the uncertainty of divergence time estimation under relaxed molecular clock models using multiple loci. Syst Biol 64:267–280

Zuckerkandl E, Pauling L (1962) Molecular disease, evolution, and genic heterogeneity. In: Kasha M, Pullman B (eds) Horizons in biochemistry. Academic, New York, pp 189–225

Zuckerkandl E, Pauling L (1965) Evolutionary divergence and convergence in proteins. In: Bryson V, Vogel HJ (eds) Evolving genes and proteins. Academic, New York, pp 97–166

Bayesian Molecular Dating

6

Tianqi Zhu

Abstract

Using methods based on the molecular clock, genetic data provide an opportunity to estimate divergence times across the tree of life. Among the most widely used methods for molecular dating are those based on the Bayesian phylogenetic approach. With major developments in phylogenetic models and computational methods in recent years, Bayesian methods allow users to estimate divergence times and evolutionary rates from multilocus data sets under complex models. In this chapter, we introduce the Bayesian phylogenetic framework and Markov chain Monte Carlo algorithms. We explain the main components of Bayesian molecular dating, including the specification of the time prior incorporating fossil calibrations, the specification of the rate prior based on relaxed-clock models of evolutionary rate drift, the likelihood model of sequence evolution for partitioned data, and approaches for summarizing the posterior estimates. We explain the infinite-sites theory, which quantifies the uncertainties in posterior time estimates due to the imprecision of fossil calibrations and the finite amount of sequence data. The chapter concludes with a list of the major Bayesian dating software packages.

Keywords

Bayesian dating · Markov chain Monte Carlo algorithm · Molecular clock models · Prior · Likelihood · Posterior · Fossil calibrations

6.1 Introduction

The tree of life is one of the most important organizing principles in biology (Hug et al. 2016). A time-tree provides much richer information than a tree without temporal information, because it allows one to correlate macroevolutionary events (species divergences and extinctions) with geological events or palaeoclimatic changes. However, resolving the timeline of the tree of life is faced with many challenges. Perhaps the fundamental difficulty is the fact that times and rates are confounded in molecular sequence data, so that the fossil data are ultimately responsible for resolving sequence distances or branch lengths into estimates of absolute times and rates.

Bayesian methods were first introduced into phylogenetics in the 1990s and have become increasingly popular (Rannala and Yang 1996; Mau and Newton 1997; Yang and Rannala 1997; Mau et al. 1999). Bayesian methods can

T. Zhu (✉)
Key Laboratory of Random Complex Structures and Data Science, National Center for Mathematics and Interdisciplinary Sciences, Academy of Mathematics and Systems Science, Chinese Academy of Sciences, Beijing, China
e-mail: zhutq@amss.ac.cn

S. Y. W. Ho (ed.), *The Molecular Evolutionary Clock*, https://doi.org/10.1007/978-3-030-60181-2_6

be used to reconstruct phylogenetic relationships, infer phylogeographic history and gene flow between species, estimate species divergence times, and delimit species boundaries. Under complex models, maximum likelihood can become intractable and frequentist statistics must rely on data summaries for inference. In these circumstances, however, it may still be possible to conduct Bayesian inference, thanks to Markov chain Monte Carlo algorithms. Indeed, advances in computing power and implementation of sophisticated computational algorithms in user-friendly software have made it possible to analyse genome-scale data sets (see Chap. 13). This chapter will focus on Bayesian dating of species divergences, which has been the subject of several detailed reviews (dos Reis et al. 2016; Bromham et al. 2018).

6.2 Bayes's Theorem

The main feature of Bayesian statistics is the use of probability distributions to describe all kinds of uncertainty, such as the occurrence of data or uncertainty in the parameters. Suppose we are interested in an unknown parameter θ, and we have collected data D. $f(\theta)$ is called the prior distribution of θ and reflects our subjective belief on the unknown parameter θ before the data are analysed. $f(D|\theta)$ is referred to as the likelihood, which is the probability of the data given the parameters under the model. Bayes's theorem states that:

$$f(\theta|D) = \frac{f(D|\theta)f(\theta)}{f(D)} = \frac{f(D|\theta)f(\theta)}{\int f(D|\theta)f(\theta)d\theta} \quad (6.1)$$

where $f(\theta|D)$, or θ given data D, is the posterior distribution of θ. Thus, the posterior combines information in the prior and in the sample data. $f(D)$ is the marginal likelihood, a normalizing constant that is used to ensure that $f(\theta|D)$ is a statistical distribution.

The three goals of statistical inference are estimation of parameter values, prediction of data outcomes, and model comparison. Bayes's theorem is used to estimate parameter θ, characterized by its posterior distribution. Equation (6.1) is derived under a certain model. If we want to compare two models M_1 and M_2, we need to focus on the posterior probability of the model given the data, that is:

$$\begin{aligned} f(M_i|D) &= \frac{\pi(M_i)f(D|M_i)}{f(D)} \\ &= \frac{\pi(M_i)\int f(D|\theta_i, M_i)f(\theta_i|M_i)d\theta_i}{f(D)} \end{aligned} \quad (6.2)$$

where $\pi(M_i)$ is the prior probability of model M_i, with $\sum_i \pi(M_i) = 1$, and $f(\theta_i|M_i)$ is the prior of θ_i under model M_i. The posterior odds is then given as:

$$\begin{aligned} \frac{f(M_1|D)}{f(M_2|D)} &= \frac{\pi(M_1)}{\pi(M_2)} \times \frac{f(D|M_1)}{f(D|M_2)} \\ &= \text{prior ratio} \times \text{Bayes factor} \end{aligned} \quad (6.3)$$

The Bayes factor is usually used for model selection problems, to choose the better model between two candidates M_1 and M_2. The Bayes factor k is defined as:

$$\begin{aligned} k &= \frac{f(D|M_1)}{f(D|M_2)} \\ &= \frac{\int f(D|\theta_1, M_1)f(\theta_1|M_1)d\theta_1}{\int f(D|\theta_2, M_2)f(\theta_2|M_2)d\theta_2} \end{aligned} \quad (6.4)$$

By averaging over the parameters θ_i in model M_i, the Bayes factor compares the marginal likelihoods $f(D|M_i)$ of the two models. The marginal likelihood $f(D|M_i)$ is the predicted probability of the data. This can be used to assess the general adequacy of the assumed model. The posterior model probabilities in Eq. (6.3) can be used to compare non-nested models as well as nested models. If we assign equal prior weight (1/2) to each of the two models, the posterior odds is the Bayes factor, which is the ratio of marginal likelihoods under the models.

6.3 The Framework of Bayesian Clock Dating

In this section, we use the program MCMCTree as an example to illustrate the framework of Bayesian molecular dating using a set of nucleotide or amino acid sequences. The details for other Bayesian dating programs are similar. MCMCTree was the first program for Bayesian phylogenetics and is part of the software package PAML (Yang 2007). PAML can read sequence data in .nuc, .aa, and .nex formats, and PHYLIP format is the 'native' format for PAML. The data can be organized in sequential format or as site pattern counts. More details on how to prepare the data can be found in the PAML documentation.

Suppose a species tree with s species is fixed, and the divergence times (the $s-1$ node ages on the rooted tree) are shared among different loci (or subsets of the data). Let $D = \{D_1, D_2, \ldots, D_L\}$ be the sequence data, where D_i is the aligned sequences at locus i. θ are the parameters in the evolutionary model. $\mathbf{t} = \{t_1, t_2, \ldots, t_{s-1}\}$ are the $s-1$ divergence times. $\mathbf{r} = \{\mathbf{r}_i\}$ are the rates, where $\mathbf{r}_i = \{r_{ij}\}$ are the rates at locus i, specified by the relaxed-clock models. According to Bayes's theorem (Eq. 6.1), the joint posterior distribution of θ, $\mathbf{t}$, and $\mathbf{r}$ is:

$$f(\theta, \mathbf{t}, \mathbf{r}|D) = \frac{f(D|\mathbf{t}, \mathbf{r}) f(\mathbf{r}|\mathbf{t}, \theta) f(\mathbf{t}|\theta) f(\theta)}{f(D)} \quad (6.5)$$

Here $f(\theta)$ is the prior for parameters in the substitution model. $f(\mathbf{t}|\theta)$ is the prior of divergence times, which is specified using a birth–death process and will be discussed in Sect. 6.4. $f(\mathbf{r}|\mathbf{t}, \theta)$ is the rate prior. The rates can vary across branches of each gene tree and the average rate can differ among loci. The strategies to assign the rate prior among loci are discussed in Sect. 6.5.3. At a given locus, the rates on branches of the gene tree are governed by the relaxed-clock models. Several commonly used clock models are introduced in Sect. 6.5, where we also discuss strategies for specifying the rate prior on branches. The likelihood is:

$$f(D|\mathbf{t}, \mathbf{r}) = \prod_i f(D_i|\mathbf{t}, \mathbf{r}_i) \quad (6.6)$$

where $f(D_i|\mathbf{t}, \mathbf{r}_i)$ is the Felsenstein's phylogenetic likelihood (Felsenstein 1981). For more details of this calculation see, for example, Chap. 4 of Yang (2014). In practice, the likelihood is expensive to calculate; approximate methods have been developed to accelerate the computation (Thorne et al. 1998; Guindon 2010; dos Reis and Yang 2011).

The denominator of Eq. (6.5), $f(D)$, is called the marginal likelihood and is in the form of a high-dimensional integral. The calculation of high-dimensional integrals with numerical methods is difficult and error prone. By using Markov chain Monte Carlo (MCMC) methods, the calculation of $f(D)$ can be avoided. This is why MCMC methods can deal with complex models. In the Metropolis–Hastings algorithm used in software such as MCMCTree, the state of the Markov chain includes substitution rate θ, divergence times $\mathbf{t}$, and evolutionary rates $\mathbf{r}$. At the current state (θ, $\mathbf{t}$, $\mathbf{r}$), a new state (θ^*, $\mathbf{t}^*$, $\mathbf{r}^*$) is proposed according to a proposal density $q(\theta^*, \mathbf{t}^*, \mathbf{r}^*|\theta, \mathbf{t}, \mathbf{r})$. The acceptance ratio is

$$\alpha = \min\left\{1, \frac{f(D|\mathbf{t}^*, \mathbf{r}^*) f(\mathbf{r}^*|\mathbf{t}^*, \theta^*) f(\mathbf{t}^*|\theta^*) f(\theta^*)}{f(D|\mathbf{t}, \mathbf{r}) f(\mathbf{r}|\mathbf{t}, \theta) f(\mathbf{t}|\theta) f(\theta)} \times \frac{q(\theta, \mathbf{t}, \mathbf{r}|\theta^*, \mathbf{t}^*, \mathbf{r}^*)}{q(\theta^*, \mathbf{t}^*, \mathbf{r}^*|\theta, \mathbf{t}, \mathbf{r})}\right\} \quad (6.7)$$

which is the product of the likelihood ratio, the prior ratio, and the proposal ratio. Each MCMC algorithm can consist of several moves that change some components of the state. For example, one might first propose $\mathbf{t}^*$, then rates $\mathbf{r}_i^*$ at each of the loci, and lastly propose θ^*. The state of the Markov chain is a vector of the parameters in the model, including species divergence times and the rates for branches at each locus. In each iteration, the MCMC chain visits a state (θ, $\mathbf{t}$, $\mathbf{r}$) of the parameter space, and the states visited during successive iterations are recorded. For an irreducible and aperiodic Markov chain, regardless of the initial state, the chain will eventually reach the stationary distribution. We require samples from the stationary distribution, so we typically discard the initial portion of the samples as the 'burn-in'.

The samples collected from the MCMC run (after the burn-in is discarded) can be summarized to characterize the posterior distribution of the parameters. For most Bayesian software packages, MCMC samples can be recorded and processed according to the user's needs. We can use the posterior mean or median and the 95% equal-tail credibility interval (CI) to give a simple characterization of the posterior distribution. To recover the marginal distribution of any parameter of interest, we simply extract the sampled values for that parameter while ignoring other parameters.

The most common credibility intervals used are the equal-tail credibility interval and the highest posterior density (HPD) interval. Given significance value α, the equal-tail credibility interval is given by $(\theta_L, \theta_U) = (F^{-1}(\alpha/2), F^{-1}(1 - \alpha/2))$, where F is the probability density function and F^{-1} its inverse (i.e., the quantile). By this construction, the probabilities of being below θ_L and above θ_U are the same, both being $\alpha/2$. For a continuous distribution, the equal-tail CI will always include the median. The HPD interval is the narrowest interval that covers $1 - \alpha$ of the probability mass.

As an example, consider an exponential distribution with mean 1. The 95% equal-tail CI is (0.025, 3.689) with length 3.664, while the HPD CI is (0, 2.996), which is the shortest CI. When prior and likelihood are in conflict or the posterior is a mixture of two distributions, the posterior distribution might have multiple modes. In such cases, the HPD CI might be two or more disconnected intervals. The 90% equal-tail CI and HPD CI of a bimodal distribution are shown in Fig. 6.1. Because there is a valley in the distribution, the HPD CI excludes the valley and consists of two disconnected intervals (θ_1, θ_2) and (θ_3, θ_4). Most molecular dating software will report the HPD CIs, and some software such as MCMCTree will calculate both equal-tail CIs and HPD CIs.

6.4 Prior on Node Times

To assign a time prior (i.e., the prior distribution of ages of nodes in the tree), we can use a birth–death process (Kendall 1948) to model the process of speciation, extinction, and species sampling (Yang and Rannala 1997). In a very small time interval with length Δt, each species existing at the time splits into two with probability $\lambda \Delta t$, and becomes extinct with probability $\mu \Delta t$. λ and μ are called the per-lineage birth rate and death rate, respectively. The probability that the number of species increases or decreases by more than 1 is of order $o(\Delta t)$.

Suppose that the process starts from an ancestor species at time t in the past. The number of present-day species S, which is a random variable, relies on this birth–death process. Let the number of species in the sample be s, then each species is sampled with probability $\rho = s/S$, and we call ρ the sampling fraction. Some important properties of this birth-death process have been described (Nee et al. 1994). The probabilities that a lineage arising at time t in the past has at least one descendant and that exactly one descendant has survived until the present time are $P(0, t)$ and $p_1(t)$, where:

$$P(0,t) = \frac{\rho(\lambda - \mu)}{\rho\lambda + (\lambda(1-\rho) - \mu)e^{(\mu-\lambda)t}} \tag{6.8}$$

and

$$p_1(t) = \frac{1}{\rho} P(0,t)^2 e^{(\mu-\lambda)t} \tag{6.9}$$

Conditional on the root age t_1, the other $s - 2$ node ages are order statistics from the kernel density:

$$g(t) = \frac{\lambda p_1(t)}{v_{t_1}} \tag{6.10}$$

where

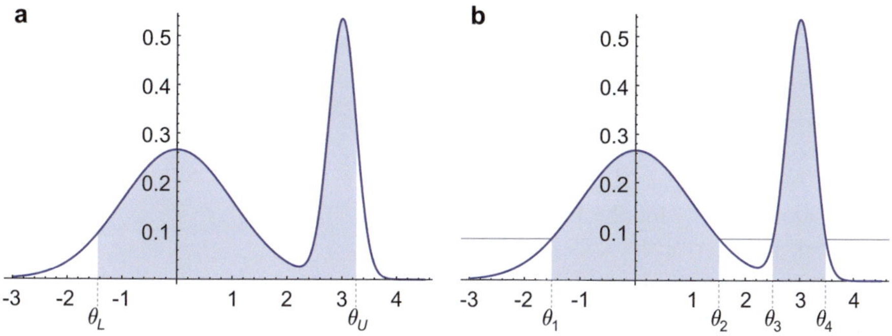

Fig. 6.1 Two kinds of credibility intervals used in Bayesian phylogenetics. (**a**) The 90% equal-tail credibility interval (θ_L, θ_U), with θ_L and θ_U as 5% and 95% quantiles. (**b**) The 90% highest posterior density (HPD) interval consists of two disconnected intervals: (θ_1, θ_2) and (θ_3, θ_4). The HPD interval is the narrowest interval that covers 90% of the probability

$$v_{t_1} = 1 - \frac{1}{\rho} P(0, t_1) e^{(\mu-\lambda)t_1} \qquad (6.11)$$

We can derive the prior density of node ages t_k, which is the $(s - k)$th order statistics of $s - 2$ random variables from the kernel:

$$f(t_k) = \frac{(s-2)!}{(s-k-1)(k-2)!} G(t_k)^{s-k-1} (1 - G(t_k))^{k-2} g(t_k) \qquad (6.12)$$

where $G(t)$ denotes the cumulative density function of $g(t)$. The joint density of the $s - 2$ node ages $t_2, t_3, \ldots, t_{s-1}$ is

$$f(t_2, t_3, \ldots, t_{s-1}) = (s-2)! \prod_{j=2}^{s-1} g(t_j) \qquad (6.13)$$

Birth rate λ and death rate μ, together with sampling fraction ρ, determine the shape of the tree. One can examine the impact of the prior on the posterior time estimation by changing these parameters. Generally, with small ρ, the tree will have long branches toward the tips. With large ρ, the tree will have long branches near the root.

On the nodes with calibrations, fossils provide important information for Bayesian molecular dating (see Chap. 8). An overview of calibration methods used in divergence dating is provided by Ho and Phillips (2009). In early dating analyses, calibrations took the form of fixed time points (Graur and Martin 2004). Nowadays, the uncertainty in the fossil data is considered in the dating process, usually by using minimum and/or maximum age bounds (e.g. Benton et al. 2009; Chap. 5). Most often the fossil evidence can provide an informative lower bound, but less informative upper bound. If the time is restricted to an interval $[t_L, t_U]$, we call such calibrations hard bounds. Sometimes the bound is one-sided, with a lower bound only $[t_L, \infty)$ or an upper bound only $(0, t_U]$.

Yang and Rannala (2006) introduced soft bounds as a new strategy to assign fossil time priors. The soft bounds $[t_L, t_U]$ allow the node age to be outside the bounds with a certain small probability, say 5%. The node age t_k then follows a general distribution $f(t_k|C)$, rather than the simple uniform distribution. Using soft bounds has a number of advantages (Yang and Rannala 2006). When the fossils conflict with each other or with the genetic data, soft bounds allow sequence data to correct poor calibrations, while it is impossible to overcome poor hard bounds regardless of the amount of genetic data. With soft bounds, it is not necessary to use 'safe' but high upper bounds, which can lead to biased posterior time estimation. In addition, soft bounds allow more reliable assessment of estimation errors, whereas hard bounds can result in misleadingly high precision when fossils and genetic data are in conflict (Yang and Rannala 2006).

Fossil calibration information is part of the time prior. Given the number of species s and the root age t_1, the remaining nodes $\mathbf{t}_{-1}$ are partitioned into two sets $\mathbf{t}_{-1} = (\mathbf{t}_C, \mathbf{t}_{-C})$ (Yang and Rannala 2006), with $\mathbf{t}_C$ for the nodes with fossil calibrations and $\mathbf{t}_{-C}$ for the nodes without fossil calibrations. In the tree shown in Fig. 6.2, $\mathbf{t}_C = \{t_2, t_4\}$ and $\mathbf{t}_{-C} = \{t_3, t_5\}$. Thus:

$$f(\mathbf{t}_{-1}|t_1, s, C) = f(\mathbf{t}_{-C}, \mathbf{t}_C|t_1, s, C) = f_{\mathrm{BD}}(\mathbf{t}_{-C}|\mathbf{t}_C, t_1, s) f(\mathbf{t}_C|C) \quad (6.14)$$

where the prior $f(\mathbf{t}_C|C)$ is specified according to the fossil calibration information. Following the definition of conditional probability,

$$f_{\mathrm{BD}}(\mathbf{t}_{-C}|\mathbf{t}_C, t_1, s) = f_{\mathrm{BD}}(\mathbf{t}_{-1}|t_1, s)/f_{\mathrm{BD}}(\mathbf{t}_C|\mathrm{t}_1, s) \quad (6.15)$$

where $f_{\mathrm{BD}}(\mathbf{t}_{-1}|t_1, s)$ is calculated according to Eq. (6.13), and $f_{\mathrm{BD}}(\mathbf{t}_C|t_1, s)$ is given by the joint distribution of order statistics:

$$f_{\mathrm{BD}}(\mathbf{t}_C|t_1, s) = \frac{(s-2)!}{(i_1-1)!(i_2-i_1-1)!\ldots(s-i_C-2)!} g(t_{i_1})g(t_{i_2})\ldots g(t_{i_C}) \times G(t_{i_1})^{i_1-1}(G(t_{i_2}) - G(t_{i_1}))^{i_2-i_1-1}\ldots(1-G(t_{i_C}))^{s-i_C-2} \quad (6.16)$$

Finally, we deal with the root age t_1. If the root has associated fossil calibration information, then $f(t_1|C)$ is assigned a prior distribution like other nodes that have fossil calibrations. Otherwise, according to the birth–death process, we assign:

$$f(t_1|s) = [P(0, t_1)(1 - v_{t_1})]^2 v_{t_1}^{s-2} \quad (6.17)$$

where $P(0, t)$ and v_{t_1} are defined in Eqs. (6.8) and (6.11).

Combining Eqs. (6.14), (6.15), and (6.17), we get the joint prior distribution of divergence times, constructed with a birth–death process with species sampling and incorporating fossil calibration information.

$$\begin{aligned} f(\mathbf{t}|s, C) &= f(t_1, t_2, \ldots, t_{s-1}|s, C) \\ &= f(t_1|s) f(\mathbf{t}_{-1}|t_1, s, C) \\ &= f(t_1|s) f_{\mathrm{BD}}(\mathbf{t}_{-C}|\mathbf{t}_C, t_1, s) f(\mathbf{t}_C|C) \\ &= \frac{f(t_1|s) f_{\mathrm{BD}}(\mathbf{t}_{-1}|t_1, s)}{f_{\mathrm{BD}}(\mathbf{t}_C|t_1, s)} f(\mathbf{t}_C|C) \end{aligned} \quad (6.18)$$

Note the distinction between $f_{\mathrm{BD}}(\mathbf{t}_C|t_1, s)$ and $f(\mathbf{t}_C|C)$. The former is the joint marginal distribution of $\mathbf{t}_C$ determined by the birth–death process, and the latter is the prior density specified according to the fossil evidence.

Stadler and colleagues (Stadler 2010; Heath et al. 2014) proposed to use the fossilized birth–death (FBD) process as the time prior (see Chap. 11). The FBD model characterizes the process of speciation, extinction, and fossilization that leads to extant and extinct (fossil) species. Parameters in the model include the birth (speciation) rate λ, death (extinction) rate μ, and sampling rate Ψ (which is the rate at which a fossil is sampled over time), as well as the sampling fraction ρ (which is the probability with which modern species are included in the data). The FBD model is used in tip-calibration approaches, in which sequence data are available for extant species and morphological measurements are available for both extant and extinct species. Compared with traditional node-dating approaches, it is unnecessary to use constraints on node ages and the FBD model allows us to include all available fossils. Some limitations of tip calibrations are discussed in Chap. 11 and by dos Reis et al. (2016).

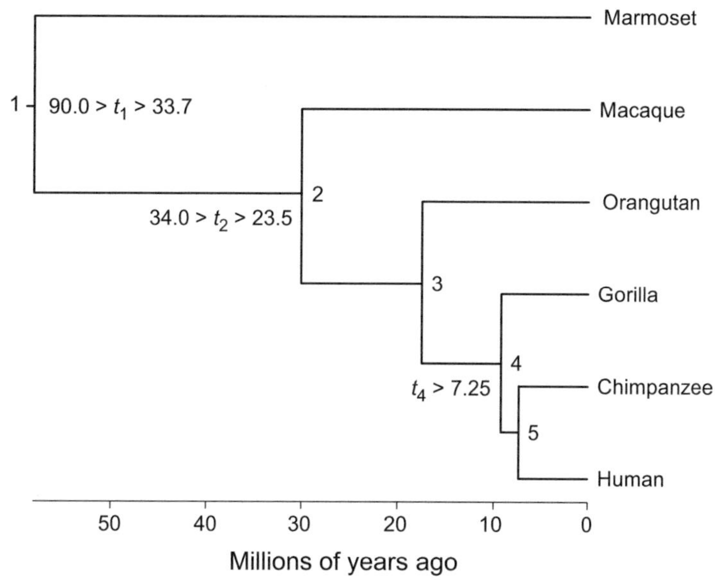

Fig. 6.2 Phylogenetic tree of six primate species, with three fossil calibrations. The five node times are partitioned into three categories: the root node t_1, the non-root nodes with fossil calibrations $\mathbf{t}_C = \{t_2, t_4\}$ and the non-root nodes without calibrations $\mathbf{t}_{-C} = \{t_3, t_5\}$

6.5 Clock Models and Rate Prior on Branches

6.5.1 Molecular Clock

The molecular clock, in its simplest form, is sometimes called the strict molecular clock and assumes that the substitution rate is constant over time (Zuckerkandl and Pauling 1965). Using genome data and under the assumptions of the substitution model, evolutionary distances can be accurately estimated. When the molecular clock is assumed, the distance between two sequences is expected to increase linearly with divergence time. Brown and Yang (2011) suggested that the strict clock models are generally appropriate in shallow phylogenies where rate variation is expected to be low. For example, if the difference between sequences is less than 5%, the molecular clock model is usually appropriate. Note that the molecular clock allows the rate to vary among loci, with the rate at locus i (r_i) treated as a parameter with a particular distribution rather than a fixed number.

6.5.2 Relaxed-Clock Models

If we assume the strict clock model, then all of the branches on the tree at each locus will share the same rate. However, the assumption of a strict clock is often violated, especially if the study species are distantly related (Langley and Fitch 1974; Yoder and Yang 2000; Hasegawa et al. 2003; see Chaps. 1 and 4). For this reason, most modern dating analyses are conducted using relaxed-clock models, which allow rates to vary throughout the tree. The two major classes of relaxed-clock models are those that assume autocorrelated rates and those that assume independent rates.

Thorne et al. (1998) and Kishino et al. (2001) introduced a method to assign a prior to the rates on branches using the geometric Brownian motion model. Let r_t denote the rate at time t, and $r_0 = r_A$ denote the rate at the root node. When there is no potential for ambiguity, r_t is abbreviated as r. In the autocorrelated-rates model, r_t follows a geometric Brownian motion process, i.e., $y_t = \log(r_t)$ follows a Brownian motion. It is natural to let $E(r_t|r_A) = r_A$. According to the property of geometric Brownian motion with drift parameter u and volatility parameter σ, $E(r_t|r_A) = r_A \times \exp(ut)$. Solving the equation leads to the drift parameter $u = 0$. Thus, the density of rate r_t given r_A is derived as

$$f(r_t|r_A) = \frac{1}{r_t\sqrt{2\pi t\sigma^2}} \exp\left\{-\frac{1}{2t\sigma^2}\left(\log(r_t/r_A) + \frac{t\sigma^2}{2}\right)^2\right\} \quad (6.19)$$

The parameter σ^2 determines how quickly the rate drifts with time, and $\mathrm{Var}(r_t|r_A) = r_A^2(\exp(\sigma^2) - 1)$.

To derive the average rate on a branch, we need to calculate an integral over time t, which can be complicated. Kishino et al. (2001) used the mean of the rates at the two ends of a branch as the average rate on the branch. Rannala and Yang (2007) used the rate at the midpoint of the branch instead. The rate at the root node is assigned a gamma prior, and the volatility parameter σ^2 is assigned another gamma prior. Then the probability density of r_1 and r_2 given r_0, where r_1, r_2, and r_0 are the rates at the midpoints of two descendent branches and the ancestral branch, respectively, can be calculated under the geometric Brownian motion model (Eq. 7 of Rannala and Yang 2007). In this way, r_1 are r_2 are autocorrelated.

Drummond et al. (2006) proposed a model with independent and identically distributed (i.i.d.) substitution rates, which incorporates rate variation among lineages without the assumption of correlation. The i.i.d. rates model is widely used in software for molecular dating, such as BEAST (Suchard et al. 2018; Bouckaert et al. 2019) and MCMCTree. For example, the rates r_1, r_2, …, r_{2s-2} for the $2s - 2$ branches might be i.i.d. from a lognormal distribution:

$$f(r_i|\mu, \sigma^2) = \frac{1}{r_i\sqrt{2\pi\sigma^2}} \exp\left\{-\frac{1}{2\sigma^2}\left(\log(r_i/\mu) + \frac{\sigma^2}{2}\right)^2\right\}, 0 < r_i < \infty \quad (6.20)$$

Note that r_i follows Lognormal($\log(\mu) - \sigma^2/2$, σ^2) instead of Lognormal(μ, σ^2), with $E(r_i) = \mu$ and $\mathrm{Var}(r_i) = \mu^2(\exp(\sigma^2) - 1)$, and furthermore, $\mathrm{Var}(\log(r_i)) = \sigma^2$. In the dating software MCMCTree, the parameters μ and σ^2 are assigned a hyperprior with a gamma distribution.

6.5.3 Prior on Rates Among Loci

When data sets with multiple loci are analysed, we also need to consider rate variation among loci. A natural way to do this is to assign a prior on the locus rate, which can be viewed as the average rate over branches at this locus, to be i.i.d. This rate prior is adopted by commonly used Bayesian dating software such as BEAST (Suchard et al. 2018; Bouckaert et al. 2019), MrBayes (Ronquist et al. 2012), and older versions of MCMCTree (Yang 2007). Consider MCMCTree as an example. Under the relaxed-clock models, i.i.d. gamma priors are used to generate parameters μ_i and σ_i^2 at locus i, allowing the average locus rate to vary among loci. At each locus, the branch rates can then be i.i.d. (independent-rates model) or specified according to the geometric Brownian motion (autocorrelated-rates) model (Rannala and Yang 2007).

In recent years, increasing attention has been paid to the time prior, but the i.i.d. rate prior among loci remained the only prior to be widely used. dos Reis et al. (2014) discovered that the i.i.d. prior is problematic in this context. In conventional Bayesian analysis, the prior will have less and less impact on the posterior as the amount of data increases. However, in divergence-time estimation with fossil calibration, as the number loci L increases, the posterior time estimates can be increasingly incorrect if the prior is not properly assigned. Furthermore, the data cannot correct errors in the prior, owing to the fact that rates and times are confounded.

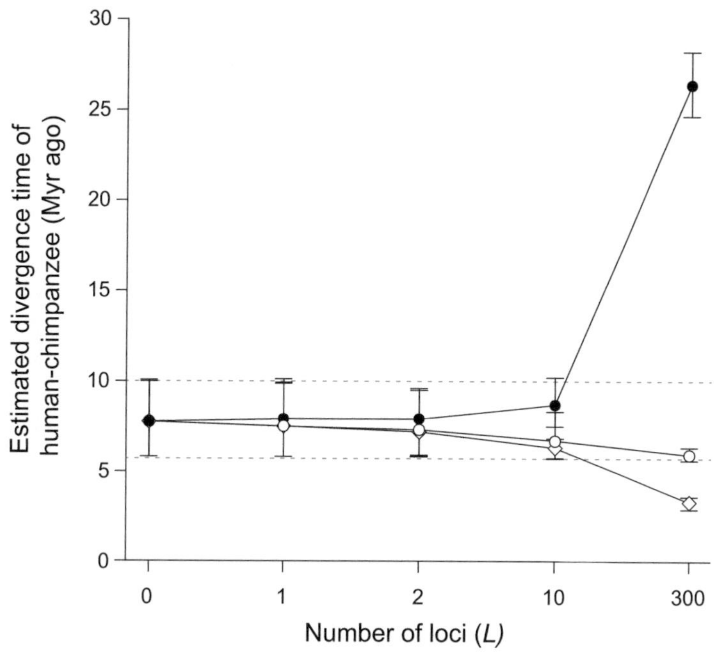

Fig. 6.3 Posterior estimates of the human–chimpanzee divergence time under the i.i.d. prior for locus rate. Three priors on locus rate were used: (1) a high rate, $\mu_i \sim$ G(2, 2) (empty diamonds) with mean 1; (2) an appropriate rate $\mu_i \sim$ G(2, 20) with mean 0.1 (empty circles); and (3) a low rate, $\mu_i \sim$ G(2, 200) with mean 0.01 (filled circles). The time unit of the rate is 100 Myr. When the high rate is used, the estimated time becomes younger as L increases. When the low rate is used, the estimated time becomes older as L increases. When more than 300 loci are used, the posterior times are outside the fossil bounds (dashed lines) in both cases

This problem is illustrated in Fig. 6.3, which shows the posterior estimates of the human–chimpanzee divergence time under the i.i.d. prior for locus rate. The fossils indicate that this divergence time should be around 5.7 to 10 Myr ago (Benton et al. 2009), and molecular studies indicate a mean rate of 10^{-9} substitutions per site per year. In an analysis of 300 loci, if a high rate prior is used, $\mu_i \sim$ G(2, 2) (with mean rate 10^{-8} substitutions per site per year), the posterior time estimate is too young, at 3.3 (2.9, 3.6) Myr ago. In contrast, if a low rate prior is used, $\mu_i \sim$ G(2, 200) (with mean rate 10^{-10} substitutions per site per year), the estimate is too old, at 26.4 (24.7, 28.2) Myr ago. In both cases, the posterior means of times are outside the fossil bounds.

The reason for this phenomenon is that as the number of loci L increases, the prior variance of the average rate across all loci goes to zero at the rate $1/L$, and thus the rate prior dominates the posterior estimates of times. If the rate prior is misspecified, the posterior divergence times will converge to incorrect values with very narrow credibility intervals. The tendency will be aggravated as L increases.

A new prior on locus rates, the compound Dirichlet prior, was developed by dos Reis et al. (2014). To avoid the average rate over loci converging to a point value as L increases, the new prior is implemented by two steps. First, we assign the average rate over loci $\bar{\mu} = \sum_{i=1}^{L} \mu_i / L$ a gamma prior G(α_μ, β_μ). Then we partition the total rate $L\bar{\mu}$ to L loci according to a symmetrical Dirichlet distribution with parameter α. The expectation and variance of locus rate μ_i, and the correlation between μ_i and μ_j, can be calculated under this construction.

$$E(\mu_i) = E(E(\mu_i|\overline{\mu})) = \alpha_\mu/\beta_\mu \tag{6.21}$$

$$\begin{aligned}\mathrm{Var}(\mu_i) &= E(\mathrm{Var}(\mu_i|\overline{\mu})) + \mathrm{Var}(E(\mu_i|\overline{\mu})) \\ &= \frac{\alpha_\mu}{\beta_\mu^2}\left(1 + \frac{\alpha_\mu + 1}{L\alpha + 1}(L-1)\right) \\ &\to \frac{\alpha_\mu}{\beta_\mu^2}\left(1 + \frac{\alpha_\mu + 1}{\alpha}\right)\end{aligned} \tag{6.22}$$

$$\begin{aligned}\mathrm{Corr}(\mu_i, \mu_j) &= \frac{L\alpha - \alpha_\mu}{L(\alpha + 1) + (L-1)\alpha_\mu} \\ &\to \frac{\alpha}{\alpha + \alpha_\mu + 1}\end{aligned} \tag{6.23}$$

The prior variances of both $\overline{\mu}$ and μ_i converge to nonzero limits, and thus the rate prior does not have an overwhelming impact on the posterior times. In the relaxed-clock models, the parameter σ_i^2, which measures the variance of the log-rate in the independent-rates model or the extent of drift in the autocorrelated-rates model, can be assigned a compound Dirichlet distribution as well. However, the prior on σ_i^2 has much less impact on posterior time estimation than that of μ_i (dos Reis et al. 2014).

The compound Dirichlet prior was evaluated by dos Reis et al. (2014), who carried out simulations for two species and analysed a data set from six primates (Fig. 6.4). Both analyses showed poor performance of the i.i.d. prior and robustness of the compound Dirichlet prior when the priors were misspecified. A primate data set with 300 loci was analysed using both locus rate priors. With the i.i.d. prior, if the prior rate is too high (or too low), the estimated divergence times are too young (or too old) (Table 6.1). In contrast, with the compound Dirichlet prior, posterior time estimates are accurate and insensitive to the rate prior.

6.6 Uncertainty of Divergence-Time Estimation

With advances in sequencing technology, more and more genome data have become available, which enables us to estimate evolutionary distances with high accuracy. However, molecular sequence data only provide information about distance, not about times and rates separately. Therefore, we need to use information in the prior and fossils to perform dating analysis. In a conventional Bayesian analysis, as the amount of sequence data increases, the posterior mean will converge to the true value of the parameter, with the CI length converging to zero. The estimate will involve less and less uncertainty, measured by the posterior variance or the square of the posterior CI length. However, owing to the confounding effect of times and rates, Bayesian divergence-time estimation with fossil calibrations is an unconventional estimation problem (Zhu et al. 2015). Even with an infinite amount of sequence data, the posterior variance will not converge to zero, and the estimates will always have uncertainties (Yang and Rannala 2006; Rannala and Yang 2007; dos Reis and Yang 2013).

6.6.1 Infinite-Sites Theory for the Molecular Clock

Yang and Rannala (2006) studied the asymptotic posterior distribution of divergence times under the strict-clock model when the number of sites approaches infinity. In this context, we consider a data set that contains only one locus because multiple loci do not contain more information than a single locus under the strict clock. The prior of rates and times are $g(r)$ and $f(t_1, t_2, \ldots, t_{s-1})$, where s is the number of species. The posterior for time t_j is:

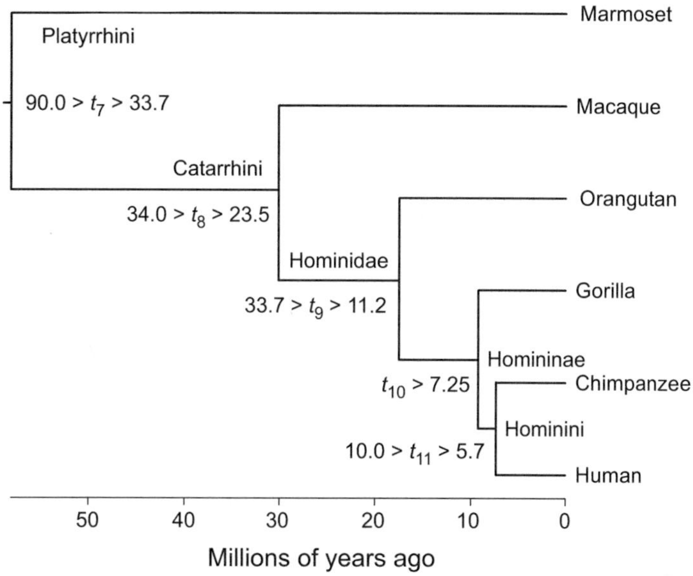

Fig. 6.4 The phylogeny of six primate species. All five nodes have fossil calibrations. The fossil bounds are soft bounds, with a 1% probability that the minimum bound is violated and 5% probability that the maximum bound is violated. The time unit on nodes is Myr

$$f\left(t_j|d_1, d_2, \ldots, d_{s-1}\right) \propto g\left(\frac{d_j}{t_j}\right) f\left(\frac{d_1}{d_j}t_j, \frac{d_2}{d_j}t_j, \ldots, \frac{d_{s-1}}{d_j}t_j\right)\left(\frac{d_j}{t_j}\right)^{2-s}\frac{1}{t_j} \quad (6.24)$$

where the d_j s are evolutionary distances that are estimated without uncertainty or error when the number of sites goes to infinity. Instead of converging to a point mass, the posterior converges to a one-dimensional distribution. If we plot the posterior means of t_j s against their true ages, percentiles, or CI lengths, the points lie on a straight line (e.g., Yang and Rannala 2006; dos Reis and Yang 2013). When analysing real data, we can plot the posterior mean times against their CI widths to assess whether the sequence data are nearly saturated. This is known as the infinite-sites theory.

6.6.2 Finite-Sites Theory Under the Strict Molecular Clock

The uncertainty in posterior time estimation has been studied by dos Reis and Yang (2013) using mathematical analysis, simulation, and analysis

Table 6.1 Posterior means of mean rate ($\bar{\mu}$) and divergence times among six primate species using the i.i.d. prior and the compound Dirichlet prior

Prior on mean rates of loci		Mean rate ($\times 10^{-8}$)	Node times (Myr ago)				
			t_7	t_8	t_9	t_{10}	t_{11}
i.i.d. prior	$\mu_i \sim$ G(2, 2)	0.199	32.9	17.2	9.2	4.2	3.3
	$\mu_i \sim$ G(2, 20)	0.098	64.4	32.5	17.5	7.8	5.9
	$\mu_i \sim$ G(2, 200)	0.019	308.8	150.4	81.8	36.3	26.4
Compound Dirichlet prior	$\bar{\mu} \sim$ G(2, 2)	0.096	65.7	33.1	17.8	8.0	6.0
	$\bar{\mu} \sim$ G(2, 20)	0.096	65.8	33.1	17.8	8.0	6.0
	$\bar{\mu} \sim$ G(2, 2)	0.096	65.8	33.1	17.8	8.0	6.0

Three priors for locus rates were used: G(2, 2) is a high rate, G(2, 20) is a medium rate, and G(2, 200) is a low rate. The mean rate is calculated by averaging locus rates over loci from the MCMC samples. The number of loci analysed is 300. Data from dos Reis et al. (2014)

of real data. Suppose that the data set contains one alignment with length N, and the sequences evolved under the strict molecular clock. dos Reis and Yang (2013) proposed a finite-sites theory which predicts that the uncertainty of the posterior approaches its infinite-data limit at the rate $1/N$. Let w be the width of the posterior CI. Because w^2 is proportional to the posterior variance, w^2 measures the uncertainty of posterior time estimation. dos Reis and Yang (2013) suggested that:

$$w^2 - w_\infty^2 \propto 1/N \quad (6.25)$$

where w_∞ is the posterior CI width for infinite data. Note that in Bayesian dating, w_∞ is not zero, as we have emphasized. Furthermore, $u_F = w_\infty^2/w^2$ is the fraction of the uncertainty in posterior time estimates that is due to uncertainties in the fossil calibrations, while $u_S = 1 - u_F$ is the fraction due to the finite amount of sequence data. As the size of the sequence data increases, u_S will go to zero and u_F will go to 1, which indicates that all uncertainty in the posterior comes from that in the fossil calibrations.

6.6.3 Finite-Sites Theory Under Relaxed-Clock Models Using Multiple Loci

In the genomic era, most analyses are conducted with multilocus data sets. Owing to violations of the strict-clock model, relaxed-clock models are used in most modern Bayesian dating analyses. Zhu et al. (2015) extended the finite-sites theory to the case of relaxed-clock analysis of multiple loci. They predicted that:

$$w^2 = \frac{b}{NL} + \frac{a}{L} + w_\infty^2 \quad (6.26)$$

where L is the number of loci, N is the sequence length at each locus, and a, b are constants that are independent of N and L. Equation (6.26) is confirmed by computer simulation and real data analysis.

According to Eq. (6.26), the sources of uncertainties in the posterior time estimation are partitioned into three parts. The first part is sampling errors in the estimates of branch lengths in the tree for each locus owing to limited sequence length, which corresponds to the term $b/(NL)$. This part of the uncertainty will be eliminated as either $N \to \infty$ or $L \to \infty$. If L is large, this component goes to zero at the rate $1/N$. The second part is due to variation in substitution rates among lineages and among loci according to the relaxed-clock model, which corresponds to the term a/L. This part of the uncertainty goes to zero at rate $1/L$ when $L \to \infty$. The last part is due to the uncertainty in fossil calibrations, which corresponds to the term w_∞^2. This uncertainty cannot be reduced by further increasing the amount of sequence data. In fact, the finite-sites theory for the strict-clock model with one locus (Eq. 6.25) is a special case of the finite-sites theory discussed in this section (Eq. 6.26), with $a = 0$ and $L = 1$.

Zhu et al. (2015) used MCMCTree to analyse a data set comprising sequences from six primate species (Fig. 6.4). The posterior uncertainties (measured by w^2) of five divergence times (t_7 to t_{11}) against $1/L$ are shown in Fig. 6.5. In all cases except t_8, when L is larger than 10, w^2 shows a strong linear relationship with $1/L$. For t_8, the linear relationship holds when L is larger than 50, which is due to the very informative fossil calibration on this node. When $L \to \infty$, the intercept of the line is the amount of uncertainty from fossil calibrations. Note that in real data sets, sequence lengths are likely to differ among loci. Although the finite-sites theory assumes that all loci have the same length, the linear relationship holds even if this assumption is incorrect.

There are 7949 genes in the primate data set. To examine the impact of the number of loci on the posterior precision (measured by the width of the 95% CI), the data set was subsampled to produce data sets with L = 1, 5, 10, 20, 50, 100, 200, and 500 loci by two strategies. Regardless of which strategy was applied, the posterior CI widths were very short with larger L. The results suggest that to improve the precision of

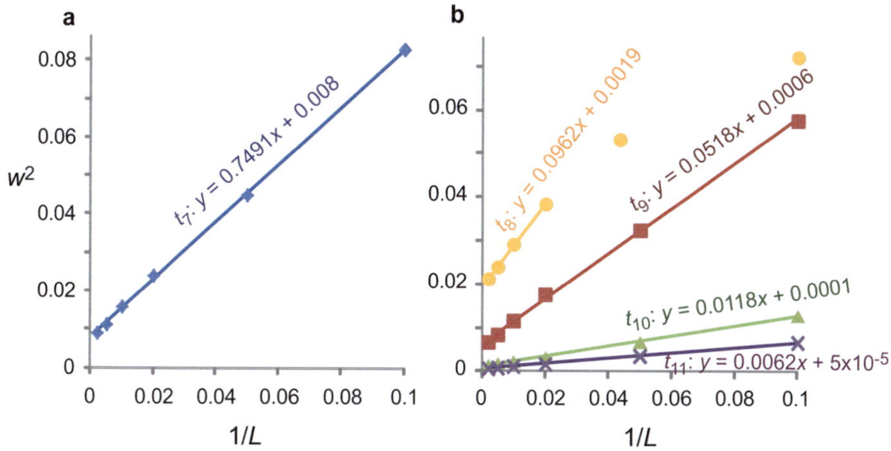

Fig. 6.5 The finite-sites theory applied to the analysis of genomic sequence data from six primate species (Fig. 6.4). The square of the 95% posterior CI widths (w^2) for (**a**) the root node t_7 and (**b**) four other node ages (t_8, t_9, t_{10}, and t_{11}) are plotted against the reciprocal of the number of loci, sampled at random from 7947 protein-coding genes (with only the third codon positions used)

posterior time estimation, increasing the number of loci is far more effective than increasing the sequence length at each locus, indicating the importance of using multilocus data in relaxed-clock dating analyses. However, even if a huge amount of sequence data is analysed, considerable uncertainty will persist in time estimates owing to the nature of the fossil calibrations.

6.7 An Example

Here we use a very simple example to illustrate the MCMC implementation of Bayesian clock dating. The data consists of 12S rRNA genes from humans and orangutans. There are $x = 90$ differences at $n = 948$ nucleotides. We assume a strict clock and the JC69 model of nucleotide substitution (Jukes and Cantor 1969). The likelihood is a function of $d = 2rt$ only, so there are no extra parameters in the evolutionary model. Thus, the only two parameters are divergence time t and the evolutionary rate on both branches r. Under these assumptions, the posterior (Eq. 6.5) is simplified as:

$$f(t, r|D) = \frac{f(D|t, r)f(r|t)f(t)}{\int f(D|t, r)f(r|t)f(t)drdt} \quad (6.27)$$

Under the JC69 substitution model, the likelihood is:

$$f(D|r, t) = \left(\frac{3}{4} - \frac{3}{4}e^{-8rt/3}\right)^x \left(\frac{1}{4} + \frac{3}{4}e^{-8rt/3}\right)^{n-x} \quad (6.28)$$

Unless specifically stated, the time unit is 100 Myr. Under the strict-clock model, the rate r does not rely on the time t. We use two prior distributions for r to examine the sensitivity to the rate prior. The first is G(1, 10), which is equivalent to the exponential distribution Exp(10), with mean 0.1 and variance 0.01. The other is inverse-gamma distribution InvG(3, 0.2), also with mean 0.1 and variance 0.01. In this case, the prior density of r is:

$$f(r) = \frac{\beta^\alpha}{\Gamma(\alpha)} r^{-\alpha-1} \exp\left(-\frac{\beta}{r}\right) \quad (6.29)$$

With $\alpha = 3$ and $\beta = 0.2$. For the root age t, the fossil information is available and we use

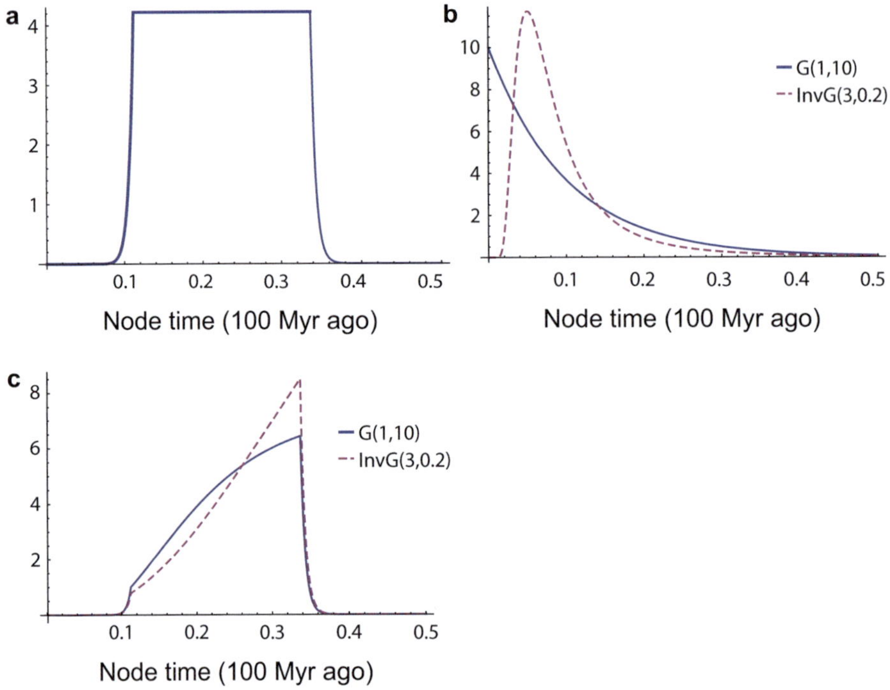

Fig. 6.6 The time prior, rate priors, and posteriors of divergence time. (**a**) The time prior with a soft bound for the node is $0.112 < t < 0.337$ (where each time unit represents 100 Myr). The probabilities that the node age is younger than 0.112 and older than 0.337 are 2.5% each. (**b**) The probability densities of two rate priors. The blue solid line is G(1, 10) and the purple dashed line is InvG (3, 0.2). Both priors have mean 0.1 (i.e., 10^{-9} substitutions per site per year) and variance 0.01. (**c**) The posterior density of the divergence time with gamma and inverse-gamma rate prior. Both posterior densities have mean 0.2637 and 95% CI (0.1325, 0.3420)

$0.112 < t < 0.337$ to calibrate in Zhu et al. (2015). The fossil bounds are implemented as soft uniform bounds, allowing 2.5% of the probability below the lower bound and 2.5% of the probability above the upper bound. In the left and right tails, there are exponential decays, and following Yang and Rannala (2006) the prior density of time t is:

$$f(t) = \begin{cases} 0.025\dfrac{\theta_1}{t_L}\left(\dfrac{t}{t_L}\right)^{\theta_1 - 1}, & t \le t_L \\ \dfrac{0.95}{t_U - t_L}, & t_L < t < t_U \\ 0.025\theta_2 \exp\{-\theta_2(t - t_U)\}, & t \ge t_U \end{cases} \tag{6.30}$$

where $\theta_1 = 0.95t_L/(0.025(t_U - t_L))$ and $\theta_2 = 0.95/(0.025(t_U - t_L))$. The time prior density is shown in Fig. 6.6a. Because there are only two parameters in the model and the data set includes only one alignment, we use Eq. (6.27) directly to calculate the joint posterior. The posterior density of divergence time t is derived by integration over r:

$$f(t|D) = \int f(t, r|D)dr \tag{6.31}$$

The gamma rate prior and inverse gamma rate prior are shown in Fig. 6.6b. The posterior distributions of t using two rate priors are shown in Fig. 6.6c. Regardless of which rate prior is used, the posterior mean of t is 0.2637 (time unit 100 Myr) and the 95% equal-tail CI is (0.1325, 0.3420). This is mainly because the two priors are similar at the tail (say, $r > 0.34$), and have the same mean and variance. At the same time, the

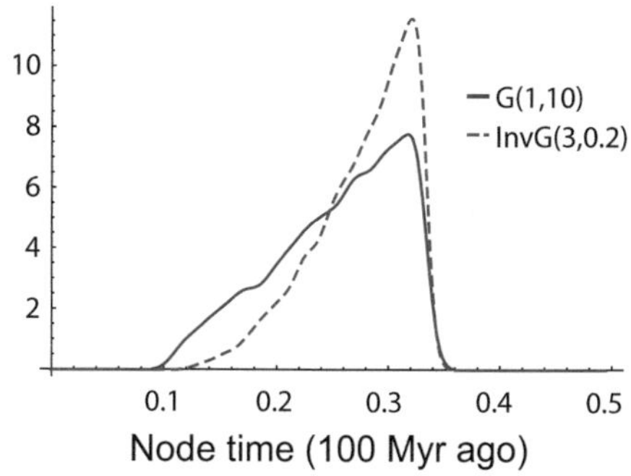

Fig. 6.7 The posterior density of the divergence time for two species with gamma rate prior G(1, 10) and inverse-gamma rate prior InvG(3,0.2). The Markov chain Monte Carlo algorithm was used to generate 10^4 posterior samples for this plot

time prior density is concentrated in the interval (0.112, 0.337). Because the size of the data set is small, the time prior and rate prior dominate the posterior. In both cases, the posterior mean is inside the interval of the fossil calibration. Note that in both cases, the posterior of time has a mode at 0.337, which is the upper bound of the fossil calibration. This is mainly due to the rate priors placing greater weight on low rates conflicting with the data. The maximum-likelihood estimate of evolutionary distance d is 0.1015 (e.g., Yang 2014). Using the posterior mean $t = 0.2637$ as an estimate of t, a simple approximate calculation leads to $r = d/(2t) = 0.19$, which is almost twice the prior mean.

A sketch of the MCMC algorithm is as follows:

1. Set initial state (θ, **t**, **r**). In the plot shown in Fig. 6.7, there is no parameter in the substitution model, and we set $t = 0.2$ and $r = 0.1$ as initial values.
2. In each iteration, do the following:
   (a) Change times **t**. In the plot shown in Fig. 6.7, a new state of t is proposed by a uniform sliding window with width $w = 0.01$. Note that divergence times should not conflict with each other.
   (b) Change $\mathbf{r}_i$ for each locus i. In the plot shown in Fig. 6.7, a new state of r is proposed by a uniform sliding window with width $w = 0.05$.
   (c) Change substitution parameters θ. (This step is skipped in the plot shown in Fig. 6.7).
   (d) Do proportional scaling if necessary. Generate multiplier c, which is a random variable near 1. Then multiply all times by c, and divide all rates by c.
   (e) Calculate the acceptance ratio α according to Eq. (6.7) and accept the new state with probability α.
   (f) Record (θ, **t**, **r**) every k iterations, where k is the sample frequency.
3. Summarize the data. Discard the first part of the samples as burn-in. In Fig. 6.7, 10,000 MCMC samples are collected to estimate the posterior density. Extract component t from vector (t, r), and the posterior mean and CI can be estimated accordingly.

6.8 Bayesian Dating Software

In this section, we briefly describe some of the Bayesian dating software packages that can be used to analyse multilocus data sets.

MCMCTree is a program in the PAML package (Yang 2007). How MCMCTree works is briefly introduced in Sect. 6.3. MCMCTree dates with soft fossil calibrations under various molecular clock models. Evolutionary rates can vary across sites, along lineages, and among loci. As a compound Dirichlet distribution is used for the rate prior among loci, the posterior time estimates are insensitive to the rate prior, making the posterior estimates more accurate and robust. MCMCTree offers an option to use a fast approximate likelihood method, which enables the analysis of genome-scale data sets (dos Reis and Yang 2011).

BEAST (Suchard et al. 2018; Bouckaert et al. 2019) is a comprehensive Bayesian MCMC analysis software package. BEAST can use MCMC to average over tree space, and thus the data are analysed without conditioning on a single tree topology. In this respect, BEAST differs from most other dating software. A variety of evolutionary models are available in BEAST. Besides the frequently used JC69, K80, HKY, and GTR models, complex substitution models can be specified. The strict-clock model, relaxed-clock models, and local-clock models can be used for dating analyses. Morphological trait models are also available.

MrBayes (Ronquist et al. 2012) is a large software package for Bayesian inference and model selection. The topological model can be fixed, constrained, or unconstrained (with all labelled trees having the same probability). The clock models include a uniform strict-clock model and three relaxed-clock models: the Thorne-Kishino 2002 (TK02) autocorrelated-rates model, the compound Poisson process (CPP) model, and the independent gamma rates (IGR) model. When calibrations are assigned to the nodes, the prior of the node ages is forced to satisfy the calibrations. In addition to dating with calibrations on node ages, MrBayes implements total-evidence dating in which fossil tips are assigned dates while internal nodes do not necessarily have any calibrations.

DPPDiv (Heath et al. 2012) is an application for estimating divergence times and substitution rates on a fixed species tree. A wide range of evolutionary models can be used for analysis, including very complex models. Clock models include the Dirichlet relaxed-clock model, independent-rates model, and strict-clock model. The priors on time include the birth–death model and the uniform distribution. Fast versions FastDPPDiv and FDPPDIV are available.

Multidivtime (Thorne et al. 1998; Kishino et al. 2001) is the first Bayesian dating program. The geometric Brownian model was introduced in Multidivtime. The prior of the divergence times is a generalization of the Dirichlet distribution to rooted tree structures. Thorne et al. (1998) approximated the likelihood surface with a multivariate normal distribution to alleviate the heavy computation of the likelihood.

PhyloBayes (Lartillot et al. 2009, 2013) is a Bayesian MCMC software package for phylogenetic reconstruction and molecular dating analysis using protein alignments. A distinguishing feature of PhyloBayes is the underlying probabilistic model CAT, which accounts for site-specific features of protein evolution. PhyloBayes implements autocorrelated as well as non-autocorrelated models of rates. Both hard and soft fossil bounds are accepted. Data augmentation methods are used to speed up the likelihood computation, so that large multilocus data sets can be analysed. Parallel computing is also allowed, making PhyloBayes especially suitable for large data sets.

6.9 Conclusions and Perspectives

Bayesian molecular clock dating has gone through rapid development in the past two decades, driven by the advancement of sequencing technologies, explosive growth of genomic data sets, rapid accumulation of morphological measurements from modern and fossil species, and development of powerful statistical models. It has now become the dominant approach for molecular dating. Through the prior for times and prior for rates, the method provides a natural framework for integrating information from different kinds of data, particularly nucleotide sequences and fossils. In contrast, non-Bayesian approaches often have difficulty in accommodating the uncertainties in the fossil data, for example, and might thus produce very precise but unreliable estimates (dos Reis et al. 2016).

Many heuristic non-Bayesian methods are computationally efficient and can be applied to very large data sets (see Chap. 12). In contrast, the MCMC algorithm used by Bayesian methods requires intensive computation, which prohibits its application to some phylogenomic data sets. Improving the mixing efficiency of MCMC algorithms so that they can handle genome-scale data sets will be a major research topic for the future. We note that MCMCTree applies approximate calculation of the likelihood function so that the program can be applied to extremely long sequences (dos Reis and Yang 2011) and faster versions of DPPDiv have also been produced.

Given the complexity and huge stochastic fluctuations of the process of fossil preservation and discovery, and the difficulty of interpretation, clock dating will remain a very inexact science for years to come. It is then essential to understand the sensitivity of time estimates to different aspects of the analysis, including the branching-process model that specifies the time prior and the model of rate drift that specifies the rate prior. For example, current rate-drift models assume that rates vary independently among genes, but it is well known that there might exist strong lineage or genome effects, with certain branches of the species phylogeny showing high (or low) rates across almost all genes. Accounting for such genome effects will be important.

Another challenging component of the dating analysis is the partitioning of sites, with the sites in the same locus or subset assumed to share the same trajectory of evolutionary rate drift. Theory predicts that time estimates will become more precise when more loci are included in the analysis or when the same set of sites are partitioned into more subsets. However high precision does not necessarily mean high accuracy. The next few years are likely to see a more systematic characterization of the limits of divergence-time estimation. Although multiple factors can cause uncertainties in a clock dating analysis, we suggest that the joint analysis of morphological and molecular data from both extant and fossil species provides the most promising approach to resolving the timescale of the tree of life.

References

Benton MJ, Donoghue PCJ, Asher RJ (2009) Calibrating and constraining molecular clocks. In: Hedges BS, Kumar S (eds) The timetree of life. Oxford University Press, Oxford, UK, pp 35–86

Bouckaert R, Vaughan TG, Barido-Sottani J, Duchêne S, Fourment M, Gavryushkina A, Heled J, Jones G, Kühnert D, De Maio N, Matschiner M, Mendes FK, Müller NF, Ogilvie HA, du Plessis L, Popinga A, Rambaut A, Rasmussen D, Siveroni I, Suchard MA, Wu C-H, Xie D, Zhang C, Stadler T, Drummond AJ (2019) BEAST 2.5: an advanced software platform for Bayesian evolutionary analysis. PLOS Comput Biol 15:e1006650

Bromham L, Duchêne S, Hua X, Ritchie AM, Duchêne DA, Ho SYW (2018) Bayesian molecular dating: opening up the black box. Biol Rev 93:1165–1191

Brown RP, Yang Z (2011) Rate variation and estimation of divergence times using strict and relaxed clocks. BMC Evol Biol 11:271

dos Reis M, Yang Z (2011) Approximate likelihood calculation on a phylogeny for Bayesian estimation of divergence times. Mol Biol Evol 28:2161–2172

dos Reis M, Yang Z (2013) The unbearable uncertainty of Bayesian divergence time estimation. J Syst Evol 51:30–43

dos Reis M, Zhu T, Yang Z (2014) The impact of the rate prior on Bayesian estimation of divergence times with multiple loci. Syst Biol 63:555–565

dos Reis M, Donoghue PCJ, Yang Z (2016) Bayesian molecular clock dating of species divergences in the genomics era. Nat Rev Genet 17:71–80

Drummond AJ, Ho SYW, Phillips MJ, Rambaut A (2006) Relaxed phylogenetics and dating with confidence. PLOS Biol 4:e88

Felsenstein J (1981) Evolutionary trees from DNA sequences: a maximum likelihood approach. J Mol Evol 17:368–376

Graur D, Martin W (2004) Reading the entrails of chickens: molecular timescales of evolution and the illusion of precision. Trends Genet 20:80–86

Guindon S (2010) Bayesian estimation of divergence times from large sequence alignments. Mol Biol Evol 27:1768–1781

Hasegawa M, Thorne JL, Kishino H (2003) Time scale of eutherian evolution estimated without assuming a constant rate of molecular evolution. Genes Genet Syst 78:267–283

Heath TA, Holder MT, Huelsenbeck JP (2012) A Dirichlet process prior for estimating lineage-specific substitution rates. Mol Biol Evol 29:939–955

Heath TA, Huelsenbeck JP, Stadler T (2014) The fossilized birth-death process for coherent calibration of divergence-time estimates. Proc Natl Acad Sci USA 111:E2957–E2966

Ho SYW, Phillips MJ (2009) Accounting for calibration uncertainty in phylogenetic estimation of evolutionary divergence times. Syst Biol 58:367–380

Hug LA, Baker BJ, Anantharaman K, Brown CT, Probst AJ, Castelle CJ, Butterfield CN, Hernsdorf AW, Amano Y, Ise K, Suzuki Y, Dudek N, Relman DA, Finstad KM, Amundson R, Thomas BC, Banfield JF (2016) A new view of the tree of life. Nat Microbiol 1:16048

Jukes TH, Cantor CR (1969) Evolution of protein molecules. In: Munro HN (ed) Mammalian protein metabolism. Academic, New York, pp 21–131

Kendall DG (1948) On the generalized birth-and-death process. Ann Math Stat 19:1–15

Kishino H, Thorne JL, Bruno WJ (2001) Performance of a divergence time estimation method under a probabilistic model of rate evolution. Mol Biol Evol 18:352–361

Langley CH, Fitch WM (1974) An examination of the constancy of the rate of molecular evolution. J Mol Evol 3:161–177
Lartillot N, Lepage T, Blanquart S (2009) PhyloBayes 3: a Bayesian software package for phylogenetic reconstruction and molecular dating. Bioinformatics 25:2286–2288
Lartillot N, Rodrigue N, Stubbs D, Richer J (2013) PhyloBayes MPI: phylogenetic reconstruction with infinite mixtures of profiles in a parallel environment. Syst Biol 62:611–615
Mau B, Newton MA (1997) Phylogenetic inference for binary data on dendrograms using Markov chain Monte Carlo. J Comput Graph Stat 6:122–131
Mau B, Newton MA, Larget B (1999) Bayesian phylogenetic inference via Markov chain Monte Carlo methods. Biometrics 55:1–12
Nee S, May RM, Harvey PH (1994) The reconstructed evolutionary process. Philos Trans R Soc B 344:305–311
Rannala B, Yang Z (1996) Probability distribution of molecular evolutionary trees: a new method of phylogenetic inference. J Mol Evol 43:304–311
Rannala B, Yang Z (2007) Inferring speciation times under an episodic molecular clock. Syst Biol 56:453–466
Ronquist F, Teslenko M, van der Mark P, Ayres DL, Darling A, Höhna S, Larget B, Liu L, Suchard MA, Huelsenbeck JP (2012) MrBayes 3.2: efficient Bayesian phylogenetic inference and model choice across a large model space. Syst Biol 61:539–542
Stadler T (2010) Sampling-through-time in birth-death trees. J Theor Biol 267:396–404
Suchard MA, Lemey P, Baele G, Ayres DL, Drummond AJ, Rambaut A (2018) Bayesian phylogenetic and phylodynamic data integration using BEAST 1.10. Virus Evol 4:vey016
Thorne JL, Kishino H, Painter IS (1998) Estimating the rate of evolution of the rate of molecular evolution. Mol Biol Evol 15:1647–1657
Yang Z (2007) PAML 4: phylogenetic analysis by maximum likelihood. Mol Biol Evol 24:1586–1591
Yang Z (2014) Molecular evolution: a statistical approach. Oxford University Press, Oxford, UK
Yang Z, Rannala B (1997) Bayesian phylogenetic inference using DNA sequences: a Markov chain Monte Carlo Method. Mol Biol Evol 14:717–724
Yang Z, Rannala B (2006) Bayesian estimation of species divergence times under a molecular clock using multiple fossil calibrations with soft bounds. Mol Biol Evol 23:212–226
Yoder AD, Yang Z (2000) Estimation of primate speciation dates using local molecular clocks. Mol Biol Evol 17:1081–1090
Zhu T, dos Reis M, Yang Z (2015) Characterization of the uncertainty of divergence time estimation under relaxed molecular clock models using multiple loci. Syst Biol 64:267–280
Zuckerkandl E, Pauling L (1965) Evolutionary divergence and convergence in proteins. In: Bryson V, Vogel HJ (eds) Evolving genes and proteins. Academic, New York, pp 97–166

7 Clock Models for Evolution of Discrete Phenotypic Characters

Michael S. Y. Lee

Abstract

Clock models, which consider rates of character change through time and across lineages, are widely used in molecular phylogenetics for inferring the pattern and timing of evolutionary divergences. However, clock models are also potentially relevant to phenotypic analyses, such as those involving morphological characters. Important hurdles need to be overcome, including biases in character sampling (e.g., changes along terminal branches are often not sampled) and strong among-lineage heterogeneity in evolutionary rates (e.g., 'living fossils' vs rapidly evolving taxa). These caveats notwithstanding, an increasing number of empirical studies have applied clock models to good effect in phylogenetic analysis, especially in combined analyses of DNA and morphological data from fossil and living taxa (total-evidence tip-dating). These studies have improved our estimates of the shape and chronology of the tree of life.

M. S. Y. Lee (✉)
College of Science and Engineering, Flinders University, Bedford Park, SA, Australia

Earth Sciences Section, South Australian Museum, Adelaide, SA, Australia
e-mail: mike.lee@samuseum.sa.gov.au

Keywords

Phenotypic characters · Evolutionary rate variation · Relaxed clock · Morphological clock · Tip-dating · Total-evidence tip-dating · Bayesian phylogenetics

7.1 Introduction

Molecular clocks have been discussed and developed extensively since the concept was first introduced (Zuckerkandl and Pauling 1962). They are now firmly established as a major area of scientific research, yielding insights into the dynamics of genetic evolution as well as the shape and timing of the tree of life, among many other things. The catalyst for the early interest in molecular clocks was the possibility that molecular change accrued at a relatively constant rate across time and across lineages, with the result that genetic divergence could be employed as a universal yardstick for relatedness (time elapsed since lineage splitting).

Subsequent work has increasingly questioned this 'strict clock' and instead revealed evidence that molecular evolutionary rates can often vary greatly (e.g., Lanfear et al. 2010). Nevertheless, there are sometimes predictable patterns to this variation, with particular groups having a relatively narrow range of typical evolutionary rates (see Chap. 1). For instance, the genomes of viruses evolve at rates up to a million times higher

S. Y. W. Ho (ed.), *The Molecular Evolutionary Clock*, https://doi.org/10.1007/978-3-030-60181-2_7

than those of mammals (Bromham and Penny 2003), while even within mammals, some groups such as rodents tend to evolve more rapidly than others. To better accommodate such variation, a range of 'relaxed' clock models have been developed, which allow rates of evolution to vary across lineages and across time (e.g., Sanderson 1997; Drummond et al. 2006).

The traits used most commonly in morphological phylogenetics are discrete phenotypic traits (e.g., presence/absence), rather than continuous (e.g., ratios and lengths) or geometric morphometric (two- or three-dimensional shape) characters. Discrete traits also arguably have the most similar evolutionary dynamics to those of molecular data (e.g., nucleotide sequences): each discrete trait flips between a small number of possible states (e.g., Lewis 2001). Thus, this chapter will focus on clock models for discrete phenotypic traits, and discrete non-molecular traits in general (e.g., certain ecological or biogeographic characters). Clock models can be readily applied to discrete phenotypic data such as morphology, and the interaction of molecular and morphological clocks can yield novel insights into evolutionary rates in both systems. However, morphological phylogenetic studies continue to be performed largely without any explicit clock model, often inferring branching order without any relative or absolute temporal dimension (e.g., most commonly using parsimony or undated Bayesian methods).

There have been comparatively few studies that have explicitly used clock models to infer phylogenies from morphological data. Several factors have contributed to this dearth of studies. The most common method of phylogenetic analysis of discrete morphological data remains parsimony (cladistics *sensu* Hennig 1966); this approach selects the tree with the smallest number of character changes and does not incorporate any temporal information. Model-based approaches, where expected changes are scaled according to relative or absolute time elapsed (such as dated Bayesian methods; Chap. 6), have only been adopted relatively recently for phylogenetic analysis of phenotypic data.

Furthermore, there are some widely appreciated issues with clock models for phenotypic data. First, unlike the case for genetic data, there was never any expectation that phenotypic evolution would follow a constant clock. Classic works from evolution's modern synthesis emphasized this rate variability across lineages, epitomized by Simpson's (1953) terms bradytely, horotely, and tachytely for low, typical, and high rates of morphological evolution. The clock models implemented in most phylogenetic packages were largely developed in the context of molecular data that had lower rate variation, and the application of these models to morphological data might be problematic (e.g., dos Reis et al. 2016). Second, phenotypic characters tend to evolve in a highly mosaic fashion, so a single common 'morphological clock' might not adequately capture biological reality (Goloboff et al. 2019). Third, the intrinsic difficulty in identifying 'unit' phenotypic characters (and thus unit changes) makes morphological clock analyses more difficult (e.g., Freudenstein 2005). Fourth, most phenotypic data sets were collected in a parsimony framework and often failed to adequately sample changes along terminal branches of the tree (Yeates 1992), making them potentially ill-suited to clock analyses (e.g., Lee and Palci 2015).

All four hurdles can be overcome, to some extent. First, even if rates of evolution of discrete phenotypic traits are much more variable than for molecular data, the ability of highly flexible relaxed-clock models to adequately model this variation should be a matter for empirical evaluation (e.g., Gavryushkina et al. 2017; Goloboff et al. 2019). Notably, the majority of comparative studies of phenotypic characters (Harvey and Pagel 1991) continue to use models where expected change is some function of time (branch duration). Most of these studies make the even stronger assumption of a strict clock, i.e., totally homogeneous rates of phenotypic evolution. Yet, few have seized upon this assumption to challenge the broad validity of the entire field of comparative biology. Only relatively recently have models in comparative studies been elaborated to incorporate rate variation across lineages (e.g., O'Meara et al. 2006; Rabosky 2014). Second, the use of multiple clocks—as already happening in analyses of genomic data (Duchêne et al. 2014)—might ameliorate issues with the lack of a single common 'morphological clock.' Third, the difficulty in identifying 'units' of phenotypic change is not unique to morphological clock studies but is critical for many other disciplines, notably morphological phylogenetics in general (e.g., Freudenstein 2005). Again, this issue has rarely been considered as a reason to reject the entire field of morphological phylogenetics. Fourth, there is nothing to deter researchers from collecting new data sets or modifying old ones so that they conform better to the assumptions of clock analyses, notably by scoring *all* variation across *all* branches, including 'autapomorphies' that change on terminal branches. Some analytical corrections to correct for biases in most earlier data sets might also work in particular situations (Matzke and Irmis 2018).

Why study phenotypic evolution using clock models? In addition to improving our understanding of the tempo and drivers of phenotypic evolution, it can help us better reconstruct ancestral states, infer divergence dates across the time-tree of life, and potentially improve our estimates of phylogenetic relationships, even in the age of genomic data. Thus, there is direct relevance for molecular clock analyses, which almost always require an absolute temporal framework. One promising new method to provide this absolute timescale is total-evidence tip-dating (Ronquist et al. 2012a), which integrates morphological clocks with molecular clocks, and combines living taxa and fossils, to simultaneously infer the tree topology and divergence dates.

7.2 Categories of Phenotypic Data

The phenotype of an organism encompasses all of its expressed morphological, ecological, and behavioural traits. When organisms extensively modify their immediate habitats, it is often difficult to draw the line where the phenotype actually ends and the environment begins (Dawkins 1982; Odling-Smee et al. 2003). However, only phenotypic traits that are genetically determined and heritable are generally appropriate for phylogenetic analyses and clock studies. Phenotypically plastic traits are environmentally determined, and there is no reason to expect differences across organisms to reflect genealogical relationships or to correlate with time elapsed since divergence.

An obvious categorization of phenotypic data can be made using their biological characteristics. The vast majority of phenotypic characters used in phylogenetic studies, including those employing clock models, are anatomical (morphological) traits, often hard parts (e.g., bones, shells, and exoskeletons) due to their accessibility in museum specimens and frequent preservation in the fossil record. However, soft anatomy, behavioural, and ecological traits can and are also regularly employed. The use of features of human language to determine population relationships and divergence times is an exciting, rapidly expanding study area (Maurits et al. 2017).

In the context of phylogenetic analysis and clock models, a different categorization of phenotypic data is potentially more useful. This is based on how they are encoded as character information and on the relevant evolutionary models (Wiens 1989):

1. Discrete characters can only take the form of a small number of states, conventionally labelled using integers starting from 0. Examples are whether two bones in the skull meet or not, whether the eyes are brown, blue, or green, and whether an organism's diet includes plant matter or not. This sort of phenotypic character is most analogous to DNA sequence data, which have four discrete states (typically A, T, C, and G).
2. Meristic characters are counts which can take any value within a large range of whole numbers. These characters share some similarities with typical discrete characters. Examples are number of dorsal fin spines in a fish and number of repeated notes in a bird's vocalization.

3. Continuous characters are measurements (or functions of multiple measurements) that vary in a single dimension and can take the value of any real number. Examples are body size, relative brain size (a ratio), or area of home range.
4. Geometric morphometric data are two- or three-dimensional data that involve identifying homologous landmarks across taxa, rescaling and aligning their shapes, and inferring the landmark displacements. Thus, unlike the first three categories, characters are not (at least potentially) independent of each other, but emerge simultaneously as displacement vectors when shapes are superimposed.

Although these definitions appear relatively clear-cut, in practice many traits can be expressed as more than one type of character. For instance, the length of a bone (a continuous character) can be reduced into a discrete character (e.g., short, medium, and long), or expressed using geometric morphometrics (e.g., via the displacement of landmarks at the proximal and distal ends). Most early versions of phylogenetic software were designed for discrete characters (e.g., PAUP*; Swofford 2003), so continuous and meristic traits were often 'discretized'. However, many modern phylogenetic programs (e.g., TNT, Goloboff and Catalano 2016; BEAST2, Bouckaert et al. 2019) can analyse all discrete, meristic, and continuous (and in the case of TNT, landmark) data simultaneously; continuous and meristic traits no longer need to be recoded into a small number of discrete states, which can result in loss of information.

This chapter will focus on discrete traits, because they remain by far the most common type of phenotypic data used in phylogenetic analyses, including those that implement clock models. Discrete phenotypic data are also more similar to DNA sequence data than are meristic, continuous, or geometric morphometric data. However, clock models can and have also been applied to other categories of phenotypic data, notably continuous and multidimensional traits (e.g., Álvarez-Carretero et al. 2019; Paterson et al. 2019).

7.3 Considerations When Working with Discrete Phenotypic Data

Phenotypic characters are usually identified in a much more idiosyncratic fashion than DNA characters. Generally, investigators survey particular phenotypic systems of interest (e.g., the skull), and then 'score' characters based on observed variation perceived to be phylogenetically informative. Invariant characters are almost never scored, and characters unique to single taxa (autapomorphies) have also historically been rarely scored. This is because autapomorphies are phylogenetically uninformative under parsimony (the dominant method for analysing morphological characters in recent decades): autapomorphies map perfectly onto any and every tree topology with no homoplasy and thus do not arbitrate between alternative trees. This bias means that existing morphological data sets typically underestimate the amount of change along terminal branches (Seligmann 2010), which raises important issues for clock analyses (see later). In contrast, DNA data are gathered much more methodically, e.g., all nucleotides for a gene or fragment can be sequenced, including invariant and autapomorphic characters. Even when there is ascertainment bias in molecular data, as when single-nucleotide polymorphisms are analysed, it is more easily corrected mathematically.

The size of phenotypic data sets (typically dozens to a few hundred characters) is also much smaller than for molecular data sets (now usually thousands to millions of characters) (Lee and Palci 2015). This means that the number of phenotypic changes along each branch is usually small. This small sample size severely limits the power of clock models to infer subtle patterns of variation in phenotypic evolutionary rates across the tree (dos Reis et al. 2016). For instance, if a branch of a certain duration that 'should' have undergone 100 nucleotide substitutions only exhibits 50 such changes, this would be rather strong evidence of a reduced rate of change. But if the expected and actual numbers of morphological changes were ten and five respectively, it is difficult to determine whether this is due to a genuine slowdown or simply due to the stochastic nature of the evolutionary process.

Discrete phenotypic data also differ from standard DNA nucleotide data in several important ways that are relevant to model-based phylogenetic and clock analyses. First, states given the same label *across* different phenotypic characters are usually not equivalent: in different characters, state 0 might denote the absence of a particular bone, or brown eye colour, or aquatic habits. For DNA sequence data, corresponding states across all characters are comparable: at all sites, an 'A' always denotes the base adenine. Second, the labelling of states *within* a particular character is also largely arbitrary: given three discrete eye-colour states, any colour could be called 0. Historically, the state that is likely to be primitive in the group under investigation (ingroup) is often allocated state 0, but this is still arbitrary: if the analysis were to consider a larger or a more restricted clade, the inferred primitive state for the 'group of interest' might change. Where character states form a continuum or morphocline (e.g., short, medium, and long process), it is often convenient to number the states in ascending or descending order, but again, the polarity of this numbering is arbitrary (e.g., either short or long could be allocated state 0).

The smaller size of phenotypic data sets, non-equivalence of states across characters, and the arbitrariness of state labels within characters severely restricts the implementation of detailed models related to aspects of morphological evolution, including clock models. These models are most useful when they can estimate evolutionary dynamics from large samples of comparable traits.

7.4 Models of Morphological Character Evolution

Model-based analyses of discrete phenotypic characters typically use the stochastic model of Lewis (2001), which is essentially the Jukes–Cantor (1969) model of molecular evolution generalized to a particular number of morphological states (e.g., see Wright and Hillis 2014). Transformations between all possible pairs of different states occur at equal rates (Fig. 7.1a); these rates are determined by (homogeneous) relative-rate parameters and (equal) equilibrium state frequencies. There are two variations of this model that are typically used, based on analogous approaches in parsimony. The unordered model is appropriate where states do not form a morphocline, and direct transformations between all states are allowed (Fig. 7.1a). The ordered model is often used where states form a clear morphocline, and direct transformations are only allowed between morphologically adjacent (similar) states; transformations from one extreme to the other are thus constrained to pass through all intermediate states (Fig. 7.1b). The distinction between unordered and ordered characters is only relevant for characters with three or more states ('multistate characters'); for binary (two-state) characters, there can be no intermediate states, and direct transformations between all (both) states must be possible.

For a data set of numerous discrete characters with a range of (2, 3, . . ., k) states, it is possible to analyse all characters using a single $k \times k$ rate matrix (or two $k \times k$ matrices, for ordered and unordered characters). Such an approach assumes that all characters, even the binary characters, can (theoretically at least) flip between all k states, and thus might be seen to be less appropriate than a more complex approach that disallows unobserved states (Gavryushkina et al. 2017). In the latter partitioned approach, all binary characters are analysed using a 2×2 matrix, three-state characters using a 3×3 matrix, and so on. Because the state labels for phenotypic characters are largely arbitrary, it generally does not make sense to have more complex substitution matrices or state frequencies analogous to the HKY or GTR models used for DNA sequence data. Across all four-state unordered characters (Fig. 7.1a), for instance, it is often meaningless to try and estimate a rate matrix where $0 \rightarrow 1$ transformations occur at a rate different from $1 \rightarrow 2$ transformations, or from $1 \rightarrow 0$ transformations. Promising attempts have been made to stochastically model asymmetry in such

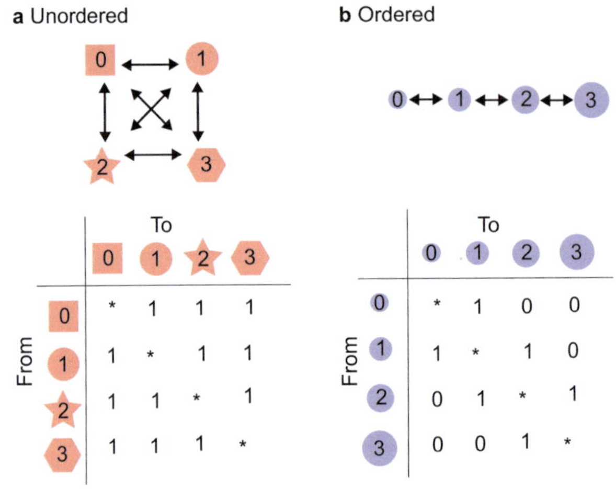

	0	1	2	3
0	*	1	1	1
1	1	*	1	1
2	1	1	*	1
3	1	1	1	*

	0	1	2	3
0	*	1	0	0
1	1	*	1	0
2	0	1	*	1
3	0	0	1	*

Fig. 7.1 Simplified substitution matrices for a four-state discrete character. In each matrix, nonzero off-diagonal rates are a function of homogeneous relative-rate parameters and equal equilibrium state frequencies (see Lewis 2001). (**a**) Unordered model, where states (e.g., coloured shapes directly representing body shape) do not form a morphocline, and each state can directly change into any of the other three states, at the same rate. (**b**) Ordered model, where states (e.g., coloured circles proportional to body size) form a linear morphocline, and each state can only directly change into adjacent states. More complex asymmetrical matrices are possible

morphological rate matrices, by allowing different characters to have different state frequencies sampled from some prior distribution (Klopfstein et al. 2015). Additionally, for comparative analyses of single characters, there is no issue regarding non-equivalent state labels, and heterogeneous rate matrices are potentially appropriate (e.g., King and Lee 2015).

Models of among-site rate variability, originally developed by geneticists, have been readily adapted for use for phenotypic characters. Thus, rate variation among phenotypic characters is often modelled using a gamma distribution (Yang 1993, 1994) widely used for molecular data, though other distributions might be more appropriate for morphological data (Harrison and Larsson 2015).

7.5 Considerations When Working with Clock Analyses of Morphological Data

The clock models developed largely for DNA data, and implemented in widely used phylogenetic packages such as MrBayes (Ronquist et al. 2012b), RevBayes (Höhna et al. 2016), BEAST (Drummond et al. 2012; Suchard et al. 2018), and BEAST 2 (Bouckaert et al. 2019), can readily accommodate discrete phenotypic data. As discussed above, the most widely used substitution model for the evolution of discrete phenotypic traits is very similar to the Jukes–Cantor model of DNA evolution. Thus, theoretically, most of the elements of molecular clock models used to characterize among-lineage rate variation in molecular data can be applied to phenotypic data, though their appropriateness might vary for several reasons. The major considerations are summarized as follows:

1. Strict clock or relaxed clock. The frequently episodic nature of phenotypic evolution means that few phenotypic data sets are likely to conform to a strict clock. In empirical studies that have compared strict and relaxed clocks using model-testing methods such as Bayes factors, some sort of relaxed clock has always been a significantly better fit (e.g., Lee and Yates 2018).
2. Shared or separate clocks. It is conceivable that different genetic loci or subsets of the

data (e.g., all mitochondrial protein-coding genes) might share correlated patterns of rate variation across branches and thus be adequately described by a single shared clock model (Duchêne et al. 2014). However, any universal genetic 'pacemaker' is unlikely to also apply to non-genetic (e.g., phenotypic) data. Instead, it is likely that on some branches genetic evolution will be fast while phenotypic evolution is slow, and vice versa. Thus, phenotypic data are usually better modelled under their own separate clock or by multiple separate clocks (e.g., Pyron 2011).

3. Uncorrelated or autocorrelated rate variation. This concept refers to whether rates of evolution vary stochastically (uncorrelated) or systematically (autocorrelated) across branches in the phylogeny (e.g., Ho et al. 2015a). The famously sporadic nature of morphological evolution means that rates of evolution on adjacent branches might be very different (Fig. 7.2a). In such cases, uncorrelated clock models might be more appropriate. The widely used uncorrelated lognormal model in BEAST is one example: it assumes that the evolutionary rate on each branch is drawn independently from a single global lognormal distribution (Drummond et al. 2006), so that expected rates on a pair of adjacent branches are no more similar than expected rates on a pair of widely separated branches. Conversely, it is possible that certain clades have characteristics that enhance or depress the rate of phenotypic evolution, resulting in adjacent branches having similar evolutionary rates. In such cases, an autocorrelated model would be more appropriate. For instance, key adaptations such as cichlid pharyngeal jaws (Liem 1973) are likely to have increased the rate of phenotypic diversification in that clade. Local clocks (Yoder and Yang 2000; Drummond and Suchard 2010) assume there are different rate 'regimes' across different regions of the tree (Fig. 7.2b). Branches in the same regime evolve at the same rate, but different regimes have different rates. The number of regimes, the shift points between regimes, and the rate in each regime can either be fixed or estimated using Markov chain Monte Carlo sampling. Another autocorrelated model is the epoch clock (Bielejec et al. 2014; Paterson et al. 2019), which assumes that rates vary across time slices, so that (for instance) all early branches might have higher rates of evolution than all later branches (Fig. 7.2c).
4. Statistical distribution of among-lineage rate variation. This refers to the question of what statistical distribution best characterizes rate variation across all the branches of a tree. It is (at least theoretically) distinct from the previous point, which refers to whether adjacent branches have similar rates, and also distinct from among-character rate variation, which relates to the distribution of rate variation across characters. Rate variation across branches can also be modelled in many different ways, and exhaustive evaluation of a large range of models has seldom been done for large and relatively homogeneous molecular data sets, let alone much smaller and idiosyncratic morphological data sets. Some common relaxed-clock models assume that rates vary in a continuous fashion across lineages, following a lognormal distribution. The uncorrelated lognormal model in BEAST and BEAST2 (see above) assumes that rates for all branches are drawn from a single global lognormal distribution. In contrast, the autocorrelated TK02 model in MrBayes assumes that the rate on a particular branch is drawn from a local lognormal distribution, where the variance is determined by length of the branch (Thorne and Kishino 2002).

7.6 Dating Evolutionary Trees Using Morphological Clocks

When taxa in a phylogenetic analysis are all from the same time slice (e.g., living species), clock models—whether used for phenotypic or molecular characters—can generally only provide relative divergence times, rather than an absolute timescale. To translate such relative ages into

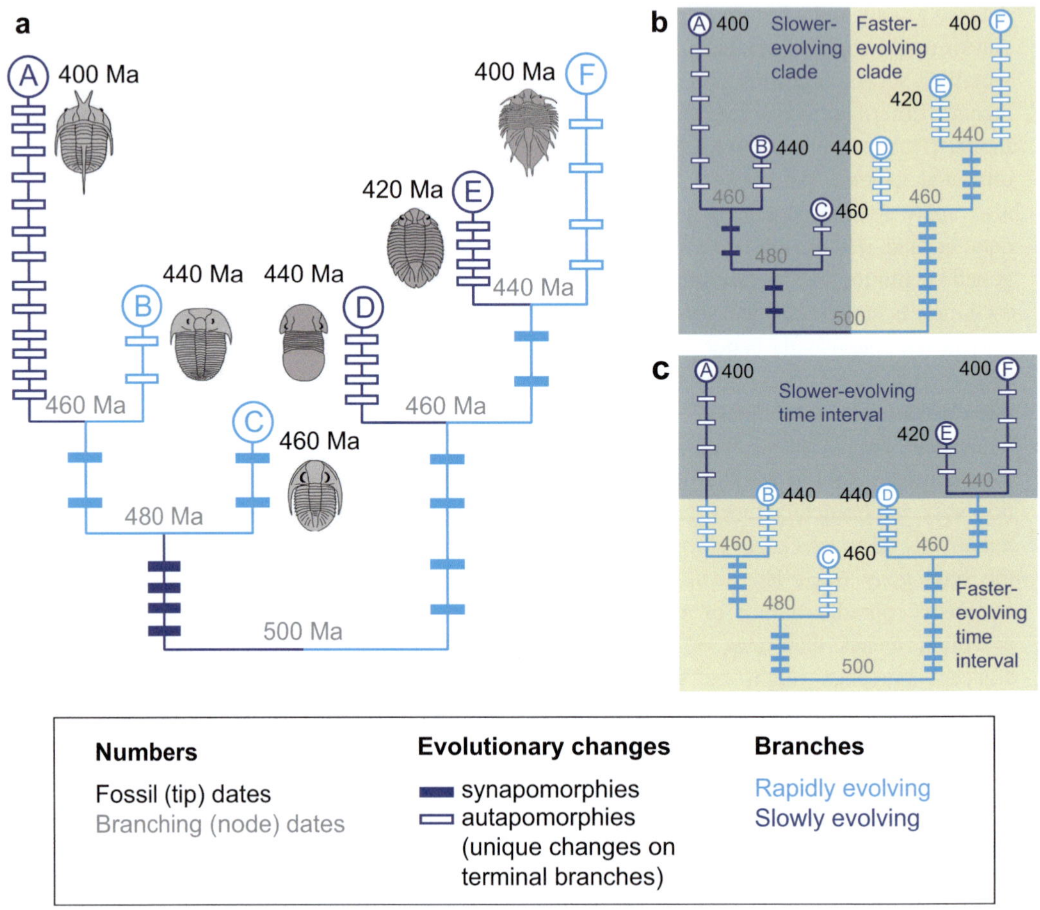

Fig. 7.2 Schematic diagram of different patterns of variation in evolutionary rate, which should be modelled using different relaxed-clock approaches. Horizontal bars denote evolutionary changes ('substitutions'). (**a**) Rates vary randomly across the tree and are thus not similar on close branches; these rates are best modelled using an uncorrelated relaxed clock. (**b**) Rates vary across the tree in a clade-specific fashion; these rates are best modelled using a local clock. (**c**) Rates vary across time slices; these rates are best modelled using an epoch clock

absolute ages, clock analyses typically have to be calibrated using external information about node dates or rates. In 'node dating', a fossil (or, less frequently, a biogeographic event) of a certain age is used to provide a minimum age constraint for the relevant divergence event. Often, multiple such calibrations are used to improve accuracy, and calibration nodes are given probabilistic age distributions rather than point ages to better accommodate potential sources of error (Parham et al. 2012; Chap. 8). Fossil-based node dating uses the following reasonable logic: the clade containing a fossil of a certain age must have evolved some time earlier, hence the age of a fossil provides a minimum age constraint on the living clade to which it belongs. Biogeographic calibrations assume that certain divergences (nodes) in the tree are temporally associated with particular dated biogeographic events (see Chap. 9). For instance, the divergence between Australian and South American marsupial clades might be assumed to be the result of vicariance related to the final breakup of Gondwana. In rate calibrations, broad estimates of average rates of molecular evolution (e.g., a 'typical' substitution rate of 1% per lineage per million years in mitochondrial DNA) can be used to translate degree of genetic divergence into absolute time.

Both node and rate calibrations for dating evolutionary trees have potentially major shortcomings. Fossil node calibrations fail to adequately accommodate uncertainty in the affinities of the calibration fossils (e.g., Rutschmann et al. 2007), and in any case only provide minimum (not maximum) age constraints. For biogeographic node calibrations, a given split might be due to dispersal (not vicariance) and thus not be associated with the putatively related tectonic event (e.g., Ho et al. 2015b). Rates of molecular evolution can vary greatly from group to group, making general 'average' values dangerous to use, though taxon-specific estimates might help reduce errors due to extrapolated rates (e.g., Arcones et al. 2019). Furthermore, with modern genomic data sets, each analysis often has a unique set of gene loci and thus unique rate dynamics (e.g., single-nucleotide polymorphisms or ultraconserved elements), making it impossible to even broadly guess the expected 'evolutionary rates'.

When taxa in a phylogenetic analysis are sampled across time, clock models for molecular and/or morphological data can, at least in principle, infer absolute divergence dates directly from the character and taxon-age data, thus providing a way to avoid the problematic assumptions of node or rate calibration. This approach, termed 'tip-dating', introduces new assumptions, notably related to the clock and diversification models (Chap. 11). Tip-dating was initially employed in the situation when viruses are sampled across real time and their nucleotide sequences analysed via molecular clocks to directly infer dated trees (see Chap. 10). An analogous situation is when taxa are sampled across geological time (e.g., living species and their long-extinct relatives), and their preserved phenotypic traits analysed via morphological clock analysis, again to infer dated trees. In both instances, the correlation between the age of a taxon and the amount of anagenetic evolution that it has undergone allows estimates of rates of change and, thus, estimates of absolute ages across the tree.

7.7 Some Case Studies

There is a rich literature on rates of morphological evolution, but phylogenetic analyses of phenotypic characters using explicit clock models remain relatively scarce. I here focus on some examples that highlight the usefulness of morphological clock methods, either for morphological analyses or for such analyses in concert with molecular data. These examples are certainly not exhaustive and focus on phylogenetic studies in which tree topology and divergence dates, as well as evolutionary rate dynamics, were all estimated. There is also a rapidly growing body of comparative studies in which relaxed-clock methods are being used to map phenotypic characters onto trees (e.g., Rabosky 2014).

7.7.1 Divergence Dating

Tip-dating approaches, initially developed in the context of molecular clock analysis of time-series samples of virus data, can also be applied to fossils and phenotypic data. Tip-dated clock analyses have been performed on morphological data alone to infer evolutionary relationships and divergence dates (e.g., Lee et al. 2014; Matzke and Wright 2016). However, when numerous terminal taxa ('tips') are still living (extant), it desirable to also include available genetic data, to allow better inference of relationships and (relative) divergence times between living taxa.

Total-evidence tip-dating refers to the simultaneous analysis of morphological and molecular data using clock-based methods for both types of data (Ronquist et al. 2012a). The method has been used to infer relationships across the tree of life, including plants (Grimm et al. 2014), insects (Ronquist et al. 2012a; Vea and Grimaldi 2016), arachnids (Sharma and Giribet 2014), fish (Arcila et al. 2015), amphibians (Pyron 2011), reptiles (Pyron 2016; Lee and Yates 2018), mammals (Herrera and Dávalos 2016; Kealy and Beck 2017), and birds (Gavryushkina et al. 2017; Crouch et al. 2019). All of these taxa have relatively complex morphologies, allowing a reasonable number of phenotypic characters to be scored and analysed using model-based methods.

The relationships of the fossil and living taxa, and divergence dates, are assessed simultaneously (usually in a Bayesian framework) to find the global solution that best fits the phenotypic, molecular, and stratigraphic data. Uncertainties in estimated variables (e.g., the positions of the fossils and the rates of molecular evolution) are fully integrated into the results (e.g., clade probabilities and node age ranges). In such cases, the ages of the fossils, along with the diversification and morphological clock models, are the primary drivers of node ages across the tree. Total-evidence tip-dating (potentially in concert with node-dating) has been argued to be theoretically superior to the traditional, sequential method of using node-dating alone, where fossils are initially analysed (typically using parsimony analysis) without recourse to their stratigraphy and often ignoring molecular data, and then their positions assumed to be known without error and used as minimum age constraints on relevant nodes in molecular clock analyses (see Chap. 5). However, tip-dating is much more computationally intensive and, as mentioned above, requires additional assumptions about the dynamics of diversification and morphological evolution.

7.7.2 Homoplasy and Tree Topology

Tip-dating approaches have usually been discussed in the context of better dating the tree of life (e.g., Ronquist et al. 2012a, b; Matzke and Wright 2016; Turner et al. 2017). However, they can also potentially improve our estimates of phylogenetic relationships, by 'nudging' the topology away from branching patterns that are highly stratigraphically incongruent, and towards phylogenies that perhaps entail slightly more homoplasy but match the fossil record much more closely (King 2021). As an example, parsimony and undated Bayesian approaches consistently unite two lineages of long-snouted crocodilians that are widely separated in time; Bayesian tip-dating approaches instead suggest (more reasonably) that these lineages are successive iterations of a similar morphotype (Lee and Yates 2018).

7.7.3 Rates of Phenotypic Evolution

Morphological evolution is widely seen as largely decoupled from molecular evolution (e.g., Lahr et al. 2014). Model testing typically suggests that morphology usually is better modelled under its own clock, rather than sharing the molecular clock (e.g., Lee and Yates 2018), even though early total-evidence tip-dating methods employed a single common clock (Ronquist et al. 2012a). Clock models have revealed the predicted high variation in rates of phenotypic evolution, to the extent that the validity of models themselves have been questioned (Puttick et al. 2016; Goloboff et al. 2019). Accordingly, when separate clocks have been applied to morphological and molecular data, the morphological data are invariably found to have larger amounts of rate variation (e.g., Pyron 2011; Lee et al. 2013; Beck and Lee 2014; Goloboff et al. 2019).

Morphological clock methods are also able to identify particular clades, lineages, or time-slices where rates of evolution are unusually high. Elevated rates of evolution were identified in some of the dinosaurian ancestors of birds, using explicit clock methods (e.g., Lee et al. 2014) and traditional time-scaling methods (Brusatte et al. 2014). Similarly, rates of morphological (as well as genetic) evolution during the Cambrian explosion (i.e., the latest Precambrian to earliest Cambrian) were estimated to be several times higher than subsequent rates (Lee et al. 2013). Surprisingly, however, after the earliest Cambrian, rates of evolution appeared to have steadied rapidly, with rates in the rest of the Lower Cambrian barely differing from rates in the Middle and Upper Cambrian, at least for trilobites (Paterson et al. 2019).

7.8 Concluding Remarks

Studies applying clock models to phenotypic data are still in their relative infancy, compared with molecular clock studies. Important concerns have been raised about the applicability of clock models to character systems such as morphology, where different characters and different taxa can exhibit highly idiosyncratic rates. However, some similar complications also manifest themselves, typically in more moderate amounts, in molecular data sets, and have been at least partly ameliorated by improved relaxed-clock models. Thus, when it comes to clock models, many of the differences between phenotypic and morphological data sets are potentially matters of degree, rather than kind. Increasing the number of empirical phenotypic studies will help identify which aspects of these clock models are adequate, and which need to be reconsidered. However, the vast majority of published phylogenetic matrices of phenotypic data are ill-suited to clock analyses, because only parsimony-informative traits were sampled. In this regard, it is vital that scientists gathering phenotypic phylogenetic data sets 'future-proof' their work by sampling all variable traits, so that their data sets are amenable to clock-based analytic approaches.

References

Álvarez-Carretero S, Goswami A, Yang Z, dos Reis M (2019) Bayesian estimation of species divergence times using correlated quantitative characters. Syst Biol 68:967–986

Arcila D, Pyron RA, Tyler JC, Ortí G, Betancur-R R (2015) An evaluation of fossil tip-dating versus node-age calibrations in tetraodontiform fishes (Teleostei: Percomorphaceae). Mol Phylogenet Evol 82:131–145

Arcones A, Ponti R, Vieites DR (2019) Mitochondrial substitution rates estimation for molecular clock analyses in modern birds based on full mitochondrial genomes. bioRxiv. https://doi.org/10.1101/855833

Beck RMD, Lee MSY (2014) Ancient dates or accelerated rates? Morphological clocks and the antiquity of placental mammals. Proc R Soc B 281:20141278

Bielejec F, Lemey P, Baele G, Rambaut A, Suchard MA (2014) Inferring heterogeneous evolutionary processes through time: from sequence substitution to phylogeography. Syst Biol 63:493–504

Bouckaert R, Vaughan TG, Barido-Sottani J, Duchêne S, Fourment M, Gavryushkina A, Heled J, Jones G, Kühnert D, De Maio N, Matschiner M, Mendes FK, Müller NF, Ogilvie HA, du Plessis L, Popinga A, Rambaut A, Rasmussen D, Siveroni I, Suchard MA, Wu C-H, Xie D, Zhang C, Stadler T, Drummond AJ (2019) BEAST 2.5: an advanced software platform for Bayesian evolutionary analysis. PLOS Comput Biol 15:e1006650

Bromham L, Penny D (2003) The modern molecular clock. Nat Rev Genet 4:216–224

Brusatte SL, Lloyd GT, Wang SC, Norell MA (2014) Gradual assembly of avian body plan culminated in rapid rates of evolution across the dinosaur-bird transition. Curr Biol 24:2386–2392

Crouch NMA, Ramanauskas K, Igić B (2019) Tip-dating and the origin of Telluraves. Mol Phylogenet Evol 131:55–63

Dawkins R (1982) The extended phenotype. Oxford University Press, Oxford, UK

dos Reis M, Donoghue PCJ, Yang Z (2016) Bayesian molecular clock dating of species divergences in the genomics era. Nat Rev Genet 17:71–80

Drummond AJ, Suchard MA (2010) Bayesian random local clocks, or one rate to rule them all. BMC Biol 8:114

Drummond AJ, Ho SYW, Phillips MJ, Rambaut A (2006) Relaxed phylogenetics and dating with confidence. PLOS Biol 4:e88

Drummond AJ, Suchard MA, Xie D, Rambaut A (2012) Bayesian phylogenetics with BEAUti and the BEAST 1.7. Mol Biol Evol 29:1969–1973

Duchêne S, Molak M, Ho SYW (2014) ClockstaR: Choosing the number of relaxed-clock models in molecular phylogenetic analysis. Bioinformatics 30:1017–1019

Freudenstein JV (2005) Characters, states and homology. Syst Biol 54:965–973

Gavryushkina A, Heath TA, Ksepka DT, Stadler T, Welch D, Drummond AJ (2017) Bayesian total-evidence dating reveals the recent crown radiation of penguins. Syst Biol 66:57–73

Goloboff PA, Catalano SA (2016) TNT version 1.5, including a full implementation of phylogenetic morphometrics. Cladistics 32:221–238

Goloboff PA, Pittman M, Pol D, Xing X (2019) Morphological data sets fit a common mechanism much more poorly than DNA sequences and call into question the Mkv model. Syst Biol 68:494–504

Grimm GW, Kapli P, Bomfleur B, McLoughlin S, Renner SS (2014) Using more than the oldest fossils: dating Osmundaceae with three Bayesian clock approaches. Syst Biol 64:396–405

Harrison LB, Larsson HC (2015) Among-character rate variation distributions in phylogenetic analysis of discrete morphological characters. Syst Biol 64:307–324

Harvey PH, Pagel M (1991) The comparative method in evolutionary biology. Oxford University Press, Oxford, UK

Hennig W (1966) Phylogenetic systematics. University of Illinois Press, Champaign, IL

Herrera JP, Dávalos LM (2016) Phylogeny and divergence times of lemurs inferred with recent and ancient fossils in the tree. Syst Biol 65:772–791

Ho SYW, Duchêne S, Duchêne D (2015a) Simulating and detecting autocorrelation of molecular evolutionary rates among lineages. Mol Ecol Resour 15:688–696
Ho SYW, Tong KJ, Foster CSP, Ritchie AM, Lo N, Crisp MD (2015b) Biogeographic calibrations for the molecular clock. Biol Lett 11:20150194
Höhna S, Landis MJ, Heath TA, Boussau B, Lartillot N, Moore BR, Huelsenbeck J, Ronquist F (2016) RevBayes: Bayesian phylogenetic inference using graphical models and an interactive model-specification language. Syst Biol 65:726–736
Jukes TH, Cantor CR (1969) Evolution of protein molecules. In: Munro HN (ed) Mammalian protein metabolism, vol III. Academic, New York, pp 21–132
Kealy S, Beck RMD (2017) Total evidence phylogeny and evolutionary timescale for Australian faunivorous marsupials (Dasyuromorphia). BMC Evol Biol 17:240
King B (2021) Bayesian tip-dated phylogenetics in paleontology: Topological effects and stratigraphic fit. Syst Biol (in press)
King B, Lee MSY (2015) Ancestral state reconstruction, rate heterogeneity, and the evolution of reptile viviparity. Syst Biol 64:532–544
Klopfstein S, Vilhelmsen L, Ronquist F (2015) A nonstationary Markov model detects directional evolution in hymenopteran morphology. Syst Biol 64:1089–1103
Lahr DJG, Laughinghouse HD, Oliverio A, Gao F, Katz LA (2014) How discordant morphological and molecular evolution among microorganisms can revise our notions of biodiversity on earth. BioEssays 36:950–959
Lanfear R, Welch JJ, Bromham L (2010) Watching the clock: studying variation in rates of molecular evolution between species. Trends Ecol Evol 25:495–503
Lee MSY, Palci A (2015) Morphological phylogenetics in the genomic age. Curr Biol 25:R922–R929
Lee MSY, Yates A (2018) Tip-dating and homoplasy: reconciling the shallow molecular divergences of modern gharials with their long fossil record. Proc R Soc B 285:20181071
Lee MSY, Soubrier J, Edgecombe GD (2013) Rates of phenotypic and genomic evolution during the Cambrian explosion. Curr Biol 23:1889–1895
Lee MSY, Cau A, Naish D, Dyke GJ (2014) Sustained miniaturization and anatomical innovation in the dinosaurian ancestors of birds. Science 345:562–566
Lewis PO (2001) A likelihood approach to estimating phylogeny from discrete morphological character data. Syst Biol 50:913–925
Liem KF (1973) Evolutionary strategies and morphological innovations: cichlid pharyngeal jaws. Syst Zool 22:425–441
Matzke NJ, Irmis RB (2018) Including autapomorphies is important for paleontological tip-dating with clocklike data, but not with non-clock data. PeerJ 6:e4553
Matzke NJ, Wright A (2016) Inferring node dates from tip dates in fossil Canidae: the importance of tree priors. Biol Lett 12:20160328
Maurits L, Forkel R, Kaiping GA, Atkinson QA (2017) BEASTling: a software tool for linguistic phylogenetics using BEAST 2. PLOS ONE 12: e0180908
Odling-Smee FJ, Laland KN, Feldman MW (2003) Niche construction: the neglected process in evolution. Princeton University Press, Princeton, NJ
O'Meara BC, Ané C, Sanderson MJ, Wainwright PC (2006) Testing for different rates of continuous trait evolution using likelihood. Evolution 60:922–933
Parham JF, Donoghue PCJ, Bell CJ, Calway TD, Head JJ, Holroyd PA, Inoue JG, Irmis RB, Joyce WG, Ksepka DT, Patané JSL, Smith ND, Tarver JE, van Tuinen M, Yang Z, Angielczyk KD, Greenwood JM, Hipsley CA, Jacobs L, Makovicky PJ, Müller J, Smith KT, Theodor JM, Warnock RCM, Benton MJ (2012) Best practices for justifying fossil calibrations. Syst Biol 61:346–359
Paterson JR, Edgecombe G, Lee MSY (2019) Trilobite evolutionary rates constrain the duration of the Cambrian explosion. Proc Natl Acad Sci USA 116:4394–4399
Puttick M, Thomas GH, Benton MJ (2016) Dating placentalia: morphological clocks fail to close the molecular fossil gap. Evolution 70:873–886
Pyron RA (2011) Divergence time estimation using fossils as terminal taxa and the origins of Lissamphibia. Syst Biol 60:466–481
Pyron RA (2016) Novel approaches for phylogenetic inference from morphological data and total-evidence dating in squamate reptiles (lizards, snakes, and amphisbaenians). Syst Biol 66:38–56
Rabosky DL (2014) Automatic detection of key innovations, rate shifts, and diversity-dependence on phylogenetic trees. PLOS ONE 9:e89543
Ronquist F, Klopfstein S, Vilhelmsen L, Schulmeister S, Murray DL, Rasnitsyn AP (2012a) A total-evidence approach to dating with fossils, applied to the early radiation of the Hymenoptera. Syst Biol 61:973–999
Ronquist F, Teslenko M, van der Mark P, Ayres DL, Darling A, Höhna S, Larget B, Liu L, Suchard MA, Huelsenbeck JP (2012b) MrBayes 3.2: efficient Bayesian phylogenetic inference and model choice across a large model space. Syst Biol 61:539–542
Rutschmann F, Eriksson T, Salim KA, Conti E (2007) Assessing calibration uncertainty in molecular dating: the assignment of fossils to alternative calibration points. Syst Biol 56:591–608
Sanderson MJ (1997) A nonparametric approach to estimating divergence times in the absence of rate constancy. Mol Biol Evol 14:1218–1231
Seligmann H (2010) Positive correlations between molecular and morphological rates of evolution. J Theor Biol 264:799–807
Sharma PP, Giribet G (2014) A revised dated phylogeny of the arachnid order Opiliones. Front Genet 5:e255
Simpson GG (1953) The major features of evolution. Columbia University Press, New York
Suchard MA, Lemey P, Baele G, Ayres DL, Drummond AJ, Rambaut A (2018) Bayesian phylogenetic and

phylodynamic data integration using BEAST 1.10. Virus Evol 4:vey016
Swofford DL (2003) PAUP*: phylogenetic analysis using parsimony (*and other methods). Version 4. Sinauer, Sunderland, MA
Thorne JL, Kishino H (2002) Divergence time and evolutionary rate estimation with multilocus data. Syst Biol 51:689–702
Turner AH, Pritchard AC, Matzke NJ (2017) Empirical and Bayesian approaches to fossil-only divergence times: a study across three reptile clades. PLOS ONE 12:e0169885
Vea IM, Grimaldi DA (2016) Putting scales into evolutionary time: the divergence of major scale insect lineages (Hemiptera) predates the radiation of modern angiosperm hosts. Sci Rep 6:e23487
Wiens JJ (1989) Phylogenetic analysis of morphological data. Smithsonian, Washington, DC
Wright A, Hillis DM (2014) Bayesian analysis using a simple likelihood model outperforms parsimony for estimation of phylogeny from discrete morphological data. PLOS ONE 9:e109210
Yang Z (1993) Maximum-likelihood estimation of phylogeny from DNA sequences when substitution rates differ over sites. Mol Biol Evol 10:1396–1401
Yang Z (1994) Maximum likelihood phylogenetic estimation from DNA sequences with variable rates over sites: approximate methods. J Mol Evol 39:306–314
Yeates DK (1992) Why remove autapomorphies? Cladistics 8:387–389
Yoder A, Yang Z (2000) Estimation of primate speciation dates using local molecular clocks. Mol Biol Evol 17:1081–1090
Zuckerkandl E, Pauling L (1962) Molecular disease, evolution, and genic heterogeneity. In: Kasha M, Pullman B (eds) Horizons in biochemistry. Academic, New York, pp 189–225

Part III

Calibrating Molecular Clocks

8 Calibrations from the Fossil Record

Jacqueline M. T. Nguyen and Simon Y. W. Ho

Abstract

Molecular clocks can be used to reconstruct evolutionary timescales based on analyses of genetic data, but these clocks need to be calibrated in order to give estimates in absolute time. Calibration is most often carried out using fossil evidence of the timing of evolutionary events, corresponding to internal nodes in phylogenetic trees. Early molecular dating studies treated fossil calibrations as point values, whereas later methods allowed calibrations to be specified as age constraints on nodes. The application of Bayesian methods to phylogenetic analysis opened up opportunities for fossil calibrations to take more complex forms. In this chapter, we trace the development and use of fossil calibrations and describe some a priori and a posteriori methods and criteria for evaluating their quality. We then present two examples of fossil calibrations from modern birds. Our chapter concludes with a discussion of the limitations of fossil calibrations, along with the changing role of the palaeontological record in molecular dating.

Keywords

Molecular clock · Molecular dating · Fossil calibration · Age constraints · Calibration prior · Phylogenetic analysis · Modern birds

8.1 Introduction

The timing of evolutionary divergences among lineages can be reconstructed from genetic data using molecular clocks, in an inference procedure known as molecular dating. This is often carried out using phylogenetic analysis of nucleotide or amino acid sequences, allowing timescales to be attached to evolutionary trees. In contrast with standard molecular phylogenetic analysis, however, molecular dating cannot be performed using the sequence data alone (see Chap. 5). This is because the sequence data only provide information about the genetic change occurring along each branch of the tree, but not about the separate contributions of evolutionary rate and absolute time to each of these branch lengths. Endless combinations of evolutionary rate (substitutions per site per year) and time duration (years) can be multiplied to give the same branch length

J. M. T. Nguyen
Australian Museum Research Institute, Australian Museum, Sydney, NSW, Australia

College of Science and Engineering, Flinders University, Adelaide, SA, Australia

PANGEA Research Centre, School of Biological, Earth and Environmental Sciences, UNSW Sydney, Sydney, NSW, Australia

S. Y. W. Ho (✉)
School of Life and Environmental Sciences, University of Sydney, Sydney, New South Wales, Australia
e-mail: simon.ho@sydney.edu.au

S. Y. W. Ho (ed.), *The Molecular Evolutionary Clock*, https://doi.org/10.1007/978-3-030-60181-2_8

(substitutions per site). Therefore, molecular dating requires external information either about the evolutionary rate across the tree or about the age of at least one of the internal nodes or some of the tips of the tree. The use of external information to constrain the ages of nodes in the tree is known as clock calibration (see Chap. 1). When the clock is calibrated, the entire phylogenetic tree can be scaled according to time to produce a 'time-tree' or 'chronogram'.

Calibrations for molecular dating can come from the fossil record, biogeography, and palaeogeography (Chap. 9), the sampling times of the sequence data (Chap. 10), or molecular date estimates from previous studies. The fossil record is the most widely used source of clock calibrations, as confirmed in a survey of nearly 700 molecular dating analyses (Hipsley and Müller 2014). Fossil calibrations can be applied to internal nodes of the tree, which represent evolutionary divergences between lineages, or to the terminal nodes or tips of the tree, which represent the fossil taxa themselves (Chap. 11).

In this chapter, we focus on the use of fossil evidence to inform the age calibrations for internal nodes in the tree. We describe how these calibrations are derived from palaeontological evidence and explain how they are used in molecular dating analyses, including Bayesian phylogenetic inference. We then discuss some a priori and a posteriori methods for evaluating fossil calibrations in molecular dating. Case studies are presented for two fossil calibrations from modern birds: the Palaeocene penguin *Waimanu manneringi* and the Miocene bristlebird *Dasyornis walterbolesi*. We conclude the chapter with a brief outline of the prospects for using fossil data in molecular dating.

8.2 Fossil Calibrations as Point Values

In the first application of molecular dating, the evolutionary rate of haemoglobin sequences was calibrated using palaeontological evidence for the age of the split between horse and modern human (see Chap. 1; Zuckerkandl and Pauling 1962). There was some uncertainty in the timing of this divergence event, which the authors believed to have occurred between 160 and 100 million years (Myr) ago. These two dates were used for calibration and the authors took the average of the resulting estimates of the evolutionary rate. Although the uncertainty in the fossil record was considered in this case, the palaeontological information was effectively distilled into a point estimate of the timing of the human–horse split.

The treatment of fossil calibrations as point estimates of node times was to continue for several decades. For example, several prominent molecular dating studies used a fossil-based point estimate of about 310 Myr for the divergence time between mammals and birds (Doolittle et al. 1996), sometimes as the sole calibration point (Kumar and Hedges 1998; Wang et al. 1999). The support for this particular fossil calibration was subsequently challenged (Lee 1999; Reisz and Müller 2004), raising serious concerns about the reliance on point calibrations that did not sufficiently acknowledge the level of uncertainty.

A fundamental problem with using point calibrations is that they make bold statements about the timing of evolutionary divergence events, when the fossil record can only provide good evidence of the earliest appearance of a lineage or clade (Fig. 8.1a; Marshall 1990; Smith and Peterson 2002). There can be a considerable time gap between the appearance of a clade and the oldest fossil that has been sampled from that clade. The time gap will be wider if the diagnostic features that allow fossil taxa to be assigned to the lineage did not appear until much later (Fig. 8.1a; Magallón 2004). For these reasons, the molecular date estimates obtained using point calibrations are often more appropriately interpreted as minimum dates (Hedges and Kumar 2004).

The fossil specimens used for calibrations have numerous sources of uncertainty relating to their identification, phylogenetic placements, and age (Benton and Donoghue 2007; Gandolfo et al. 2008). The fossils need to preserve features that allow them to be assigned to a specific lineage or clade, but their relationships to the taxa included

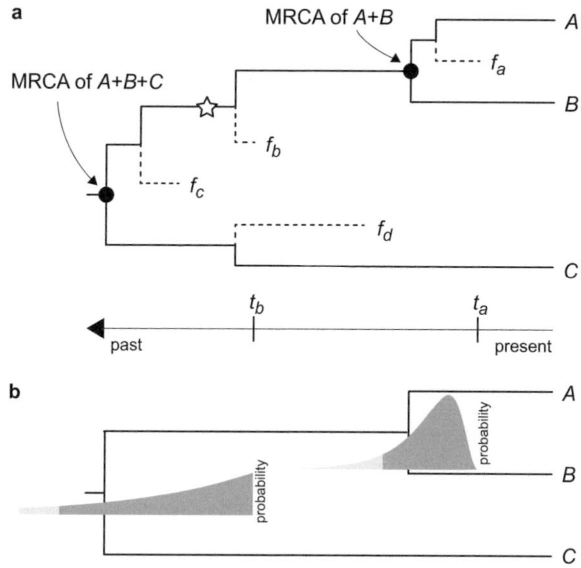

Fig. 8.1 (**a**) Phylogenetic tree showing the use of fossil evidence to provide age constraints for calibrating the molecular clock. Fossil f_a, which has age t_a, is placed in the crown group of $A + B$ and puts a minimum bound of t_a on the age of the most recent common ancestor (MRCA) of A and B. The fossils f_b and f_c are on the stem lineage leading to $A + B$. Although f_c is older than f_b, a diagnostic morphological feature (star) of A and B only appeared later, along the sister lineage of f_c that led to $A + B$. Therefore, f_b is the oldest fossil that can be confidently assigned to the lineage leading to $A + B$. The age of f_b places a minimum bound of t_b on the age of the MRCA of $A + B + C$. The oldest fossil in the sister group, f_d, can also be used to put a minimum bound on the MRCA of $A + B + C$. (**b**) Calibration densities for Bayesian molecular dating. A lognormal distribution has been selected for the age of the MRCA of $A + B$, with an offset (minimum bound) of t_a. The mean and standard deviation of this distribution must also be chosen. An exponential distribution has been selected for the age of the MRCA of $A + B + C$, with an offset of t_b. The rate parameter of the distribution must also be chosen. For both distributions, the light grey tail contains 5% of the probability density and the beginning of this tail is sometimes referred to as a 'soft' maximum bound

in the molecular dating analysis might not be resolved with confidence. It can also be difficult to identify which specific node in the tree should be calibrated using that fossil because this depends on whether the fossil can be placed in a crown group or on a stem lineage (Fig. 8.1a; Wilson et al. 1977; Magallón 2004).

Although molecular dating can be performed with just a single calibration point, the use of multiple calibrations is preferred because it can minimize the impact of any erroneous calibrations (Lee 1999; Lukoschek et al. 2012). Calibrations close to the root of the tree are particularly effective because they can help to counteract the underestimation of deep divergence times (Duchêne et al. 2014), which can be caused by poor modelling of the molecular evolutionary process across long timeframes. The introduction of molecular dating methods that allowed evolutionary rate variation across branches, known as relaxed molecular clocks, increased the need for employing multiple calibrations to enable better estimation of among-lineage rate variation (Thorne et al. 1998; Sanderson 2003). These methods also provided statistical frameworks that allowed the field to move away from using point calibrations.

8.3 Fossil Calibrations as Age Constraints

The fossil record is most effective in supplying minimum age constraints, which can be identified for many groups of organisms across the tree of life (Benton and Donoghue 2007). In molecular dating, these minimum age constraints are usually implemented in the form of 'hard' bounds, which exclude the possibility of the node being younger than the specified value. Thus, minimum bounds represent strong statements about the possible ages of nodes in the phylogeny. On their own, however, minimum bounds are insufficient for molecular dating because they do not impose any upper limits on node ages.

An important aspect of using age constraints for molecular dating is that at least one maximum constraint or point calibration needs to be included. Maximum age constraints are difficult to establish because they are equivalent to postulating the absence of a particular clade at some point in the past. If the maximum bound is too small, then it brings the risk of incorrectly claiming the absence of a lineage at a point in time when it was actually present. If the bound is too large, then it might be too uninformative for molecular dating. The choice of maximum age constraints can have a considerable impact on the molecular date estimates (Hug and Roger 2007; Warnock et al. 2015), as seen in a recent debate concerning the evolutionary timescale of modern birds (Jarvis et al. 2014; Cracraft et al. 2015; Mitchell et al. 2015).

Several methods have been proposed for deriving maximum age constraints from fossil evidence. One approach is to interpret the absence of evidence of a clade at some time point as being (weak) evidence of absence. If the clade of interest and its putative stem taxa are absent from a fossil assemblage that has been well sampled, then we might consider this to be an indication that the clade had not yet evolved. A more careful approach can involve the use of a taphonomic control group comprising taxa that have biological and ecological features similar to those of the clade of interest (Bottjer and Jablonski 1988). If the clade of interest is absent from older, well-sampled strata that contain the taphonomic control group, then this provides support for the absence of the clade of interest at that point in time.

Another approach to setting a maximum bound is to bracket the age of a node by constructing a 95% confidence interval, which can be calculated by a method that uses the distribution of the ages of the oldest fossil on each branch of the tree (Marshall 2008). This bracketing method uses an ultrametric tree: one in which the branch lengths are proportional to time, such that all extant tips are equally distant from the root. The method then identifies the fossil that covers the largest time duration of its corresponding branch and adds a confidence interval to the age of that fossil. The construction of the confidence interval is based on the branch-coverage proportions for all of the available fossils, which are assumed to follow a uniform distribution. The age-bracketing method involves a number of assumptions, such as random fossilization and the correct phylogenetic placement of fossils (Strauss and Sadler 1989; Marshall 2008).

Phylogenetic bracketing can be used to set a maximum bound on the age of a node (e.g., Reisz and Müller 2004; Benton and Donoghue 2007). In this approach, the age range of a node is constrained by the ages of its neighbouring nodes (i.e., those that are immediately ancestral and descendent), which can be based on fossil evidence or on an independent molecular date estimate. Phylogenetic bracketing requires prior knowledge of the evolutionary relationships, along with an age estimate for at least one of the neighbouring nodes. The method still carries the risk of setting a maximum age constraint that is too young. The need to specify maximum bounds can be partly avoided by taking a Bayesian approach to molecular dating, as described below.

8.4 Fossil Calibrations in Bayesian Molecular Dating

8.4.1 Calibration Priors

The rich information of the fossil record is used most effectively in Bayesian molecular dating methods, which provide a framework that naturally integrates multiple sources of information and their uncertainty (see Chap. 6; dos Reis et al. 2016; Bromham et al. 2018). The Bayesian statistical framework was applied to phylogenetic analysis in the mid-1990s, with the first Bayesian molecular dating analyses following soon afterwards (Thorne et al. 1998; Kishino et al. 2001; Aris-Brosou and Yang 2002). In Bayesian molecular dating, prior probability distributions need to be specified for all of the parameters in the model, including the node times. To obtain a dated phylogenetic tree in which the nodes are scaled in absolute time units (e.g., Myr), an informative prior distribution must be specified for the evolutionary rate or for at least one of the internal node ages. The prior distribution of relative node times also needs to be specified, although this is sometimes combined with the prior on the tree topology (e.g., Stadler 2009).

An early Bayesian dating program, multidivtime, required the user to specify a gamma prior distribution for the age of the root node, in addition to allowing the user to set age constraints on any of the other internal nodes (Thorne et al. 1998; Kishino et al. 2001). Minimum age constraints are analogous to uniform prior densities between the specified minimum age and infinity, whereas maximum age constraints are analogous to uniform prior densities between zero and the specified maximum age. Soon after the initial applications of Bayesian molecular dating, Huelsenbeck et al. (2000) suggested using fossil information to construct non-uniform prior distributions for internal node times. New Bayesian phylogenetic software packages enabled users to select parametric distributions for the calibration priors on any internal nodes in the tree (Drummond et al. 2006; Yang and Rannala 2006). The first such use of a fossil calibration prior was in an analysis of equid evolution in the Americas, where a normal prior with a mean of 55 Myr and standard deviation of 5 Myr was specified for the divergence time between rhinoceroses and horses (Weinstock et al. 2005).

A calibration prior describes the prior information about the age of a node in the tree, so it can be based on a single fossil specimen or on an interpretation of a body of palaeontological evidence. In the former case, the calibration prior represents the probability distribution of the time gap between the appearance of a clade (as defined by its most recent common ancestor) and the age of the oldest fossil in that clade to be preserved, sampled, and identified. This time gap depends on a number of factors, including the rate of morphological evolution (which affects the confidence with which we can assign fossils to the clade of interest), the preservation probability of the fossil, the geological processes that affect the survival of the fossil through time, and the sampling efforts and sampling biases of present-day researchers and collectors (Magallón 2004; Reisz and Müller 2004; Donoghue and Benton 2007).

The choice of parametric distribution for the calibration prior is largely subjective. Unfortunately, palaeontological evidence is rarely detailed enough to allow a well-justified choice of non-uniform probability density and its parameters; these choices are often made without explicit justification (Warnock et al. 2012). In many cases, there is no prior information except for the minimum age of the node. Intuitively, the distribution should have a declining probability for values that are increasingly distant from the age of the fossil (Hedges and Kumar 2004; Donoghue and Benton 2007). Common choices for the calibration prior include lognormal, exponential, and gamma distributions (Ho 2007), all of which have a declining tail of probability towards greater ages (Fig. 8.1b). Thus, these distributions have 'soft' maxima at the values that mark the 5% tail of the probability. The lognormal, exponential, and gamma distributions are bounded at zero, but when they are used for calibration priors they are often offset so that the minimum bound is fixed to some nonzero value based on fossil evidence (Fig. 8.1b).

A different approach to calibration priors was taken with the development of uniform priors with soft bounds (Yang and Rannala 2006), implemented in the software MCMCTree in the PAML package (Yang 2007). Uniform calibration priors continued the tradition of using the fossil record to set age constraints on node times, but soft bounds allowed a declining tail of probability (e.g., 2.5 or 5%) of the node age being outside the chosen time interval. In principle, this enables the sequence data to overrule erroneous calibration constraints.

Using different prior distributions for node ages can have a strong influence on the posterior date estimates (Inoue et al. 2010; Warnock et al. 2012, 2015), so the choice of distribution needs to be considered carefully. One way of obviating the responsibility of choosing the parameters of the calibration priors is to construct them using a hierarchical approach. For example, the calibrations can be modelled using exponential priors, with offsets being based on the fossil evidence and with the rate parameters (of the exponential distributions) being distributed according to a Dirichlet process (Heath 2012). The number of distinct rate parameters, and their assignment to the exponential priors, are random variables in the analysis.

There have been efforts to develop methods that can construct the calibration densities in an objective manner, typically by modelling the time gap between the oldest fossil within a clade and the most recent common ancestor of that clade. The CladeAge method generates a probability density for the origin time of a clade based on the oldest known fossil in that clade, an estimate of the fossil sampling rate, and estimates of the speciation and extinction rates (Matschiner et al. 2017). Another class of methods combines estimates of fossil preservation rates with a model of lineage diversification that is fitted to the stratigraphic ranges of the fossil taxa within the clade (Wilkinson et al. 2011; Nowak et al. 2013). The number of fossils and the distribution of their ages across chosen time periods then inform the estimates of the rates of speciation, fossil preservation, and fossil discovery. This procedure allows a probability density to be generated for the time gap between the age of a clade and the appearance of its oldest known fossil. The probability density can then be used as a calibration prior in a Bayesian molecular dating analysis. This approach is limited to groups of organisms that have a sufficient fossil record to inform the parameters of the lineage diversification model, such as primates (Wilkinson et al. 2011). However, these sequential methods have largely been rendered obsolete by the development of the fossilized birth–death model, which allows the fossil occurrences and molecular data to be analysed jointly in a single framework (see Chap. 11; Heath et al. 2014).

8.4.2 The Induced Calibration Prior

An important challenge to using fossil calibrations in Bayesian molecular dating is that the calibration prior specified by the user is not necessarily preserved when it is combined with the other priors in the analysis. For this reason, some researchers prefer the term 'calibration density' when referring to the probability density defined by the user (Heled and Drummond 2012). If the calibration densities overlap between ancestral and descendent nodes, then they are necessarily truncated so that the order of the nodes is preserved. In some implementations of Bayesian molecular dating, such as those in BEAST (Bouckaert et al. 2019) and MrBayes (Ronquist et al. 2012), the priors on node times are constructed multiplicatively by combining the calibration densities with the prior on the relative node times. For example, some model of diversification might be used to generate the prior on the tree topology and relative node times (e.g., Stadler 2009). The induced or effective prior distributions of the node ages might then differ from the calibration densities that were initially specified by the user (Kishino et al. 2001; Ho and Phillips 2009; Heled and Drummond 2012; Warnock et al. 2012). This problem is particularly noticeable when the calibration densities are diffuse, because this involves the largest extent of overlap between the ancestral and descendent nodes (Warnock et al. 2015; Barba-Montoya et al. 2017).

The discrepancy between the specified calibration density and the induced prior distribution can be reduced by enforcing monophyly for the calibrated nodes, or partly avoided by applying the diversification model to the uncalibrated nodes while conditioning on the nodes that have specified calibration densities (Yang and Rannala

2006; Heled and Drummond 2012, 2015). However, the calibration densities can also interact with each other and influence the node probabilities, owing to the rank ordering of the internal nodes imposed by the calibrations (Ho and Phillips 2009; Heled and Drummond 2012; Rannala 2016). An alternative is to examine the induced priors by sampling from the joint prior distribution, to check whether they are consistent with the intended, user-specified calibration densities (e.g., Warnock et al. 2015). This can be done, for example, by running the Bayesian analysis without any molecular sequence data. If any disparities are observed, a potential solution is to adjust the calibration priors to ensure that the induced priors reflect the age information supported by the fossil evidence (Kishino et al. 2001; Warnock et al. 2015).

8.5 A Priori Evaluation of Fossil Calibrations

The accuracy of molecular date estimates depends critically on the reliability of the calibrations that are employed for the molecular clock. There is growing recognition of the need for clear and convincing justifications for fossil calibrations. Accordingly, various groups of researchers have proposed methods for evaluating the quality and influence of fossil calibrations. Some have put forward specific criteria for selecting fossils for molecular dating (Gandolfo et al. 2008; Parham et al. 2012), whereas others have described approaches for investigating the influence of different calibrations and calibration schemes on estimates of node times (see Sect. 8.6). In this section, we describe some of the key considerations when selecting fossil calibrations, focusing on the criteria described by Gandolfo et al. (2008) and Parham et al. (2012).

8.5.1 Phylogenetic Placement

Fossil calibrations are usually applied to specific internal nodes (evolutionary divergence events) in the tree, meaning that the phylogenetic placement of each fossil should be determined as precisely as possible. The fossil calibration should preferably be based on a single specimen that preserves characters that allow its assignment to a specific branch or clade in the tree. The phylogenetic position of the fossil should be supported by an apomorphy-based diagnosis or a phylogenetic analysis including the specimen, although most of the fossils used as calibrations in molecular dating have not been explicitly analysed in a phylogenetic framework (Magallón 2004; Gandolfo et al. 2008). Any uncertainty in the placement of the fossil can be taken into account by considering the age constraint that is imposed by the fossil across a set of candidate tree topologies or a set of bootstrap replicates (e.g., Sterli et al. 2013). If molecular data are also available for the extant taxa in the data set, a joint analysis of morphological and molecular data can be performed. In this case, there should be some degree of congruence between the relationships supported by morphological and molecular data, because any discrepancies can lead to uncertainty in the placement of the fossil calibration.

Some fragmentary fossils might not preserve a sufficient number of features to allow many characters to be coded, but might still preserve diagnostic apomorphies that allow the fossil to be assigned unambiguously to a particular lineage or clade. For example, the Oligo-Miocene logrunner *Orthonyx kaldowinyeri* is known only from fragmentary leg bones, but these bones exhibit apomorphies that unambiguously place the species in the family Orthonychidae (Nguyen et al. 2014). This bird is the oldest known crown oscine passerine and has been used in molecular dating studies to provide a minimum age for Orthonychidae (Moyle et al. 2016; Oliveros et al. 2019).

The identification and phylogenetic placements of fossils can have a large impact on molecular date estimates. These can change in light of subsequent fossil discoveries, reinterpretation of available fossils, and new knowledge about phylogenetic relationships. For example, the Cretaceous bird *Vegavis iaai* was initially placed in crown Anseriformes (ducks, geese, and swans) in a phylogenetic analysis (Clarke et al. 2005) and was used as a fossil calibration for this group in several molecular dating studies (e.g., Pacheco et al. 2011; Ksepka and Phillips 2015). Subsequent phylogenetic analyses placed *Vegavis iaai* outside crown Anseriformes (e.g., Lee et al. 2014; Agnolín et al. 2017; Worthy et al. 2017), and a re-evaluation of this species questioned its affinities to Galloanseres (a clade comprising Galliformes and Anseriformes) altogether (Mayr et al. 2018b).

When selecting fossil calibrations, input from palaeontologists is essential for verifying the identification of fossils and investigating any major taxonomic and phylogenetic revisions since their original description (Gandolfo et al. 2008; Parham et al. 2012). Therefore, it is important that the fossils are well documented and accessible to researchers, for example by being registered in the collections of a public museum or other institution.

8.5.2 Fossil Age

A fossil specimen can only be useful for calibration if there is an estimate of its geological age. Fossils are rarely dated directly and radiometric dating of their associated strata is not always possible, so their ages are often inferred using other methods including stratigraphy and correlation. The stratigraphic placement of a fossil provides a relative age that can be used to establish a numerical age. When translating a relative age into a numerical age, a published geological timescale (e.g., Raine et al. 2015; Cohen et al. 2020) or geochronological literature should be referenced to explain how the numerical age was established. The estimated age of the fossil can be corroborated or constrained with more than one method of dating or with evidence from multiple studies. For fossil calibrations taking the form of minimum age constraints, the youngest possible age of the fossil should be used (Gandolfo et al. 2008). This involves using the minimum value of the age range of the fossil-bearing stratum, as well as the lower (minimum) limit of the error interval in radiometric dates.

Ongoing revisions in geochronology and stratigraphy are continually improving the accuracy and precision of these age estimates. After the initial publication of a fossil description, there can be changes to the stratigraphic interpretations and geochronology of the fossil deposit, as well as changes in stratigraphic classifications and regional and global geological timescales. For example, the Gelasian stage was moved from the Pliocene to the Pleistocene in 2009, leading to a 43% increase in the age of the Pliocene–Pleistocene boundary from 1.806 to 2.588 Myr (Gibbard et al. 2010). In terms of magnitude, there has been an even larger change in the age of the lower boundary of the Norian stage (Upper Triassic). The base of the Norian was defined as 216.5 Myr in the 2004 Geologic Time Scale (Gradstein et al. 2004) but is estimated to be 227 Myr in the 2020 International Chronostratigraphic Chart (v2020/3, Cohen et al. 2020).

Therefore, best practice dictates that the description of a fossil calibration should include explicit details of the precise locality and stratigraphic context of the specimen, and a numerical date with reference to published radiometric dates or a geological timescale. This information allows researchers to check whether there have been any revisions to the stratigraphic placement and estimated age of the fossil, prior to using the fossil as a calibration.

8.6 A Posteriori Evaluation of Fossil Calibrations

Once a set of fossil calibrations has been selected, they can be employed in a molecular dating analysis. Inevitably, however, some fossil calibrations are less reliable than others, and using different subsets of the calibrations might lead to different molecular date estimates. A simple way to address the problem of calibration choice is to repeat the molecular dating analysis using different candidate sets of calibrations. For example, molecular date estimates might be compared for a set of conservatively chosen 'safe but late' calibrations and for a set of more assertive 'early but risky' fossil calibrations (Sauquet et al. 2012).

This approach allows the impacts of various calibration schemes to be compared, although it does not necessarily give any insight into which set of date estimates is more accurate.

A number of methods have been developed to investigate the quality of fossil calibrations by employing them in molecular dating analyses and then using some criterion to evaluate them. These methods can be used to choose the node placements of the calibrations, to choose from among candidate sets of calibrations, or to exclude poor calibrations. Methods for a posteriori evaluation of fossil calibrations are based on the principle that any erroneous calibrations will lead to inconsistencies in the molecular date estimates and should be revised or discarded. They differ from a priori methods (described above in Sect. 8.5) in that the assessment of fossil calibrations is based primarily or exclusively on their age and phylogenetic placement, rather than on qualitative features of the fossils themselves.

The internal consistency of a set of calibrations can be evaluated using a cross-validation approach. The first implementation of such a method involved performing a molecular dating analysis using each fossil individually, then calculating the sum of squared differences between the molecular and fossil-based estimates of the remaining nodes considered for calibration (Near et al. 2005). Then, the calibrations with the greatest sum of squared differences are progressively removed until there is no further significant reduction in the variance of the differences between molecular and fossil-based estimates of node ages. The application of this approach to a data set from turtles led to seven of 17 candidate calibrations being discarded, with large shifts resulting in the molecular date estimates for some nodes (Near et al. 2005). The cross-validation approach can also be used to evaluate different node placements of fossil calibrations (Rutschmann et al. 2007). In this case, the sum of squared (relative) differences between estimated and fossil ages is calculated for all possible combinations of the candidate placements of the fossil calibrations.

There are several potential problems associated with using cross-validation to select a set of calibrations on the basis of internal consistency. A notable weakness of the approach is that it treats all of the fossil calibrations as point estimates of node ages, rather than as age constraints (Parham and Irmis 2008). Furthermore, the cross-validation approach might call for the removal of the calibrations that are actually the most informative (Marshall 2008). For example, if a large majority of the fossil calibrations are substantial underestimates of the corresponding node ages, then the cross-validation approach would support the retention of these calibrations while erroneously excluding any fossil calibrations that are actually much closer to the true node ages. These shortcomings can be circumvented by implementing the fossil calibrations as age constraints rather than as point values (Marshall 2008).

Candidate fossil calibrations can be evaluated by using reliably dated evolutionary divergences as reference points, as in the likelihood-checkpoint approach (Pyron 2010). In this method, a number of nodes with well-constrained ages are chosen as checkpoints. A molecular dating analysis is then performed using each set of fossil calibrations that is being considered. The preferred set of calibrations is the one that produces molecular date estimates with the highest likelihood for the checkpoint nodes. The likelihood is calculated on the basis of probability distributions, either lognormal or exponential, that are selected for each checkpoint node (Pyron 2010). Other probability distributions are likely to be more appropriate, because they provide more realistic penalties for molecular date estimates that contradict the fossil evidence (Lee and Skinner 2011). A weakness of the likelihood-checkpoint approach is that the potentially most reliable fossil calibrations are sequestered as checkpoints rather than being used as calibrations for the molecular dating analysis itself (Lee and Skinner 2011).

In Bayesian molecular dating, conflicting signals among calibrations can potentially be identified through comparison of the prior and posterior distributions of the node ages (Sanders and Lee 2007). The reasoning behind this approach is that if a calibration has a temporal signal that differs from the rest of the calibrations, its posterior distribution will tend to be shifted from its prior distribution. Bayesian methods have also been used to identify sets of consistent calibrations. One method builds on the age-bracketing method used to select maximum age constraints (described in Sect. 8.3) while accounting for phylogenetic uncertainty (Dornburg et al. 2011). Another method, Bayes factor cluster analysis, involves the inference of dated trees from all possible pairs of calibrations, then retains the calibration pairs that yield similar marginal likelihoods (Andújar et al. 2014). However, the effectiveness of Bayes factor cluster analysis is reduced when the calibrations are distant from each other on the tree or when there is large variation in evolutionary rates. In general, any Bayesian phylogenetic method of evaluating fossil calibrations is susceptible to being misled by discrepancies between the user-specified calibration densities and the induced priors on node times, as described in Sect. 8.4.2 (Warnock et al. 2015).

Methods for the a posteriori evaluation of fossil calibrations have not been used widely, despite the importance of calibrations in molecular dating. The low uptake of these methods might partly be due to the difficulty in identifying good fossil calibrations for most groups of organisms in the first place. Another reason might be that Bayesian methods offer varied and flexible ways to incorporate fossil information into molecular dating analyses, which has removed the need to treat fossil calibrations as simple point values or hard bounds on node ages.

8.7 Case Studies

Here we present two examples to illustrate the use of fossil specimens to set minimum age constraints for molecular dating analyses of modern birds. In these examples, we refer to the a priori criteria described in the previous section of this chapter (Gandolfo et al. 2008; Parham et al. 2012). However, we do not attempt to construct maximum age constraints. Similar treatments of fossil calibrations have been presented for turtles (Joyce et al. 2013), insects (Kohli et al. 2016; Evangelista et al. 2017), and other groups of organisms.

8.7.1 An Ancient Penguin from the Early Palaeocene

Penguins (Sphenisciformes) are well represented in the fossil record. These birds have a high fossilization potential because they have robust bones and live in shallow marine environments. One of the oldest known representatives of Sphenisciformes is the early Palaeocene penguin *Waimanu manneringi* Jones, Ando, and Fordyce, 2006 (in Slack et al. 2006) from the Waipara Greensand in the Waipara River in Canterbury, New Zealand. The holotype and only known specimen of *W. manneringi*, CM (Canterbury Museum) zfa 35, is a partial skeleton of one individual, including the pelvis, leg bones, and vertebrae (Fig. 8.2a).

Waimanu manneringi provides a reliable minimum age constraint for the divergence between Sphenisciformes and Procellariiformes (albatrosses, petrels, and allies). It exhibits several apomorphies that unambiguously place *W. manneringi* in Sphenisciformes (Slack et al. 2006). The phylogenetic position of *W. manneringi* has consistently been shown to be outside crown-group penguins (Fig. 8.2a), based on analyses of morphological data (e.g., Mayr et al. 2017, 2018a) and combined morphological and molecular data (e.g., Ksepka and Clarke 2010; Ksepka et al. 2012; Blokland et al. 2019).

A clade comprising *W. manneringi* and *Muriwaimanu* (formerly in *Waimanu*) *tuatahi* has been resolved as the sister lineage to all other fossil and extant penguins in some phylogenetic analyses (e.g., Ksepka et al. 2012; Mayr et al. 2017; Blokland et al. 2019). Although *M. tuatahi* also derives from the Waipara Greensand and is represented by several skeletons, it was found higher up in the strata from that of

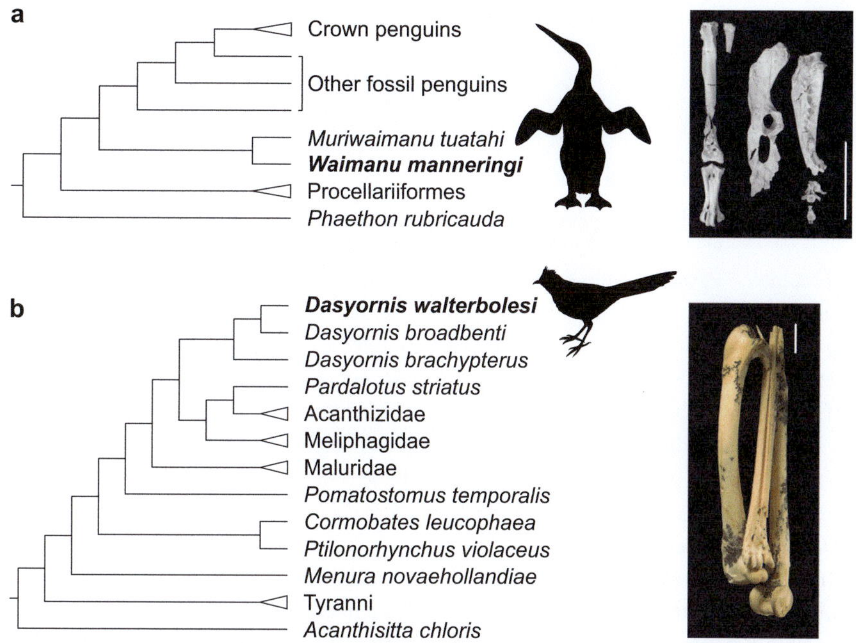

Fig. 8.2 Examples of two fossil specimens that provide calibrations for molecular dating analyses of modern birds. (**a**) Holotype specimen (CM zfa 35) of the Palaeocene stem penguin *Waimanu manneringi*, comprising a partial skeleton (scale bar = 100 mm), from Canterbury, New Zealand (Slack et al. 2006). An analysis of combined morphological and molecular data grouped the fossil species with *Muriwaimanu tuatahi* as the sister lineage to all other fossil and extant penguins (Blokland et al. 2019). Silhouette based on artwork by Chris Gaskin. (**b**) Holotype specimen (QM F50580) of the Miocene crown-group bristlebird *Dasyornis walterbolesi*, comprising hindlimb bones of one individual (scale bar = 2 mm), from Riversleigh, Australia (Nguyen 2019). An analysis of morphological data grouped the fossil species with the extant species of *Dasyornis*. Silhouette based on artwork by Peter Schouten

W. manneringi and is slightly younger in age (60–58 Myr; Slack et al. 2006).

A sister relationship between Sphenisciformes and Procellariiformes is supported by several phylogenetic analyses of genomic data (e.g., Jarvis et al. 2014; Prum et al. 2015) and morphological data (Livezey and Zusi 2007). Other phylogenetic analyses of morphological data inferred a sister relationship between Sphenisciformes and a clade comprising Podicipediformes (grebes) and Gaviiformes (loons) (Smith 2010), and proposed affinities of Sphenisciformes with Suliformes (cormorants and allies) and Pelecaniformes (pelicans and allies) (Mayr 2005). The phylogenetic placement of *W. manneringi* in stem-group Sphenisciformes does not conflict with current hypotheses regarding the sister relationships of Sphenisciformes, so does not affect its reliability as a fossil calibration.

The Waipara Greensand yields microfossils that indicate that it falls within the Teurian stage of the New Zealand geological timescale (e.g., Strong 1984; Hollis and Strong 2003). The top of the Waipara Greensand marks the Teurian–Waipawan local stage boundary, which correlates with the Palaeocene–Eocene international stage boundary (Cooper 2004; Slack et al. 2006; Raine et al. 2015). *Waimanu manneringi* derives from the basal part of the Waipara Greensand. On the basis of calcareous nannofossils, including two age-diagnostic taxa, the age of *W. manneringi* is constrained to late early Palaeocene (61.6–60.5 Myr; Slack et al. 2006). Based on these data, an age of 60.5 Myr is recommended as a minimum constraint for the split between Sphenisciformes and Procellariiformes.

8.7.2 Walter's Bristlebird from the Early Miocene

Songbirds (oscines, Passeriformes) are enormously diverse and make up nearly half of all living birds, but have a relatively patchy fossil record. Recent years have seen the description of a number of fossil songbirds that are potentially useful for calibrating the molecular dating analyses of modern birds (Worthy and Nguyen 2020). These include the early Miocene bristlebird *Dasyornis walterbolesi* Nguyen, 2019 from the Riversleigh World Heritage Area in Queensland, Australia. This species is the oldest known representative of the bristlebird family Dasyornithidae, which has a deep divergence from most of the remaining songbirds. The holotype specimen of *D. walterbolesi*, QM (Queensland Museum) F50580, comprises the major hindlimb bones (femur, tibiotarsus, and tarsometatarsus) of one individual bird (Fig. 8.2b).

The holotype specimen of *Dasyornis walterbolesi* provides a firm calibration for the split between Dasyornithidae and its sister group. It possesses several apomorphies that unambiguously place it in Dasyornithidae (Nguyen 2019). Phylogenetic analyses of morphological data also provide robust support for this placement (Fig. 8.2b), regardless of whether the analysis includes topological constraints based on genetic evidence (a 'molecular scaffold').

A sister relationship between Dasyornithidae and a clade comprising Acanthizidae (thornbills and gerygones), Pardalotidae (pardalotes), and Meliphagidae (honeyeaters) was supported in analyses of molecular data (Gardner et al. 2010; Marki et al. 2017; Oliveros et al. 2019). However, a phylogenetic analysis of morphological data found a sister relationship between *D. broadbenti* and a clade comprising Maluridae, Meliphagidae, and Acanthizidae (Worthy et al. 2010). The differences between the interfamilial relationships supported by these molecular and morphological studies do not affect the phylogenetic placement of *D. walterbolesi* in Dasyornithidae.

The type locality of *Dasyornis walterbolesi*, Camel Sputum Site, is part of Godthelp's Hill Sequence from D-Site Plateau in Riversleigh. Based on biocorrelation of mammalian faunas, this site is allocated to Riversleigh Faunal Zone B and is inferred to be early Miocene in age (Archer et al. 1989, 1997). Arena et al. (2016) refined the biostratigraphy and biochronology of Riversleigh using its mammalian faunas and placed Camel Sputum Site in Interval B3 of Faunal Zone B, which is the youngest zone interval. The age of this site is further constrained by uranium–lead radiometric dating of flowstone associated with fossil bone in the deposit, which yielded an estimated age of 17.75 ± 0.78 Myr (Woodhead et al. 2016). Based on these data, a minimum age of 16.97 Myr is recommended for the divergence between Dasyornithidae and its sister lineage.

8.8 Concluding Remarks

Fossil calibrations have been a pivotal component of the majority of molecular dating studies throughout the history of the molecular clock. They have undergone an impressive amount of expansion and development, but have also been subject to considerable scrutiny. The use of fossil calibrations flourished with the development of Bayesian phylogenetic methods, which provide a natural means of incorporating various sources of information and uncertainty. Even with genome-scale data sets, the precision of molecular date estimates ultimately depends on the precision of the calibrations (see Chap. 13; Thorne and Kishino 2002; Yang and Rannala 2006; dos Reis and Yang 2013). Therefore, further refinements to fossil calibrations will lead to better estimates of evolutionary timescales. Collaboration between geneticists and palaeontologists will be of great value to this endeavour.

Revisions of the taxonomy and phylogenetic relationships of fossil taxa, as well as ongoing improvements in geochronology and stratigraphic classifications, can have a strong influence on fossil calibrations and molecular date estimates. Because of these continual revisions, it is important to document specimen and provenance data in detail so that researchers can readily determine whether any changes have occurred since the initial description of the fossil (Gandolfo et al. 2008; Parham et al. 2012). These and other aspects of molecular dating studies can benefit substantially from the input of palaeontologists.

A range of methods are now available for implementing fossil calibrations in molecular dating analyses, but one unsatisfying aspect is that they use the palaeontological evidence in a limited and often indirect way. For example, minimum age constraints are based only on the age of the single oldest fossil in the descendent lineage or clade; other fossils are uninformative for this approach and are not taken into account. This shortcoming has motivated the development of methods that allow greater incorporation of the information from the fossil record. The fossilized birth–death model allows fossil taxa to be included as tips in the phylogenetic tree, with their placement inferred as part of a total-evidence dating analysis of molecular and morphological data (Stadler 2010; Didier et al. 2012). Alternatively, the unresolved fossilized birth–death model can use the full set of known fossil occurrences, with phylogenetic constraints on these occurrences being specified by the user (see Chap. 11; Heath et al. 2014). These methods are best suited to groups of organisms with richly preserved morphological features or with extensive fossil records. Therefore, we expect that fossil calibrations will continue to be an important component of molecular dating analyses of most branches across the tree of life.

References

Agnolín FL, Egli FB, Chatterjee S, Marsà JAG, Novas FE (2017) Vegaviidae, a new clade of southern diving birds that survived the K/T boundary. Sci Nat 104:87

Andújar C, Soria-Carrasco V, Serrano J, Gómez-Zurita J (2014) Congruence test of molecular clock calibration hypotheses based on Bayes factor comparisons. Meth Ecol Evol 5:226–242

Archer M, Godthelp H, Hand SJ, Megirian D (1989) Fossil mammals of Riversleigh, northwestern Queensland: preliminary overview of biostratigraphy, correlation and environmental change. Aust Zool 25:29–65

Archer M, Hand SJ, Godthelp H, Creaser P (1997) Correlation of the Cainozoic sediments of the Riversleigh World Heritage fossil property, Queensland, Australia. In: Aguilar J-P, Legendre S, Michaux J (eds) Actes du Congrès BiochroM'97. École Pratique des Hautes Études, Institut de Montpellier, Montpellier, pp 131–152

Arena DA, Travouillon KJ, Beck RMD, Black KH, Gillespie AK, Myers TJ, Archer M, Hand SJ (2016) Mammalian lineages and the biostratigraphy and biochronology of Cenozoic faunas from the Riversleigh World Heritage Area, Australia. Lethaia 49:43–60

Aris-Brosou S, Yang Z (2002) Effects of models of rate evolution on estimation of divergence dates with special reference to the metazoan 18S ribosomal RNA phylogeny. Syst Biol 51:703–714

Barba-Montoya J, dos Reis M, Yang Z (2017) Comparison of different strategies for using fossil calibrations to generate the time prior in Bayesian molecular clock dating. Mol Phylogenet Evol 114:386–400

Benton MJ, Donoghue PCJ (2007) Paleontological evidence to date the tree of life. Mol Biol Evol 24:26–53

Blokland JC, Reid CM, Worthy TH, Tennyson AJD, Clarke JA, Scofield RP (2019) Chatham Island Paleocene fossils provide insight into the palaeobiology, evolution, and diversity of early penguins (Aves, Sphenisciformes). Palaeontol Electron 22(3):78

Bottjer DJ, Jablonski D (1988) Paleoenvironmental patterns in the evolution of post-Paleozoic benthic marine invertebrates. Palaios 3:540–560

Bouckaert R, Vaughan TG, Barido-Sottani J, Duchêne S, Fourment M, Gavryushkina A, Heled J, Jones G, Kühnert D, De Maio N, Matschiner M, Mendes FK, Müller NF, Ogilvie HA, du Plessis L, Popinga A, Rambaut A, Rasmussen D, Siveroni I, Suchard MA, Wu C-H, Xie D, Zhang C, Stadler T, Drummond AJ (2019) BEAST 2.5: an advanced software platform for Bayesian evolutionary analysis. PLOS Comput Biol 15:e1006650

Bromham L, Duchêne S, Hua X, Ritchie AM, Duchêne DA, Ho SYW (2018) Bayesian molecular dating: opening up the black box. Biol Rev 93:1165–1191

Clarke JA, Tambussi CP, Noriega JI, Erickson GM, Ketcham RA (2005) Definitive fossil evidence for the extant avian radiation in the Cretaceous. Nature 433:305–308

Cohen KM, Harper DAT, Gibbard PL, Fan J-X (2020) The ICS International Chronostratigraphic Chart v2020/03. International Commission on Stratigraphy, International Union of Geological Sciences, https://stratigraphy.org/chart

Cooper RA (ed) (2004) The New Zealand geological timescale. Institute of Geological and Nuclear Sciences, Lower Hutt

Cracraft J, Houde P, Ho SYW, Mindell DP, Fjeldså J, Lindow B, Edwards SV, Rahbek C, Mirarab S, Warnow T, Gilbert MTP, Zhang G, Braun EL, Jarvis ED (2015) Response to comment on "Whole-genome analyses resolve early branches in the tree of life of modern birds". Science 349:1460b

Didier G, Royer-Carenzi M, Laurin M (2012) The reconstructed evolutionary process with the fossil record. J Theor Biol 315:26–37

Donoghue PCJ, Benton MJ (2007) Rocks and clocks: calibrating the Tree of Life using fossils and molecules. Trends Ecol Evol 22:424–431

Doolittle RF, Feng D-F, Tsang S, Cho G, Little E (1996) Determining divergence times of the major kingdoms of living organisms with a protein clock. Science 271:470–477

Dornburg A, Beaulieu JM, Oliver JC, Near TJ (2011) Integrating fossil preservation biases in the selection of calibrations for molecular divergence time estimation. Syst Biol 60:519–527

dos Reis M, Yang Z (2013) The unbearable uncertainty of Bayesian divergence time estimation. J Syst Evol 51:30–43

dos Reis M, Donoghue PCJ, Yang Z (2016) Bayesian molecular clock dating of species divergences in the genomics era. Nat Rev Genet 17:71–80

Drummond AJ, Ho SYW, Phillips MJ, Rambaut A (2006) Relaxed phylogenetics and dating with confidence. PLOS Biol 4:e88

Duchêne S, Lanfear R, Ho SYW (2014) The impact of calibration and clock-model choice on molecular estimates of divergence times. Mol Phylogenet Evol 78:277–289

Evangelista DA, Djernaes M, Kohli MK (2017) Fossil calibrations for the cockroach phylogeny (Insecta, Dictyoptera, Blattodea), comments on the use of wings for their identification, and a redescription of the oldest Blaberidae. Palaeontol Electron 20(3):1FC

Gandolfo M, Nixon KC, Crepet WL (2008) Selection of fossils for calibration of molecular dating models. Ann Missouri Bot Gard 95:34–42

Gardner JL, Trueman JWH, Ebert D, Joseph L, Magrath RD (2010) Phylogeny and evolution of the Meliphagoidea, the largest radiation of Australasian songbirds. Mol Phylogenet Evol 55:1087–1102

Gibbard PL, Head MJ, Walker MJC, Subcommission on Quaternary Stratigraphy (2010) Formal ratification of the Quaternary System/Period and the Pleistocene Series/Epoch with a base at 2.58 Ma. J Quat Sci 25:96–102

Gradstein FM, Ogg JG, Smith AG (eds) (2004) A geologic time scale 2004. Cambridge University Press, Cambridge, UK

Heath TA (2012) A hierarchical Bayesian model for calibrating estimates of species divergence times. Syst Biol 61:793–809

Heath TA, Huelsenbeck JP, Stadler T (2014) The fossilized birth-death process for coherent calibration of divergence-time estimates. Proc Natl Acad Sci USA 111:E2957–E2966

Hedges SB, Kumar S (2004) Precision of molecular time estimates. Trends Genet 20:242–247

Heled J, Drummond AJ (2012) Calibrated tree priors for relaxed phylogenetics and divergence time estimation. Syst Biol 61:138–149

Heled J, Drummond AJ (2015) Calibrated birth-death phylogenetic time-tree priors for Bayesian inference. Syst Biol 64:369–383

Hipsley CA, Müller J (2014) Beyond fossil calibrations: realities of molecular clock practices in evolutionary biology. Front Genet 5:138

Ho SYW (2007) Calibrating molecular estimates of substitution rates and divergence times in birds. J Avian Biol 38:409–414

Ho SYW, Phillips MJ (2009) Accounting for calibration uncertainty in phylogenetic estimation of evolutionary divergence times. Syst Biol 58:367–380

Hollis CJ, Strong CP (2003) Biostratigraphic review of the Cretaceous/Tertiary boundary transition, mid-Waipara River section, North Canterbury, New Zealand. New Zeal J Geol Geophys 46:243–253

Huelsenbeck JP, Larget B, Swofford D (2000) A compound Poisson process for relaxing the molecular clock. Genetics 154:1879–1892

Hug LA, Roger AJ (2007) The impact of fossils and taxon sampling on ancient molecular dating analyses. Mol Biol Evol 24:1889–1897

Inoue J, Donoghue PCJ, Yang Z (2010) The impact of the representation of fossil calibrations on Bayesian estimation of species divergence times. Syst Biol 59:74–89

Jarvis ED, Mirarab S, Aberer AJ, Li B, Houde P, Li C, Ho SYW, Faircloth BC, Nabholz B, Howard JT, Suh A, Weber CC, da Fonseca RR, Li J, Zhang F, Li H, Zhou L, Narula N, Liu L, Ganapathy G, Boussau B, Bayzid MS, Zavidovych V, Subramanian S, Gabaldon T, Capella-Gutierrez S, Huerta-Cepas J, Rekepalli B, Munch K, Schierup M, Lindow B, Warren WC, Ray D, Green RE, Bruford MW, Zhan X, Dixon A, Li S, Li N, Huang Y, Derryberry EP, Bertelsen MF, Sheldon FH, Brumfield RT, Mello CV, Lovell PV, Wirthlin M, Schneider MPC, Prosdocimi F, Samaniego JA, Velazquez AMV, Alfaro-Nunez A, Campos PF, Petersen B, Sicheritz-Ponten T, Pas A, Bailey T, Scofield P, Bunce M, Lambert DM, Zhou Q, Perelman P, Driskell AC, Shapiro B, Xiong Z, Zeng Y, Liu S, Li Z, Liu B, Wu K, Xiao J, Xiong Y, Zheng Q, Zhang Y, Yang H, Wang J, Smeds L, Rheindt FE, Braun M, Fjeldsa J, Orlando L, Barker FK, Jonsson KA, Johnson W, Koepfli K-P, O'Brien S, Haussler D, Ryder OA, Rahbek C, Willerslev E, Graves GR, Glenn TC, McCormack J, Burt D, Ellegren H, Alstrom P, Edwards SV, Stamatakis A, Mindell DP, Cracraft J, Braun EL, Warnow T, Jun W, Gilbert MTP, Zhang G

(2014) Whole-genome analyses resolve early branches in the Tree of Life of modern birds. Science 346:1320–1331

Joyce WG, Parham JF, Lyson TR, Warnock RCM, Donoghue PCJ (2013) A divergence dating analysis of turtles using fossil calibrations: an example of best practices. J Paleontol 87:612–634

Kishino H, Thorne JL, Bruno WJ (2001) Performance of a divergence time estimation method under a probabilistic model of rate evolution. Mol Biol Evol 18:352–361

Kohli MK, Ware JL, Bechly G (2016) How to date a dragonfly: fossil calibrations for odonates. Palaeontol Electron 19:1.1FC

Ksepka DT, Clarke JA (2010) The basal penguin (Aves: Sphenisciformes) *Perudyptes devriesi* and a phylogenetic evaluation of the penguin fossil record. Bull Am Mus Nat Hist 337:1–77

Ksepka D, Phillips MJ (2015) Avian diversification patterns across the K-Pg boundary: influence of calibrations, datasets, and model misspecification. Ann Missouri Bot Gard 100:300–328

Ksepka DT, Fordyce RE, Ando T, Jones CM (2012) New fossil penguins (Aves, Sphenisciformes) from the Oligocene of New Zealand reveal the skeletal plan of stem penguins. J Vert Paleontol 32:235–254

Kumar S, Hedges SB (1998) A molecular timescale for vertebrate evolution. Nature 392:917–920

Lee MSY (1999) Molecular clock calibrations and metazoan divergence dates. J Mol Evol 49:385–391

Lee MSY, Skinner A (2011) Testing fossil calibrations for vertebrate molecular trees. Zool Scr 40:538–543

Lee MSY, Cau A, Naish D, Dyke GJ (2014) Morphological clocks in paleontology, and a mid-Cretaceous origin of crown Aves. Syst Biol 63:442–449

Livezey BC, Zusi RL (2007) Higher-order phylogeny of modern birds (Theropoda, Aves: Neornithes) based on comparative anatomy. II. Analysis and discussion. Zool J Linn Soc 149:1–95

Lukoschek V, Keogh JS, Avise JC (2012) Evaluating fossil calibrations for dating phylogenies in light of rates of molecular evolution: a comparison of three approaches. Syst Biol 61:22–43

Magallón S (2004) Dating lineages: Molecular and paleontological approaches to the temporal framework of clades. Int J Plant Sci 165:S7–S21

Marki PZ, Jønsson KA, Irestedt M, Nguyen JMT, Rahbek C, Fjeldså J (2017) Supermatrix phylogeny and biogeography of the Australasian Meliphagides radiation (Aves: Passeriformes). Mol Phylogenet Evol 107:516–529

Marshall CR (1990) The fossil record and estimating divergence times between lineages: maximum divergence times and the importance of reliable phylogenies. J Mol Evol 30:400–408

Marshall CR (2008) A simple method for bracketing absolute divergence times on molecular phylogenies using multiple fossil calibration points. Am Nat 171:726–742

Matschiner M, Musilová Z, Barth JMI, Starostová Z, Salzburger W, Steel M, Bouckaert R (2017) Bayesian phylogenetic estimation of clade ages supports trans-Atlantic dispersal of cichlid fishes. Syst Biol 66:3–22

Mayr G (2005) Tertiary plotopterids (Aves, Plotopteridae) and a novel hypothesis on the phylogenetic relationships of penguins (Spheniscidae). J Zool Syst Evol Res 43:61–71

Mayr G, De Pietri VL, Scofield RP (2017) A new fossil from the mid-Paleocene of New Zealand reveals an unexpected diversity of world's oldest penguins. Sci Nat 104:9

Mayr G, De Pietri VL, Love L, Mannering AA, Scofield RP (2018a) A well-preserved new mid-Paleocene penguin (Aves, Sphenisciformes) from the Waipara Greensand in New Zealand. J Vert Paleontol 37: e1398169

Mayr G, De Pietri VL, Scofield RP, Worthy TH (2018b) On the taxonomic composition and phylogenetic affinities of the recently proposed clade Vegaviidae Agnolín et al., 2017 – neornithine birds from the Upper Cretaceous of the Southern Hemisphere. Cretac Res 86:178–185

Mitchell KJ, Cooper A, Phillips MJ (2015) Comment on "Whole-genome analyses resolve early branches in the tree of life of modern birds". Science 349:1460a

Moyle RG, Oliveros CH, Andersen MJ, Hosner PA, Benz BW, Manthey JD, Travers SL, Brown RM, Faircloth BC (2016) Tectonic collision and uplift of Wallacea triggered the global songbird radiation. Nat Commun 7:12709

Near TJ, Meylan PA, Shaffer HB (2005) Assessing concordance of fossil calibration points in molecular clock studies: an example using turtles. Am Nat 165:137–146

Nguyen JMT (2019) A new species of bristlebird (Passeriformes, Dasyornithidae) from the early Miocene of Australia. J Vert Paleontol 39:e1575838

Nguyen JMT, Boles WE, Worthy TH, Hand SJ, Archer M (2014) New specimens of the logrunner *Orthonyx kaldowinyeri* (Passeriformes: Orthonychidae) from the Oligo-Miocene of Australia. Alcheringa 38:245–255

Nowak MD, Smith AB, Simpson C, Zwickl DJ (2013) A simple method for estimating informative node age priors for the fossil calibration of molecular divergence time analyses. PLOS ONE 8:e66245

Oliveros CH, Field DJ, Ksepka DT, Barker FK, Aleixo A, Andersen MJ, Alström P, Benz BW, Braun EL, Braun MJ, Bravo GA, Brumfield RT, Chesser RT, Claramunt S, Cracraft J, Cuervo AM, Derryberry EP, Glenn TC, Harvey MG, Hosner PA, Joseph L, Kimball RT, Mack AL, Miskelly CM, Peterson AT, Robbins MB, Sheldon FH, Silveira LF, Smith BT, White ND, Moyle RG, Faircloth BC (2019) Earth history and the passerine superradiation. Proc Natl Acad Sci USA 116:7916–7925

Pacheco MA, Battistuzzi FU, Lentino M, Aguilar RF, Kumar S, Escalante AA (2011) Evolution of modern birds revealed by mitogenomics: timing the radiation and origin of major orders. Mol Biol Evol 28:1927–1942

Parham JF, Irmis RB (2008) Caveats on the use of fossil calibrations for molecular dating: a comment on Near et al. Am Nat 171:132–136

Parham JF, Donogue PCJ, Bell CJ, Calway TD, Head JJ, Holroyd PA, Inoue JG, Irmis RB, Joyce WG, Ksepka DT, Patané JSL, Smith ND, Tarver JE, van Tuinen M, Yang Z, Angielczyk KD, Greenwood JM, Hipsley CA, Jacobs L, Makovicky PJ, Müller J, Smith KT, Theodor JM, Warnock RCM, Benton MJ (2012) Best practices for justifying fossil calibrations. Syst Biol 61:346–359

Prum RO, Berv JS, Dornburg A, Field DJ, Townsend JP, Lemmon EM, Lemmon AR (2015) A comprehensive phylogeny of birds (Aves) using targeted next-generation DNA sequencing. Nature 526:569–573

Pyron RA (2010) A likelihood method for assessing the molecular divergence time estimates and the placement of fossil calibrations. Syst Biol 59:185–194

Raine JI, Beu AG, Boyes AF, Campbell HJ, Cooper RA, Crampton JS, Crundwell MP, Hollis CJ, Morgans HEG, Mortimer N (2015) New Zealand geological timescale NZGT 2015/1. New Zeal J Geol Geophys 58:398–403

Rannala B (2016) Conceptual issues in Bayesian divergence time estimation. Philos Trans R Soc B 371:20150134

Reisz RR, Müller J (2004) Molecular timescales and the fossil record: a paleontological perspective. Trends Genet 20:237–241

Ronquist F, Teslenko M, van der Mark P, Ayres DL, Darling A, Höhna S, Larget B, Liu L, Suchard MA, Huelsenbeck JP (2012) MrBayes 3.2: efficient Bayesian phylogenetic inference and model choice across a large model space. Syst Biol 61:539–542

Rutschmann F, Eriksson T, Abu Salim K, Conti E (2007) Assessing calibration uncertainty in molecular dating: the assignment of fossils to alternative calibration points. Syst Biol 56:591–608

Sanders KL, Lee MSY (2007) Evaluating molecular clock calibrations using Bayesian analyses with soft and hard bounds. Biol Lett 3:275–279

Sanderson MJ (2003) r8s: inferring absolute rates of molecular evolution and divergence times in the absence of a molecular clock. Bioinformatics 19:301–302

Sauquet H, Ho SYW, Gandolfo M, Jordan GJ, Wilf P, Cantrill DJ, Bayly MJ, Bromham L, Brown GK, Carpenter RJ, Lee DM, Murphy DJ, Sniderman JMK, Udovicic F (2012) Testing the impact of calibration on molecular divergence times using a fossil-rich group: the case of *Nothofagus* (Fagales). Syst Biol 61:289–313

Slack KE, Jones CM, Ando T, Harrison GL, Fordyce RE, Arnason U, Penny D (2006) Early penguin fossils, plus mitochondrial genomes, calibrate avian evolution. Mol Biol Evol 23:1144–1155

Smith ND (2010) Phylogenetic analysis of Pelecaniformes (Aves) based on osteological data: implications for waterbird phylogeny and fossil calibration studies. PLOS ONE 5:e13354

Smith AB, Peterson KJ (2002) Dating the time of origin of major clades: molecular clocks and the fossil record. Annu Rev Earth Planet Sci 30:65–88

Stadler T (2009) On incomplete sampling under birth-death models and connections to the sampling-based coalescent. J Theor Biol 261:58–66

Stadler T (2010) Sampling-through-time in birth-death trees. J Theor Biol 267:396–404

Sterli J, Pol D, Laurin M (2013) Incorporating phylogenetic uncertainty on phylogeny-based palaeontological dating and the timing of turtle diversification. Cladistics 29:233–246

Strauss D, Sadler PM (1989) Classical confidence intervals and Bayesian probability estimates for the ends of local taxon ranges. Math Geol 21:411–427

Strong CP (1984) Cretaceous tertiary boundary, mid-Waipara River Section, North Canterbury, New-Zealand. New Zeal J Geol Geophys 27:231–234

Thorne JL, Kishino H (2002) Divergence time and evolutionary rate estimation with multilocus data. Syst Biol 51:689–702

Thorne JL, Kishino H, Painter IS (1998) Estimating the rate of evolution of the rate of molecular evolution. Mol Biol Evol 15:1647–1657

Wang DY-C, Kumar S, Hedges SB (1999) Divergence time estimates for the early history of animal phyla and the origin of plants, animals and fungi. Proc R Soc B 266:163–171

Warnock RCM, Yang Z, Donoghue PCJ (2012) Exploring uncertainty in the calibration of the molecular clock. Biol Lett 8:156–159

Warnock RCM, Parham JF, Joyce WG, Lyson TR, Donoghue PCJ (2015) Calibration uncertainty in molecular dating analyses: there is no substitute for the prior evaluation of time priors. Proc R Soc B 282:20141013

Weinstock J, Willerslev E, Sher A, Tong W, Ho SYW, Rubenstein D, Storer J, Burns J, Martin L, Bravi C, Prieto A, Froese D, Scott E, Lai X, Cooper A (2005) Evolution, systematics, and phylogeography of Pleistocene horses in the New World: a molecular perspective. PLOS Biol 3:e241

Wilkinson RD, Steiper ME, Soligo C, Martin RD, Yang Z, Tavaré S (2011) Dating primate divergences through an integrated analysis of palaeontological and molecular data. Syst Biol 60:16–31

Wilson AC, Carlson SS, White TJ (1977) Biochemical evolution. Annu Rev Biochem 46:573–639

Woodhead J, Hand SJ, Archer M, Graham I, Sniderman K, Arena DA, Black KH, Godthelp H, Creaser P, Price E (2016) Developing a radiometrically-dated chronologic sequence for Neogene biotic change in Australia, from the Riversleigh World Heritage Area of Queensland. Gondwana Res 29:153–167

Worthy TH, Nguyen JMT (2020) An annotated checklist of the fossil birds of Australia. Trans R Soc South Aust 144:66–108

Worthy TH, Hand SJ, Nguyen JMT, Tennyson AJ, Worthy JP, Scofield RP, Boles WE, Archer M (2010) Biogeographical and phylogenetic implications of an early Miocene wren (Aves: Passeriformes: Acanthisittidae) from New Zealand. J Vert Paleontol 30:479–498

Worthy TH, Degrange FJ, Handley WD, Lee MSY (2017) The evolution of giant flightless birds and novel phylogenetic relationships for extinct fowl (Aves, Galloanseres). R Soc Open Sci 4:170975

Yang Z (2007) PAML 4: phylogenetic analysis by maximum likelihood. Mol Biol Evol 24:1586–1591

Yang Z, Rannala B (2006) Bayesian estimation of species divergence times under a molecular clock using multiple fossil calibrations with soft bounds. Mol Biol Evol 23:212–226

Zuckerkandl E, Pauling L (1962) Molecular disease, evolution, and genic heterogeneity. In: Kasha M, Pullman B (eds) Horizons in biochemistry. Academic, New York, pp 189–225

9 Biogeographic Dating of Phylogenetic Divergence Times Using Priors and Processes

Michael J. Landis

Abstract

Historical biogeographic processes shaped the distribution of life throughout space and time. A species range might expand, contract, or subdivide throughout its evolutionary history, during which extrinsic factors such as the palaeogeographic arrangement of land masses can influence how a species range evolves. Phylogenetic studies can therefore benefit from incorporating biogeographic and palaeogeographic evidence into their analyses in order to better estimate species divergence times and species relationships. This chapter begins by outlining a conceptual framework for using biogeography to date phylogenies, with some emphasis on the inherent uncertainty of reconstructing past events. Following this, the chapter explores two methods (prior- and process-based methods) for estimating divergence times using biogeographic evidence and discusses their applications and merits.

Keywords

Phylogenetic inference · Time-calibration · Dating · Biogeography · Palaeogeography

M. J. Landis (✉)
Department of Biology, Rebstock Hall, Washington University in St. Louis, St. Louis, MO, USA
e-mail: michael.landis@wustl.edu

9.1 Introduction

How have the evolutionary lineages of life diversified during Earth's history? Phylogenetic inference seeks to answer this question in two principal ways: with topological estimates that explain relationships among lineages and with dating estimates to bracket when lineages diverged in geological time (in millions of years). Today, tree topology is readily estimated from molecular sequence data thanks to nearly 50 years of parallel advancements in the fields of molecular evolution, genetic sequencing, and phylogenetic inference. But even with this bounty of molecular sequences, those data alone are incapable of dating geological divergence times under existing models of molecular evolution. For now and for the foreseeable future, extrinsic information is needed to time-calibrate (or date) the ages of phylogenetic lineages, information such as that from the fossil record or from the lasting impression of palaeogeographical scenarios upon biogeographic patterns (see Chap. 5). Fossils have been, and remain, indispensable in how biologists understand evolution. In the current era of phylogenetic inference, the fossil record is unquestionably the preferred source of evidence for time-calibrating molecular phylogenies (see Chap. 8). Under the best conditions, a fossil specimen preserves the ancient presence of an evolutionary lineage through its morphological features and its age. With palaeontological expertise, that morphology can be used to diagnose the fossil

S. Y. W. Ho (ed.), *The Molecular Evolutionary Clock*, https://doi.org/10.1007/978-3-030-60181-2_9

taxon's phylogenetic relationship to extinct and extant taxa. If the phylogenetic hypothesis places the fossil within a clade of extant taxa (a crown group), then that clade must be at least as old as the fossil, i.e., a crown-group fossil constrains the minimum age for that clade (Marshall 1990).

For many, this line of reasoning is quite direct and intuitive. But, in practice, fossil dating is complicated by a range of challenges, spanning the theoretical to the empirical. Researchers are actively advancing how we understand fossil dating in topics as varied as its application (Donoghue and Moore 2003; Parham et al. 2012), its robustness (Warnock et al. 2015; Brown and Smith 2017), and how to explicitly incorporate fossil morphology into the model of divergence-time estimation (Pyron 2011; Ronquist et al. 2012), with diversification processes that allow for the preservation of fossil-taxon occurrences (Heath et al. 2014) and fossil-taxon time series (Stadler et al. 2018), with fossilization processes under the multispecies coalescent (Ogilvie et al. 2018), and more (see Chap. 11).

While many theoretical obstacles in fossil dating are likely to surrender to human creativity (and computational muscle) over time, some empirical obstacles appear to be immutable properties of our natural world. Perhaps the largest empirical complication is the paucity of known fossils for many groups of organisms. To make this concern concrete, take the turtles (order Testudines), a clade with a fossil record so exquisite that turtles serve as a model clade for experimenting with fossil dating strategies (Joyce et al. 2013; Warnock et al. 2015). In contrast, consider the daisy family (Asteraceae), where new fossil discoveries are rare and when found rewrite our understanding of their diversification (Barreda et al. 2015). While the Paleobiology Database is not a perfect reflection of the fossil record itself, Asteraceae is represented by far less than one-thousandth as many described fossil specimens per living species when compared with Testudines ($\frac{100}{26000} \div \frac{8000}{350} \approx 0.00017$) on the database (Turtle Taxonomy Working Group 2017; Paleobiology Database 2018; Angiosperm Phylogeny Website 2018). Many plant, fungus, and insect clades lack fossils useful for dating, creating widespread demand for dating methods that do not directly depend on the fossil record.

Biogeographic dating is one possible alternative. Biogeographic evidence, like the fossil record, is fundamentally linked to our understanding of how evolution has generated and maintained biodiversity over geological time. Since the earliest days of evolutionary thinking, biogeographic disjunctions have been viewed as both intriguing and perplexing patterns (Wallace 1855; Darwin 1859; Wallace 1876). In particular, how is it that closely related lineages come to inhabit distinct regions that are separated today by a geographical barrier? One lineage must have either dispersed over the barrier after it was formed, or dispersed into the new region before the barrier existed. Studying disjunctions from an evolutionary perspective makes it clear that dispersal opportunities are shaped by palaeogeographical dynamics that play a central role in shaping biogeographic processes and patterns: the opening and closing of ancient seaways, the formation of mountains, and the surfacing of volcanic islands.

It follows that palaeogeography and biogeography, together, can serve as a valuable source of dating information. Take, for instance, a clade of species that is endemic to a young oceanic island, far from its mainland relatives (Fleischer et al. 1998). The endemics did not spontaneously originate on the island, so what sequence of biogeographic events can explain the clade's geographical distribution? At one extreme, all lineages within the clade might have first originated on the mainland before the new island formed, then independently colonized the island only after its origination, followed by any necessary extirpation and/or extinction of mainland lineages. In this case, the clade might be older than the island. A second, more plausible scenario is that one lineage colonized the new island and then radiated upon it (Baldwin and Sanderson 1998). If this second scenario were true, the clade would be younger than the island. Time-calibrating phylogenies with biogeography and palaeogeography operates by this reasoning: that some biogeographic scenarios are more likely than alternatives for a given palaeogeographical context, and that likelihood should influence what ages we estimate for the lineages involved (Ho et al. 2015; de Baets et al. 2016).

Although historical biogeography and palaeogeography are each fascinating in isolation, I will only focus on how they serve to inform divergence times in this chapter. Translating possible biogeographic histories into statistical information, specifically, to inform or date divergence times in a tree, is the exercise of dating phylogenies using historical biogeography. I will begin this chapter by reviewing several examples of how palaeogeography, biogeography, and diversification together generate information to date phylogenies. Then, I will discuss two Bayesian frameworks for biogeographic time-calibration, and illustrate the utility of both with empirical examples. First, I will consider biogeographic node calibrations that date key divergence events using expert-defined prior node densities. Second, I will explore a newer class of likelihood-based biogeographic dating methods, which explicitly model and probabilistically weight alternative biogeographic histories. This chapter concludes by characterizing general views of the utility of biogeographic dating in the field, its uses, its shortcomings, and its future.

9.2 Linking Phylogeny, Biogeography, and Palaeogeography

In its simplest form, biogeographic dating relies on combined evidence from palaeogeographic, biogeographic, and molecular phylogenetic data (Renner 2005; Ho et al. 2015; de Baets et al. 2016). What dating information can be extracted from these data depends entirely on what clade, what regions, and what timescales are under study. Rather than speaking in generalities, this section will examine several idealized scenarios portrayed in Fig. 9.1 to develop guidelines for identifying what biogeographic scenarios might enrich a phylogenetic dating analysis.

For clarity, we will consider geography in a discrete setting, with two regions, a northern region (N) and a southern region (S). The palaeogeographic context, which describes features such as the availability of and connectivity between regions, changes at the time labelled T. Figure 9.1 shows three alternative palaeogeographic scenarios: the new availability of a region where the initial existence of region S begins only after time T (Fig. 9.1a); the new connectivity between the regions N and S after time T (Fig. 9.1b); and the lost connectivity between the regions N and S after time T (Fig. 9.1c).

Three biogeographic and phylogenetic scenarios are shown in Fig. 9.1 and these are explained in more detail below. In each biogeographic scenario, I assume that we know precisely when and how the palaeogeographic context changed, how that influenced species range evolution, and how that relates to lineage diversification. Note that the examples assume identical statements of phylogenetic topology and species ranges across scenarios, highlighting how node-age distributions (in red) should respond to alternative palaeogeographic histories. All scenarios require at least one dispersal event from region N into region S. Multiple dispersals could also explain the observed biogeographic patterns, but those explanations would generally be eliminated by parsimony or penalized by probability: all else being equal, the probability of one dispersal, p, is greater than or equal to the probability of two dispersals, $0 \leq p^2 \leq p \leq 1$.

The first biogeographic scenario describes the recent colonization of and radiation within a newly accessible region (Fig. 9.1d). One example of a newly available region is the birth of an oceanic island through volcanic activity (Fig. 9.1a; Clague and Sherrod 2014). Regional availability provides a strong maximum age constraint for divergence times, since the region could not have been inhabited before it existed. New interregional connectivity generates a related, but weaker, source of dating information. Connectivity between regions increases following events such as the merging of two continents or the erosion of an intermediate mountain range

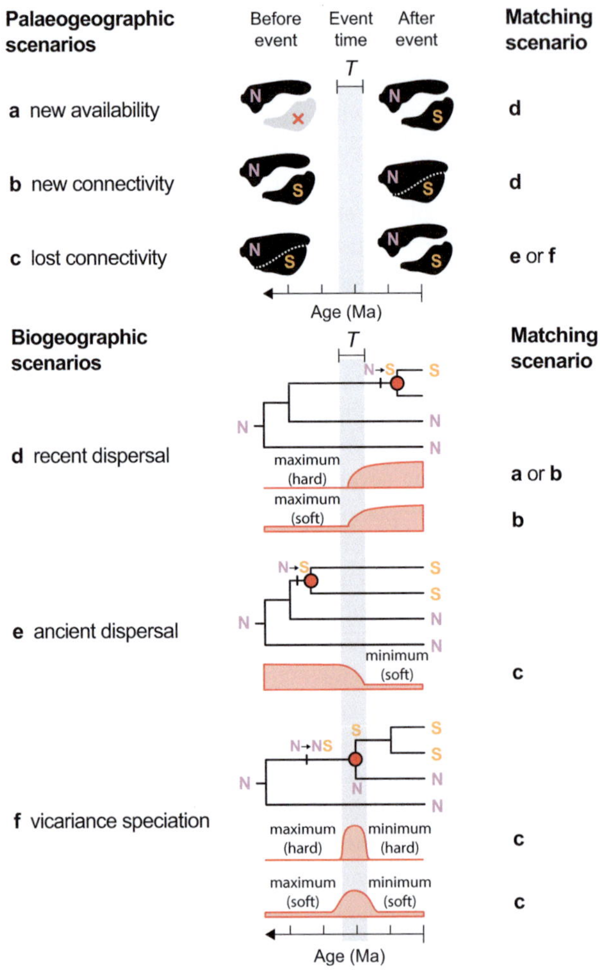

Fig. 9.1 Cartoon of biogeographic dating scenarios. Node-age distributions (red) by combining evidence that supports particular palaeogeographic scenarios (top) and biogeographic scenarios (bottom), either with prior- or process-based methods (further details are given in Sect. 9.3 of the main text). The three palaeogeographic scenarios show how connectivity and availability between two regions, N and S, changed during the time interval *T* (grey bar): (**a**) region S originates; (**b**) regions N and S merge; and (**c**) regions N and S split. The three biogeographic scenarios show various timings for biogeographic and lineage splitting events and whether those timings are consistent or not with the proposed palaeogeographic scenarios (**a–c**): (**d**) a recent dispersal scenario; (**e**) an ancient dispersal scenario; and (**f**) a vicariance scenario. Panels (**d–f**) all show the same topology and species ranges, differing only in the ages of biogeographic and lineage splitting events

(Fig. 9.1b). In this scenario, colonization events can precede the formation of the barrier, but they occur at a much lower rate (i.e., with lower probability per unit time).

The second biogeographic scenario describes ancient dispersal (Fig. 9.1e). Ancient dispersal involves a lineage colonizing a region before a geographical barrier is formed. Interregional connectivity is lost with the emergence of mountains, the splitting of continents, the submergence of a land bridge (Fig. 9.1c), or the loss of an intermediate connecting region (absent in Fig. 9.1 since it requires >2 regions). Lineages can freely disperse between regions before the barrier forms, but they do so at a much lower rate afterwards (i.e., with lower probability per unit time). For the example shown in Fig. 9.1e, dispersal into region S occurs before the barrier forms, with the lineages diverging for unspecified reasons well before regions N and S separate.

The third biogeographic scenario describes vicariance (Fig. 9.1f). Vicariance speciation involves two phases. First, a new geographical barrier interrupts the gene flow within a widespread species range. Widespread is a relative term, which operationally means that the species range spans multiple regions; in Fig. 9.1f, one lineage becomes widespread when it expands from region N into both regions N and S. Second, while this new barrier stands, subdivided populations in the species range develop heritable incompatibilities that establish one or both isolates as new lineages. Note that vicariance requires that the split lineage is ancestrally widespread, implying that vicariance scenarios involve an ancient dispersal event at some point in the clade's ancestry. While an ancient dispersal scenario prefers clade ages that predate the age of the new geographical barrier (Fig. 9.1e), a vicariance scenario constrains both the minimum and maximum clade ages to follow the age of the new barrier.

To reiterate, the biogeographic scenarios presented in Fig. 9.1 are simplified in order to introduce three categories of biogeographic dating evidence. When appropriate, each biogeographic scenario generates its own set of divergence-time constraints: recent dispersal scenarios are used to constrain the maximum age of a divergence, ancient dispersal scenarios provide a minimum age constraint, and vicariance scenarios provide both maximum and minimum age constraints. But how do we know when it is appropriate to invoke biogeographic evidence to date a phylogeny? And how much influence should biogeography and palaeogeography have on clade-age estimates? After all, in standard phylogenetic analyses, we do not precisely know the relationships among lineages, their divergence times, and the biogeographic history of the clade (which depends on phylogenetic knowledge), not to mention our poor knowledge of the exact sequence and timing of many palaeogeographic events. The next section discusses these matters in more detail.

9.3 Time-Calibrating Trees with Biogeography

A central premise of biogeographic dating is that palaeogeography informs phylogenetic node ages through the clade's biogeographic history. But how do we translate palaeogeographic events and biogeographic patterns into information about clade ages in practice? To begin, I will reintroduce several familiar modelling components used in molecular phylogenetics that were detailed in Chaps. 5 and 6. Here, and for the rest of the chapter, we will assume that we are interested in estimating phylogeny by modelling a molecular substitution process (Felsenstein 1981) where lineages diversify following a branching process (Nee et al. 1994) and molecular rates along branches vary according to a relaxed-clock model (Thorne et al. 1998). By fitting such a phylogenetic model to molecular data, we can simultaneously estimate the parameters for the tree topology, the node ages, the substitution process, the relaxed molecular clock, and the diversification process.

Without calibrations, the node ages can be estimated in units of relative time at best (Thorne et al. 1998). To estimate geological divergence times, researchers have typically relied on fossil evidence, secondary calibrations from backbone phylogenies, or biogeographic hypotheses. Regardless of the dating method, the exact relationship between any line of extrinsic evidence and the timing of one (or several) divergence events is not known with absolute certainty. Because they readily accommodate this inherent source of uncertainty, Bayesian phylogenetic approaches have proven extremely effective for divergence-time estimation (Drummond et al. 2006; Yang and Rannala 2006; Ronquist et al. 2012; Chaps. 6 and 13).

Briefly reviewing Bayesian phylogenetics will help frame how we survey biogeographic dating methods. The chief aim of Bayesian phylogenetics is to estimate the distribution of evolutionary parameters that have a high probability of generating the data that we observe in nature, such as the observed data being the molecular sequences and geographic ranges of the species being compared. This estimated distribution is called the posterior distribution, and it is defined as being proportional to the product of the likelihood function and the prior distribution.

$$\underbrace{P(\tau, a, \theta|X)}_{\text{posterior}} = \frac{1}{Z} \times \underbrace{P(X|\tau, a, \theta)}_{\text{likelihood}} \times \underbrace{P(\tau, a, \theta)}_{\text{prior}} \quad (9.1)$$

In the above formulation, the posterior distribution defines the joint probability over the possible tree topologies, τ, divergence times, a, and other model parameters, θ, such as various rates of evolution and diversification-process parameters, conditional upon the observed data collected from nature, X. The normalization term, $\frac{1}{Z}$, is the reciprocal of the marginal likelihood, which is not directly relevant to the topic of dating discussed here. What is important to recognize here is that the likelihood function, $P(X \mid \tau, a, \theta)$, is a function of the observed data, X, while the prior distribution, $P(\tau, a, \theta)$, is not. The likelihood function defines a model of evolution that can generate our observed data, X, that we use to fit the model parameters, θ. This means that if our observations of the natural world for X changed, so would our parameter estimates for θ. In contrast, the prior distribution θ remains constant regardless of the value of X. In this sense, the prior density is data-independent and the likelihood function is data-dependent, which has consequences for the dating estimates that are discussed later. The next two sections introduce Bayesian strategies for time-calibrating phylogenies: prior-based node calibration methods and process-based dating methods. In these sections, I will discuss some of the strengths and sensitivities of each approach.

9.3.1 Prior-Based Node Calibrations

Prior-based node calibrations, often called node calibrations or node priors for short, are used to constrain the range of divergence times for targeted nodes in phylogenies. During inference, the prior probability of the calibrated node's age is taken into account when computing the joint probability of all node ages in the phylogeny. Dated phylogenies with node ages that do not conform to all specified node calibrations are scored with low probabilities, and are thus disfavoured during estimation. In practice, node calibrations are most often applied using fossil evidence. Over decades, palaeontologists and evolutionary biologists have developed a rich literature of techniques and best practices (Parham and Irmis 2007; Parham et al. 2012; Joyce et al. 2013; Warnock et al. 2015) that we can extend to frame principles for biogeographic node calibrations here.

Applying fossil-based node calibrations involves two major steps: the calibration must first be justified, then the age constraints must be specified. Justification involves determining that a fossil specimen is a valid representative of early-diverging stem or crown lineages of the target node that is present in an explicitly stated phylogenetic hypothesis. Parham et al. (2012) advocate for the placement of fossils through the cladistic analysis of morphology of fossil and extant taxa. During justification, the biologist defines the prior probability for the distribution of possible ages relating to the calibrated node. Some aspects of specification are considered standard practice, particularly that fossil specimens represent the minimum age of the split, so the calibrated lineage must be at least as old as the fossil representative (Marshall 1990). Yet other aspects of the prior are not so easily specified. In particular, when did a lineage first originate, i.e., what is its maximum age? The clade could lack early fossil representation because of taphonomy or simply because the lineage had not yet originated (Jaanusson 1976), rendering the true maximum age unknowable. While models exist to estimate origin times under fossil sampling distributions (Strauss and Sadler 1989; Marshall 2008), assigning maximum age constraints to calibrations is, to some, a dubious exercise (Heads 2012). Nonetheless, explicitly or implicitly, all sources of calibration uncertainty are encoded in the phylogenetic position and the prior age density.

Superficially, biogeographic node calibrations resemble fossil node calibrations in that they both assert an evolutionary hypothesis to justify the prior preference for certain node-age estimates. To justify, specify, and validate a biogeographic node calibration, however, requires principles that are distinct from what fossil-based methods use (Kodandaramaiah 2011; Ho et al. 2015; de Baets et al. 2016). I have outlined these principles below and diagrammed them in Fig. 9.2.

An overview of prior-based biogeographic dating

(I) Justification
  (a) Declare the phylogenetic hypothesis of lineage relationships.
  (b) Record the biogeographic distributions for the taxa.
  (c) Indicate which node will be calibrated by identifying a biogeographic disjunction between one clade and its close relatives.
  (d) Assert the hypothesis that a (dated) palaeogeographic event facilitated or maintained the biogeographic disjunction that is represented by the calibrated node.

(II) Specification
  (a) Record the range of possible dates for the palaeogeographic event named in Step I-d.
  (b) Define a prior to constrain the range of plausible ages for the node specified in Step I-c as influenced by the dated palaeogeographic event of Step II-a.
  (c) Define standard models of molecular evolution and lineage diversification.

(III) Estimation
  (a) Estimate the posterior of dated phylogenies under the chosen biogeographic prior.
  (b) Confirm that dated phylogenies do not imply internal inconsistencies, i.e., no resulting biogeographic scenarios that contradict premises for justifying the calibration.
  (c) Assess prior sensitivity of posterior estimates.

Justifying a biogeographic node calibration leverages a combination of palaeogeographic, biogeographic, and phylogenetic evidence. Typically, the researcher begins with a phylogenetic hypothesis, such as the topology estimated from a molecular phylogeny. Next, biogeographic disjunctions are identified from species range data mapped to the tips of the phylogenetic hypothesis. Lastly, the researcher identifies and asserts that a particular palaeogeographic event influenced the age of the proposed biogeographic scenario. For example, take the phylogenetic hypothesis (Step I-a) and range data (Step I-b) of the biogeographic scenario in Fig. 9.1d. In this case, we assert that a dispersal event into region S followed by in situ speciation in region S explains the biogeographic disjunction between regions N and S today (Step I-c), where the dispersal event must have occurred after region S came into existence (Step I-d).

Justifying biogeographic node calibrations is often challenging for purely practical reasons: small and recent radiations often lack sufficient molecular variation to claim strong phylogenetic hypotheses, while backbone phylogenies built from, for example, genera as taxa generally represent deeper timescales, making it difficult to assert specific historical biogeographic scenarios with certainty. Perhaps more troublingly, several researchers (Renner 2005; Kodandaramaiah 2011) hold that circular reasoning is sometimes required to justify biogeographic node calibrations: to assert that a biogeographic event should favour a certain phylogenetic hypothesis (i.e., a range of node ages for a divergence event), the practitioner must first assume a phylogenetic hypothesis (i.e., that biogeographic change induced the divergence event).

Once the calibrated node has been justified, specifying the calibration density requires two main considerations: When was the palaeogeographic event and, relative to it, when might the calibrated divergence event have occurred? Depending on the event, the palaeogeographic event might have occurred

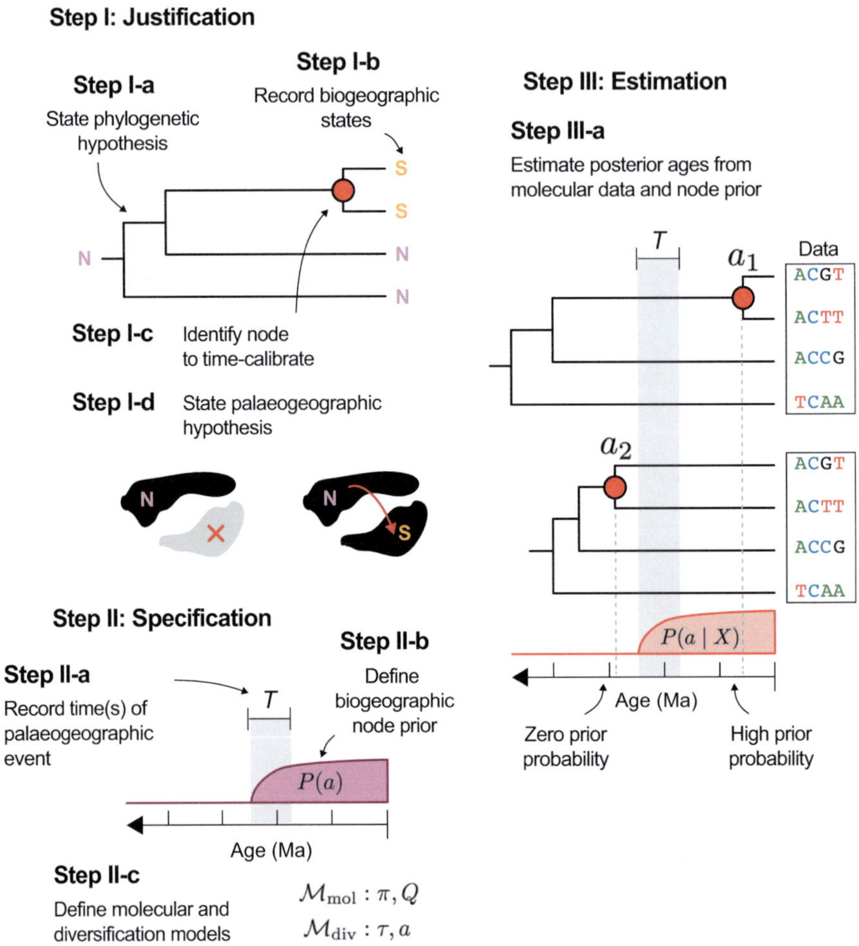

Fig. 9.2 Diagram of prior-based biogeographic dating. Node calibration involves three major steps. Justification (Step I) asserts a divergence scenario by interpreting evidence from a phylogenetic hypothesis (I-a) and the biogeographic states of the taxa (I-b) to identify a divergence event (I-c) whose biogeographic disjunction was hypothetically caused by a palaeogeographic event (I-d). Specification (Step II) designs a model to estimate divergence times by recording the time of the divergence-causing palaeogeographic event (II-a) then assigning a node prior of divergence times relative to the palaeogeographic event time (II-b), along with specifying standard molecular and diversification model components (II-c). Estimation (Step III) fits the model with the node prior to the molecular data to estimate a posterior density of dated phylogenies. The prior density (purple; Step II-b) has zero probability for ages older than T, which is reflected in the posterior density (red). Thus, the posterior density prefers young ages (a_1) rather than old ages (a_2). Steps III-b and III-c are not shown. See Sect. 9.3.1 for additional details

instantaneously or gradually, and its date (or duration) might be known precisely or not (Step II-a). When in doubt, conservative times that minimize the dating information are preferred, e.g., using an island age that is slightly too old reduces the influence of a young-dispersal node calibration. To define the offset between the calibrated divergence time and the palaeogeographic event, the shape of the calibration density should reflect whatever scenario was used for its justification (Step II-b). The divergence event might have preceded, coincided with, or followed the palaeogeographic event (Fig. 9.1). For example, the young-dispersal scenario acts as a hard maximum age constraint for the crown node of a subclade's radiation, so the prior probability should be zero for all subclade ages that are older than the newly available region (Fig. 9.1d). Ancient-dispersal scenarios include only a minimum age constraint (Fig. 9.1e). Vicariance speciation scenarios apply both minimum and maximum age constraints (Fig. 9.1f). Node ages that do not conform to the asserted biogeographic scenario can be assigned zero prior probability (hard constraints) or small nonzero probabilities (soft constraints). Typical node prior densities are simple univariate densities (e.g., uniform, exponential, gamma, and lognormal) with shapes that reflect the asserted biogeographic hypothesis. In addition to the node prior, a standard model of molecular evolution and lineage diversification is specified (Step II-c).

The final step, estimation, first estimates the posterior of time-calibrated phylogenies (Step III-a), then validates those estimates in two ways. One type of validation re-examines whether the initial justification for the calibration is compatible with the subsequent estimates for all nodes in the phylogeny (Step III-b). A vicariance-based node calibration, for example, will assert that a conspicuous biogeographic disjunction resulted from, say, continental rifting that subdivided an ancestor's range. This justification excludes long-distance dispersal as an explanation, but if a vicariance-based calibration inadvertently induces long-distance dispersal events elsewhere in the phylogeny, then the justification itself is questionable (discussed by Kodandaramaiah 2011). It is also crucial to validate that the specified calibrations do not interact negatively with other model priors and with the data (Step IV-b). Though choosing well-behaved priors in Bayesian phylogenetics is often difficult (Alfaro and Holder 2006), misbehaviour can be detected through prior-sensitivity analyses in which the prior and posterior divergence-time estimates are compared under a variety of calibration hypotheses (Warnock et al. 2015; Brown and Smith 2017).

Practical details aside, hundreds of studies have employed prior-based node calibrations based on biogeographic evidence over the past two decades. Listed below is a sample of biogeographic node-calibration analyses applied to a diversity of clades and historical scenarios (Fig. 9.1). Pioneering examples include the work of Fleischer et al. (1998), who calibrated the Hawaiian honeycreeper radiation using the ages of the modern High Islands while being careful to explicate what they assumed to justify the calibrations. That same year, Baldwin and Sanderson (1998) dated the Hawaiian silversword radiation, not using the island ages themselves, but rather the date of dry-summer conditions that appeared in the western North American continent during the mid-Miocene (15 Myr ago). Under the assumption that the arid climate predated the radiation of the silverswords' arid-adapted ancestors (crown Madiinae), this palaeoclimatic shift was used to impose maximum age constraints to calibrate the tree. Hawaiian palaeogeography has been subsequently used to date numerous clades, including *Megalagrion* damselflies (Jordan et al. 2003) and *Hyposmocoma* moths (Haines et al. 2014).

Island palaeogeography outside of Hawaii has also proven useful for node calibration. Studying Sapotaceae, Swenson et al. (2014) used the age when New Caledonia last surfaced from the sea to date five independent colonizations of New Caledonia, shifting crown node ages to be younger by 15–20% (up to 5 million years) relative to when palaeogeography was ignored. Plana et al. (2004) took the formation of two volcanic islands to time-calibrate the radiation of African begonias, a clade with no described fossils. From 16 possible biogeographic calibrations dependent on volcanic and continental island ages, Andújar et al. (2014) computed Bayes factors to select eight biogeographic scenarios that yielded congruent divergence times for *Carabus* beetle diversification.

To marine organisms, the appearance of new land in the ocean acts as a geographical barrier that potentially limits, rather than facilitates, new dispersal opportunities. To this end, many biogeographic calibration studies have focused on the closure of the Isthmus of Panama during the Pliocene: Cowman and Bellwood (2011) used this event, along with fossil calibrations, to date four families of coral reef fishes; Thacker (2017) dated two fish families, Elotridae and

Apogonidae, and showed that using biogeographic calibrations reduced mean divergence times by roughly 30–35%; Swart et al. (2015) also used the isthmus to date Carangidae fishes, which, during validation, revealed that an ancient divergence event and biogeographic disjunction was incidentally consistent with the closure of the Tethys Sea. Yet, as a land bridge, the isthmus simultaneously facilitates terrestrial dispersal: Fuchs et al. (2007) calibrated the basal node of a North-South American disjunction in woodpeckers (Picidae), arguing that dispersal was improbable before the isthmus began to form.

Gondwanan vicariance has been proposed extensively to date clades distributed throughout the Southern Hemisphere, with mixed results in studies of terrestrial invertebrates. Allwood et al. (2010) invoked Gondwanan vicariance to date the evolution of velvet worms (Onychophora), a monophyletic phylum with no inferred transoceanic dispersals. If it is true that New Zealand was entirely below sea level during the late Oligocene (Mildenhall et al. 2014), then this vicariance calibration would imply that velvet worms somehow survived the island's submersion. Allegrucci et al. (2010) examined the Gondwanan distribution of cave crickets (Raphidophoridae), finding that while many vicariance calibrations appear justified, other biogeographic disjunctions can only be explained by invoking long-distance dispersal or with exceedingly ancient (Precambrian) crown-node ages (also see Beasley-Hall et al. 2018). McCulloch et al. (2016) studied the effect of various fossil and tectonic calibrations on the date estimates of Gondwanan stoneflies (Plecoptera). Examining 17 uncalibrated nodes that corresponded to Gondwanan disjunctions, McCulloch et al. found that five of those nodes had age estimates consistent with tectonic events: vicariance sufficiently explained some, but not all, divergence events.

Mountain-building episodes have also been used to calibrate divergence times. On one hand, mountains can serve as geographical barriers between regions: Mansion and Zeltner (2004) used the dates of uplift of the Sierra mountains and various Mexican mountains to calibrate nodes of the *Zeltnera* (Gentianaceae) phylogeny. Mountains also resemble islands to high-altitude specialists: Chaves et al. (2011) dated the serial expansion of *Adelomyia* hummingbirds, a clade of cloud forest endemics, using calibrations based on the south-to-north uplift of the Andean mountains.

Like many fossil-based node calibrations, biogeographic node calibrations are often contentious. Node calibrations should be, and are, subject to measured scientific scrutiny. For example, efforts to date the plant clade Crypteroniaceae stoked discourse among phytogeographers about if, when, and how one might use biogeographic calibrations. Conti et al. (2002) applied tectonic calibrations for Crypteroniaceae, including its migration from the African to Asian continent aboard India as it drifted northwards following the breakup of Gondwana. Moyle (2004) proposed an alternative phylogenetic hypothesis that undermined the justification of Africa-India vicariance calibrations and questioned the range of dates used to represent Gondwanan rifting (see response by Conti et al. 2004), but a comprehensive effort to validate the calibrations by Rutschmann et al. (2004) supported the original conclusions of Conti et al. (2002).

There are important cases where the fossil record undermines the justification of key calibrations. Gibb et al. (2015) questioned the use of Gondwanan vicariance to time-calibrate several passerine phylogenies in the literature. The calibration depended on the rifting of Zealandia from Australia throughout the Late Cretaceous (~82 Myr ago) to date the split between a clade composed of two species of flightless wrens that are endemic to New Zealand from all remaining passerines. Gibb et al. (2015) took two issues with the justification of this calibration. First, the wrens might have arrived in Zealandia after the rifting event if their ancestors could fly. Second, flightless wrens would have been extirpated from New Zealand if and when the island was inundated in the late Oligocene (22–25 Myr ago). Gibb et al. (2015) also took one issue with the specification: that the Australia-Zealandia rift might not have completed until much later (~55 Myr ago). In another example, Goswami and Upchurch

(2010) found that the vicariance calibrations used by Heads (2010) were unjustified in explaining the disjunction between New World monkeys and Old World monkeys. The calibrations require that early primate lineages were globally distributed, first diverging during the breakup of Pangaea in the Jurassic (~160 Myr ago). It would be unprecedented, Goswami and Upchurch argued, for primates to be sufficiently ancient and widely distributed, yet leave no appearances in the fossil record until the late Paleocene (~56 Myr ago). Moreover, no eutherian mammal fossils were known prior to 125 Myr ago (Goswami and Upchurch 2010). If primates were so ancient, per the Pangaean vicariance hypothesis, that fact would radically alter our understanding of evolutionary processes, vicariance speciation, and the fossil record.

In other cases, sole reliance on fossils for dating can result in node-age estimates that are unusual when viewed in light of biogeographic and palaeogeographic evidence. Ali (2020) identified that the fossil-calibrated phylogeny of *Kurixalus* frogs (Lv et al. 2018) implies that Taiwan was colonized at least 10 Myr before the island had originated. Román-Palacios et al. (2018) detected a similar issue in a fossil-dated phylogeny of Caribbean anoles that they estimated. In it, a clade of island endemics appeared to predate the emergence of the island itself. The quality of a dating analysis is improved by examining where complementary lines of evidence agree or disagree.

To summarize this section, biogeographic node calibrations encode expert knowledge and diversification hypotheses as prior probabilities to constrain when key lineage-splitting events occurred. Biogeographers and evolutionary biologists are developing new conceptual frameworks for how to justify and specify palaeogeographic node calibrations (Kodandaramaiah 2011; Ho et al. 2015; de Baets et al. 2016). Two dominant themes emerge from the few examples of biogeographic node calibration listed above. First, biogeographic node calibrations tend to be used to date clades with little or no representation in the fossil record, including plants, insects, fish, and birds. And, second, the justification for some biogeographic calibrations, particularly vicariance calibrations, wither away when assessed critically, while others remain firm.

Beyond their application in the literature, it is important to recognize that node priors are only as good as their justification and specification. With justification, how certain is it that a particular biogeographic event influenced the phylogenetic split of interest? If this scenario is unlikely, then any dates estimated under its premise are equally questionable. If one could confidently state the probability, p, of a key biogeographic scenario informing the age of the calibrated node, a_n, then the node prior could be treated as a mixture of priors where a_n follows the calibration prior with probability p and an uninformative uniform prior with probability $1 - p$. This is a calibration with soft bounds (Yang and Rannala 2006). What complicates this strategy is that the value of p should depend on the biogeographic states at the tips of the tree, the tree topology and distribution of divergence times, the tempo and mode of the biogeographic process, and the unknown interactions between biogeography and palaeogeography: that is, p needs to be inferred from the data with a biogeographic model.

Specifying node prior calibration densities is also challenging. What probability density represents all of the uncertainty concerning the age of the divergence event relative to the palaeogeographic event? This depends a great deal on the palaeogeography, biogeography, and phylogeny of the system in question. For example, a prior density for a lineage from the mainland colonizing and radiating in an island system might be constructed as a compound prior

$$a_n = a_\mathrm{i} - t_\mathrm{c} - t_\mathrm{d} \qquad (9.2)$$

where a_n is the age of the first divergence event among a clade of island endemics, a_i is the age of the island, t_c is the waiting time for a mainland lineage to colonize the new island, and t_d is the waiting time until divergence following genetic isolation. Decomposing the prior in this manner clarifies some of the assumptions made to specify it. For example, we might define a_n as the age of a uniformly distributed island age minus two exponentially distributed waiting times: this would be fairly consistent with the definition of a gamma distribution centred on a random age away from zero. While it might be straightforward to acquire a radiometric date (with error terms) for the island age component, a_i, it is less clear what the expected waiting times until colonization, t_c, and until lineage splitting, t_d, should be. Like the justification probability, p, the values of t_c and t_d are quantities that one typically estimates from data phylogenetically, rather than what one asserts a priori.

Despite the many conceptual challenges, prior-based calibration methods are easy to apply, computationally speaking. Most prior densities are standard univariate parametric distributions, so it is simple and fast to compute the node-age probability under the calibration. That said, multiple node calibrations can induce a joint prior density that behaves in unpredictable ways, making it imperative to assess the prior sensitivity of the calibration model: see Warnock et al. (2015) for an excellent overview of this subject. Biogeographic node calibrations can be specified in any phylogenetic software that supports node priors, such as MrBayes (Ronquist et al. 2012), RevBayes (Höhna et al. 2016), BEAST (Bouckaert et al. 2014), and MCMCTree (Yang and Rannala 2006).

9.3.2 Process-Based Biogeographic Dating

The previous section explored how prior-based methods date phylogenies that reflect hypothetical biogeographic scenarios. This section introduces a second class of dating methods, called process-based methods, that instead rely on data-dependent biogeographic inference. Process-based dating methods estimate posterior node-age densities that are shaped through the likelihood function rather than through the prior. The distinction between prior-and process-based approaches is important to many researchers, because process-based inference means that node-age estimates result from the palaeogeographic, biogeographic, and phylogenetic evidence provided, rather than as an interpretation of the evidence that one asserts on the inference problem through the prior. The two biogeographic dating strategies are directly analogous to prior-based (Donoghue and Moore 2003; Yang and Rannala 2006; Parham et al. 2012) and process-based (Pyron 2011; Ronquist et al. 2012; Heath et al. 2014; Gavryushkina et al. 2017) fossil dating strategies.

As the name suggests, process-based biogeographic dating methods require that we define a biogeographic process of range evolution. Much like any process of molecular evolution, biogeographic processes are used to compute transition probabilities for biogeographic change over time, for example to compute the probability that a lineage dispersed from a continent into an island within 1 million years of the lineage's existence. To compute this probability, again similar to molecular processes, biogeographic processes probabilistically average (integrate) over all possible histories that are compatible with the biogeographical data observed at the tips of the tree, weighing each history by its probability. Biogeographic models come in a variety of forms, with special attention paid to discrete regions (Ree et al. 2005; Sanmartín et al. 2008), continuous regions (Lemmon and Lemmon 2008; Lemey et al. 2009; Quintero et al. 2015), range-dependent speciation-extinction rate variation (Goldberg et al. 2011; Caetano et al. 2018), cladogenetic range-inheritance events (Matzke 2014), and factors including geography (Landis et al. 2013; Tagliacollo et al. 2015), ecology (Meseguer et al. 2015; Landis et al. 2021), morphology (Sukumaran et al. 2016), and interspecific competition (Quintero and Landis 2020).

Unlike molecular processes, however, biogeographic processes can readily incorporate palaeogeographic information to structure their transition rates (and thus their transition probabilities) between regions so that they depend on the geological age of a possible range-evolution event. These models are often called time-stratified (Ree et al. 2005; Ree and Smith 2008) or epoch models (Bielejec et al. 2014). As an example, a dispersal event from a continent into a 5-million-year-old island at any time during a 1 Myr interval would have zero probability before the island surfaced (e.g., 10 Myr ago) and nonzero probability only after it originated (e.g., just 1 Myr ago). Although evolutionary rate and geological time are non-identifiable from one another under standard molecular models (Zuckerkandl and Pauling 1962; Thorne et al. 1998), the parameters can be separately identified under biogeographical models that are palaeogeographically structured (Landis 2017).

A major strength of process-based dating methods is their ability to model the uncertainty that is inherent to biogeographic dating approaches. Rather than justifying and specifying individual biogeographic node calibrations as one does in a prior-based approach, process-based methods assess the likelihood of observed biogeographic disjunctions by fitting models with historical interactions between palaeogeographic, biogeographic, and phylogenetic components. By integrating over hypothetical scenarios involving palaeogeographic, biogeographic, and phylogenetic interactions that compete to explain the observed molecular and biogeographic variation, process-based dating methods generate posterior distributions of node-age densities through the model's likelihood function. A brief overview of how to design a process-based analysis is given below (Fig. 9.3).

An overview of process-based biogeographic dating

(I) Specification
  (a) Define a set of discrete regions.
  (b) Define a model of palaeogeographical dynamics.
  (c) Define a palaeogeography-dependent model of biogeographic evolution.
  (d) Define standard models of molecular evolution and lineage diversification.

(II) Estimation
  (a) Estimate the posterior of divergence times from molecular and biogeographic data.
  (b) Assess prior sensitivity of posterior estimates.
  (c) Conduct simulation experiments to assess confidence in empirical results.

Process-based dating relies on two major steps: specification and estimation. Unlike prior-based dating, there is no justification step because the specified model averages over possible biogeographic and palaeogeographic justifications (hypotheses). Specification, however, is more involved. The first step is to define the set of regions that will adequately portray the relevant palaeogeographic dynamics and biogeographic disjunctions in the analysis (Step I-a). For example, two regions are sufficient to capture the dynamics in Fig. 9.1. Further dividing region S into regions SW and SE would not expose any new dating information. Next, characterize the palaeogeographic dynamics in terms of the availability and connectivity between regions with respect to time (Step I-b). In practice, this can be done by defining a vector of adjacency graphs, where nodes correspond to regions, edges correspond to dispersal routes, and each graph in the vector is indexed by the timing of a palaeogeographic event (Buerki et al. 2011; Landis 2017). The time-stratified biogeographic model of Step I-c will then be defined to condition on the palaeogeographic structure defined in Step I-b by, for example, declaring that dispersal rates between two regions are greater than zero only if the regions are connected. To estimate the topology and branch-length parameters, standard molecular phylogenetic models are used (Step I-d).

Once the molecular, biogeographic, and palaeogeographical models are specified, all model components are simultaneously fitted to the molecular and biogeographic data in a joint Bayesian analysis (Step II-a). The estimates are then validated in two ways. Prior sensitivity experiments or simulation experiments can be used to validate process-derived age estimates. Prior sensitivity analyses for process-based methods behave similarly to those for prior-based methods (see previous section), except that the priors applied to the biogeographic process parameters should also be tested for sensitivity. Sensitivity analyses are useful to assess the

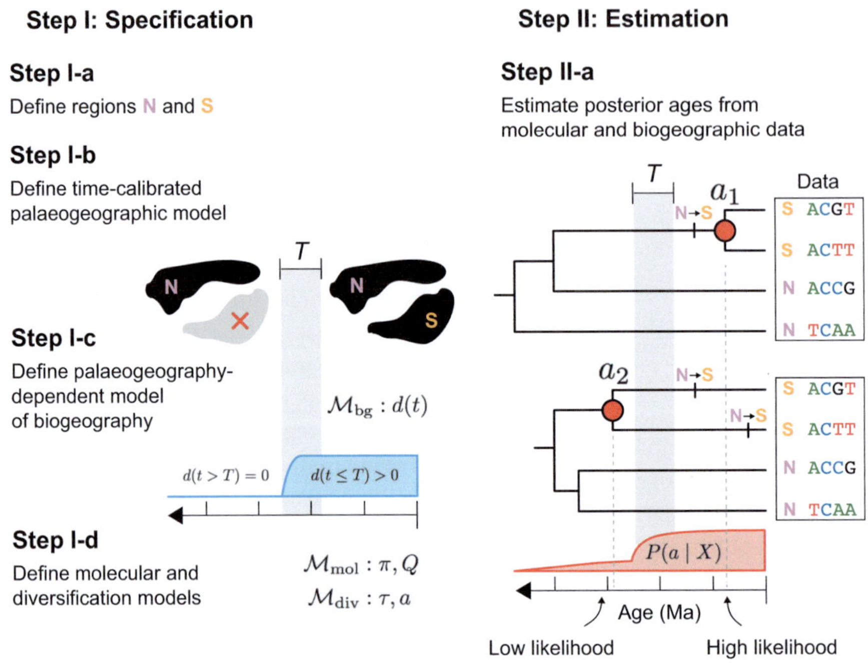

Fig. 9.3 Diagram of process-based biogeographic dating. Node calibration involves two major steps. Specification (Step I) first defines a set of biogeographic regions (I-a), then defines a dated palaeogeographic model for the availability and connectivity of those regions (I-b) that structures a time-stratified biogeographic model (I-c). Here, the biogeographic model, $\mathcal{M}_{\text{bg}}$, defines palaeogeography-dependent dispersal rates, $d(t)$, that are zero before region S appears at time T and positive afterwards (blue density). Simple molecular and diversification models ($\mathcal{M}_{\text{mol}}$ and $\mathcal{M}_{\text{div}}$) are also defined (I-d). Estimation (Step II) fits the model to the molecular and biogeographic data to estimate a posterior density of dated phylogenies (II-a). The model likelihood favours few, young dispersal events over many, (young or ancient) dispersal events, which is reflected in the posterior density (red). Thus, the posterior density prefers young ages (a_1) rather than old ages (a_2). Steps II-b and II-c are not shown. See main text for details

accuracy of divergence-time estimates under a variety of controlled settings, such as under the assignment of alternative prior densities to model parameters. Simulation experiments involve simulating many (≥100) phylogenetic data sets under the model defined in Steps I-b, I-c, and I-d, then estimating divergence times for those simulated data. If the estimated ages agree with the true simulated ages, that instils confidence that the process-based method behaves well under controlled conditions. But if the method grossly misestimates divergence times, then it is likely that any empirical estimates are incorrect or worse. Simulation experiments cannot be conducted naturally using prior-based dating methods, since priors do not generate biogeographic data, making this form of validation somewhat unique to process-based methods.

Process-based biogeographic dating methods are quite young relative to prior-based methods, with only two examples in the literature at this time. In the study that introduced the method, Landis (2017) used the process-based techniques to date the phylogeny of crown turtles (Testudines). Testudines has received excellent taxonomic and phylogenetic treatment for decades (Crawford et al. 2015), and possesses a superbly documented fossil record that spans the Mesozoic and Cenozoic. These data have been used to generate numerous fossil-based estimates of the clade's age (Near et al. 2004; Hugall et al. 2007; Joyce et al. 2013), which are valuable to validate whether alternative dating methods find similar ages. Biogeographically speaking, Testudines exhibits an 'out-of-Gondwana' distribution (Joyce et al. 2016) that remains imprinted in extant ranges thanks to their tendency for slow-and-steady dispersal. Together, these natural and scientific conditions made Testudines ideal for studying the behaviour of process-based biogeographic dating methods.

Landis (2017) dated the global expansion of Testudines lineages by jointly modelling interactions between processes of diversification, molecular evolution, biogeographic change, and palaeogeography while assuming a fixed topology. To do this, I introduced a global empirical model of continental drift for 25 regions and 26 time intervals, stretching from the Cambrian (540 Myr ago) to the present. The continental drift model defined connectivity between regions as strong, weak, or absent with three sets of graphs, defining $\sim$200 connections per time interval, with over 1000 changes to interregional connectivity in total. It is not obvious how important geographic connectivity is to any particular clade: birds might disperse freely across mountains, but turtles might not. Rather than asserting the importance of connectivity a priori, as one must do when justifying a prior-based node calibration, the combined importance of strong, weak, and/or absent connectivity on biogeographic change was estimated from the data. Fitting the model to the Testudines data set estimated a mean posterior root age of 205 Myr (95% highest posterior density of 135–358 Myr), a result that was congruent with the fossil-based estimates from the literature, which reported estimates as low as 150 Myr (Joyce 2007) and as high as 325 Myr (Dornburg et al. 2011).

In the second study, Landis et al. (2018) estimated divergence times for the silversword alliance, an iconic adaptive radiation of plants that dispersed throughout the Hawaiian Islands (Carlquist 1966). Like many island plant endemics, silverswords lack any described fossils, so estimating the age of this clade has been consequently difficult (Baldwin and Sanderson 1998). To complicate matters, the topology within the silversword clade is not completely resolved, making it difficult to justify calibrations for a specific phylogenetic hypothesis. Although there was likely to have been only one long-distance dispersal event from the North American continent into the Hawaiian archipelago (Baldwin et al. 1991), silverswords are found only throughout the High Islands, the set of younger islands that are not more than 6 million years old (Clague and Sherrod 2014). The exact sequence and timing of island colonization events within the archipelago must inform the divergence times, but no single biogeographic scenario is suitable for justifying or specifying node calibrations, further limiting the applicability of prior-based methods.

Similar to the approach taken by Landis (2017), Landis et al. (2018) estimated what combination of evolutionary histories for silverswords with divergence times, tree topologies, biogeographic histories, and island origination times had a high probability of generating the observed molecular and biogeographical data. The silversword crown age was dated to be roughly 3.5 Myr, and at most 5.1 Myr, a date that is consistent with the maximum age estimate of Baldwin and Sanderson (1998) that disregarded island ages to date the clade. Standard practice for using prior-based methods would assign the maximum age of the clade as equal to the oldest inhabited island (Kauai), while potential dating information from younger islands would be disregarded because it cannot easily be justified. But process-based dating applied to the silverswords extracted additional information for subclade ages when using all islands' ages rather than only the oldest island's age. For example, when testing prior sensitivity of the age estimates, one subclade (*Argyroxiphium*) age was estimated to be twice as old when only using the age of Kauai in comparison with an analysis using all islands' ages.

Keep in mind that evolutionary biologists and biogeographers are often interested in dating phylogenies for the purposes of estimating ancestral states. Dated phylogenies that rely on biogeographic node calibrations generally cannot be used to later reconstruct ancestral species ranges: prior-based calibrations are justified through the assertion of a historical biogeographic scenario, so subsequent ancestral range estimates would be biased towards scenarios that conform to the prior hypothesis (de Jong 2007). As part of the process-based dating analysis, Landis et al. (2018) estimated distributions of possible biogeographic scenarios, then used those estimated histories to test various biogeographic hypotheses, such as which island silverswords first colonized and whether dispersal and speciation processes favoured young or old islands. Because process-based dating methods do not involve a justification step, they avoid many such forms of circular reasoning ascribed to prior-based methods (Renner 2005; Kodandaramaiah 2011; de Jong 2007).

In summary, process-based biogeographic dating methods measure the probabilities of competing node-age distributions by averaging (integrating) over all possible palaeogeographic, biogeographic, and phylogenetic histories that are defined by the likelihood model. Because process-based dating methods fit models to palaeogeographic and biogeographic data, they are more data-intensive than prior-based methods. This makes process-based methods more sensitive to data errors, such as errors in coding species ranges or island ages, which could skew first arrival times to islands. In addition, process-based methods are more computationally intensive than prior-based methods. Estimating divergence times using a process-based approach requires repeatedly computing the biogeography model's likelihood function, which can be as slow as computing the molecular likelihood function or worse. Most biogeographic models scale abysmally with increasing numbers of regions (Ree and Sanmartín 2009), and current methods designed to circumvent this issue cannot treat phylogenies as random variables (Landis et al. 2013).

Biogeographic model adequacy is also a major concern. Models that neglect major features of range evolution will assign inaccurate probabilities in support of alternative biogeographic scenarios. For example, an extremely inadequate model might treat all geographical barriers as entirely impermeable, assigning zero probability to any dated phylogeny that requires long-distance dispersal to explain biogeographic disjunctions. But, in reality, long-distance dispersal always has a nonzero probability, however small. At the moment, phylogenetic models of biogeography are still in their infancy (Albert and Antonelli 2017), but are maturing steadily (Sanmartín et al. 2008; Webb and Ree 2012; Matzke 2014; Meseguer et al. 2015; Quintero et al. 2015; Tagliacollo et al. 2015; Sukumaran et al. 2016). Not only are the biogeographic models young, but so are the methods that apply them to divergence-time estimation problems (Landis 2017; Landis et al. 2018), meaning the properties of those methods are poorly understood relative to prior-based methods. Further limiting the method's use, the computational framework needed to jointly model phylogenetic, biogeographic, and palaeogeographic interactions is specialized and only currently available in RevBayes (Höhna et al. 2016).

9.4 Conclusions

Recounting the biological history of Earth requires knowledge of the order and timing of the constituent events. Our record of historical events is incomplete, meaning that we rely on inference to retrace the past. But it is no trivial matter to locate key phylogenetic events, such as when or where lineages diverged, throughout most of the tree of life. Advances in phylogenetic inference let us establish a geological timescale for lineage divergences by recruiting extrinsic evidence: for example, the age and morphology of a fossil specimen can indicate the early origins of a particular clade. Palaeogeographic events, such as the birth of islands, the building of mountains, or the separation of continents, have also proven useful for dating evolutionary lineages, particularly lineages in clades with poor fossil representation. That framework, biogeographic dating, works by adopting a phylogenetic perspective to disentangle how a clade's age and biogeographic pattern might have been influenced by one or several palaeogeographic events.

In this chapter, I reviewed the conceptual basis of biogeographic dating under two methodological frameworks: prior-based biogeographic node-calibration methods and process-based biogeographic dating methods. Although prior- and process-based methods both convert biogeographic hypotheses into information to estimate divergence times, they do so in different ways. Prior-based methods require that the researcher first justifies that a particular biogeographic scenario resulted from a palaeogeographic event and, second, specifies a range of plausible origination times for the newly diverged lineage(s). Process-based methods specify a model of palaeogeography-dependent biogeographic evolution that is fitted to the observed species ranges by probabilistically averaging over the distribution of possible historical scenarios.

Prior-based dating is extremely flexible. Part of this flexibility emerges from the conceptual foundation of biogeographic node calibrations, which has been regarded as somewhat murky (Renner 2005; Kodandaramaiah 2011; de Baets et al. 2016). But being murky comes with advantages and disadvantages. Prior-based dating affords the biologist complete control to set the precision and accuracy of the dating estimates in such a way that comports with their expert description of the system's evolutionary history. From the computational perspective, prior-based node calibrations tend to be fairly easy to design and evaluate in phylogenetic analyses. But it is not uncommon to hear experts disagree about whether a calibration has a correct justification or specification. While those disagreements are not easily settled quantitatively because the justification and specification are, ultimately, related to prior model design, they can be settled logically (Goswami and Upchurch 2010; Kodandaramaiah 2011). Several researchers have also voiced concern that justifying a biogeographic node calibration often requires circular reasoning, or at least the assertion of an unfalsifiable hypothesis (Renner 2005; Kodandaramaiah 2011; de Baets et al. 2016). One last consequence of prior-based dating is that the biogeographically dated trees cannot later be used to estimate ancestral species ranges without improperly 'double counting' the biogeographic evidence (de Jong 2007). Despite any complications, prior-based methods are still the most popular and widely used biogeographic dating strategy.

Process-based methods extract calibration information from the joint distribution of palaeogeographic data, biogeographic data, and molecular data. When compared with prior-based methods, the biologist is required to make fewer strong assertions about specific biogeographic scenarios, such as exactly how species are related, how completely a geographic barrier disrupts geodispersal, and the exact choreography of a clade's spatial radiation. Rather than supposing a particular scenario a priori, process-based dating approaches average over all historical biogeographic scenarios that are defined by the model, weighing each scenario by its probability of having occurred. While process-based approaches have some features that are theoretically appealing, they depend entirely on the adequacy of our biogeographic models, which are still quite simple despite ongoing developments. Process-based methods are fundamentally more complex than prior-based methods, with a computational burden that might limit their application in practice.

Regardless of whether one uses prior-based or process-based methods, the most satisfying divergence-time estimates are those that are correct (high accuracy) and confident (high precision). Precise results without accuracy are especially disturbing, leading one to draw the wrong conclusions from the evidence at hand. Seeking highly precise dating estimates should not be a goal at the expense of all else (Graur and Martin 2004). How do we extract that dating information from biogeography without introducing bias? When applying biogeographic prior-based calibrations, first, be critical of whether the calibration is truly justified and defensible. What empirical evidence or computational experiment could render the justification invalid? Second, avoid recycling densities that were previously published in the literature. Instead, the biologist should define a new density that represents her prior beliefs for when a divergence event occurred after reviewing the phylogenetic, biogeographic, and palaeogeographic evidence. Experiment with densities with higher variance or soft tails during validation. For example, if it is not certain that a vicariance scenario occurred, consider the use of a 'soft' vicariance prior (Fig. 9.1f, bottom), or instead use the less constrained ancient dispersal prior (Fig. 9.1e). Process-based methods are also prone to giving biased estimates when the biogeographic data contain coding errors, when the palaeogeographic model is too restrictive or contains inaccurate dates, and when the biogeographic model is severely inadequate. Simulation experiments might help to detect and protect against such biases.

We do not know the exact limits for when we can accurately or precisely date nodes using biogeography. It is worth mentioning that fossil-based estimates are generally deemed superior to biogeography-based estimates for many good reasons. The timing of biogeographic scenarios contains many uncertainties: the timing of many palaeogeological events is not known precisely, the actual palaeogeographical event might be prolonged over many millions of years, or the relationship between the palaeogeographic event and the biogeographic event is itself unclear. Fossil specimens are often dated with fairly high precision when compared with biogeographic scenarios, and fossilization and divergence scenarios involve fewer contingent events to describe patterns than do biogeographic scenarios. This ultimately makes it easier to model fossilization scenarios (with a prior or a process). Despite the superiority of fossil-based dating in these respects, the fossil record varies in reliability: some groups (such as carnivorans and cetaceans) have exceptional records, while many groups are poorly represented (bacteria, plants, fungi, soft-bodied invertebrates, and small vertebrates). Fossil and biogeographic dating methods, however, should be regarded as complementary, not competing, strategies. Applying the two strategies simultaneously will, in principle, improve dating estimates beyond what might be learned using only one half of the available evidence.

Our understanding of evolution draws from diverse lines of evidence, and so our attempts to chronicle evolutionary history will require similarly diverse evidence and methods. No matter what method is used to time-calibrate a phylogeny, it is a delicate and often difficult practice. But biologists, in exercising care, curiosity, and patience, are making steady progress towards a richer portrait of when, where, and how life diversified.

Acknowledgements Feedback from Nate Upham, Rachel C. Warnock, Edgar Benevides, and Luis Palazessi helped improve an early draft of the chapter. I am also grateful to an anonymous reviewer and to the editor of this book, Simon Y. W. Ho, for their remarks.

Funding

MJL was supported by a NSF Postdoctoral Fellowship (DBI-1612153) to MJL and a Gaylord Donnelley Environmental Fellowship through the Yale Institute of Biospheric Studies.

References

Albert JS, Antonelli A (2017) Society for the study of systematic biology symposium: frontiers in parametric biogeography. Syst Biol 66:125–127

Alfaro ME, Holder MT (2006) The posterior and the prior in Bayesian phylogenetics. Annu Rev Ecol Evol Syst 37:19–42

Ali JR (2020) Geological data indicate that the interpretation for the age-calibrated phylogeny for the *Kurixalus*-genus frogs of South, South-east and East Asia (Lv et al., 2018) needs to be rethought. Mol Phylogenet Evol 145:106053

Allegrucci G, Trewick SA, Fortunato A, Carchini G, Sbordoni V (2010) Cave crickets and cave weta (Orthoptera, Rhaphidophoridae) from the southern end of the world: a molecular phylogeny test of biogeographical hypotheses. J Orthoptera Res 19:121–130

Allwood J, Gleeson D, Mayer G, Daniels S, Beggs JR, Buckley TR (2010) Support for vicariant origins of the New Zealand Onychophora. J Biogeogr 37:669–681

Andújar C, Soria-Carrasco V, Serrano J, Gómez-Zurita J (2014) Congruence test of molecular clock calibration hypotheses based on Bayes factor comparisons. Methods Ecol Evol 5:226–242

Angiosperm Phylogeny Website (2018) The angiosperm phylogeny website. http://www.mobot.org/MOBOT/research/APweb/

Baldwin BG, Sanderson MJ (1998) Age and rate of diversification of the Hawaiian silversword alliance (Compositae). Proc Natl Acad Sci USA 95:9402–9406

Baldwin BG, Kyhos DW, Dvorak J, Carr GD (1991) Chloroplast DNA evidence for a North American origin of the Hawaiian silversword alliance (Asteraceae). Proc Natl Acad Sci USA 88:1840–1843

Barreda VD, Palazzesi L, Tellería MC, Olivero EB, Raine JI, Forest F (2015) Early evolution of the angiosperm clade Asteraceae in the Cretaceous of Antarctica. Proc Natl Acad Sci USA 112:10989–10994

Beasley-Hall PG, Tierney SM, Weinstein P, Austin AD (2018) A revised phylogeny of macropathine cave crickets (Orthoptera: Rhaphidophoridae) uncovers a paraphyletic Australian fauna. Mol Phylogenet Evol 126:153–161

Bielejec F, Lemey P, Baele G, Rambaut A, Suchard MA (2014) Inferring heterogeneous evolutionary processes through time: from sequence substitution to phylogeography. Syst Biol 63:493–504

Bouckaert R, Heled J, Kühnert D, Vaughan T, Wu C-H, Xie D, Suchard MA, Rambaut A, Drummond AJ (2014) BEAST 2: a software platform for Bayesian evolutionary analysis. PLOS Comput Biol 10: e1003537

Brown JW, Smith SA (2017) The past sure is tense: on interpreting phylogenetic divergence time estimates. Syst Biol 67:340–353

Buerki S, Forest F, Alvarez N, Nylander JAA, Arrigo N, Sanmartín I (2011) An evaluation of new parsimony-based versus parametric inference methods in biogeography: a case study using the globally distributed plant family Sapindaceae. J Biogeogr 38:531–550

Caetano DS, O'Meara BC, Beaulieu JM (2018) Hidden state models improve state-dependent diversification approaches, including biogeographical models. Evolution 72:2308–2324

Carlquist S (1966) The biota of long-distance dispersal. I. Principles of dispersal and evolution. Q Rev Biol 41:247–270

Chaves JA, Weir JT, Smith TB (2011) Diversification in *Adelomyia* hummingbirds follows Andean uplift. Mol Ecol 20:4564–4576

Clague DA, Sherrod DR (2014) Growth and degradation of Hawaiian volcanoes. US Geol Surv Prof Pap 1801:97–146

Conti EA, Eriksson T, Schönenberger J, Sytsma KJ, Baum DA (2002) Early Tertiary out-of-India dispersal of Crypteroniaceae: evidence from phylogeny and molecular dating. Evolution 56:1931–1942

Conti EA, Rutschmann F, Eriksson T, Sytsma KJ, Baum DA (2004) Calibration of molecular clocks and the biogeographic history of Crypteroniaceae: a reply to Moyle. Evolution 58:1871–1876

Cowman PF, Bellwood DR (2011) Coral reefs as drivers of cladogenesis: expanding coral reefs, cryptic extinction events, and the development of biodiversity hotspots. J Evol Biol 24:2543–2562

Crawford NG, Parham JF, Sellas AB, Faircloth BC, Glenn TC, Papenfuss TJ, Henderson JB, Hansen MH, Simison BW (2015) A phylogenomic analysis of turtles. Mol Phylogenet Evol 83:250–257

Darwin C (1859) On the origin of species. John Murray, London

de Baets K, Antonelli A, Donoghue PCJ (2016) Tectonic blocks and molecular clocks. Philos Trans R Soc B 371:20160098

de Jong R (2007) Estimating time and space in the evolution of the Lepidoptera. Tijdschr Entomol 150:319–346

Donoghue MJ, Moore BR (2003) Toward an integrative historical biogeography. Integr Comp Biol 43:261–270

Dornburg A, Beaulieu JM, Oliver JC, Near TJ (2011) Integrating fossil preservation biases in the selection of calibrations for molecular divergence time estimation. Syst Biol 60:519–527

Drummond AJ, Ho SYW, Phillips MJ, Rambaut A (2006) Relaxed phylogenetics and dating with confidence. PLOS Biol 4:e88

Felsenstein J (1981) Evolutionary trees from DNA sequences: a maximum likelihood approach. J Mol Evol 17:368–376

Fleischer RC, McIntosh CE, Tarr CL (1998) Evolution on a volcanic conveyor belt: using phylogeographic reconstructions and K-Ar-based ages of the Hawaiian Islands to estimate molecular evolutionary rates. Mol Ecol 7:533–545

Fuchs J, Ohlson JI, Ericson PGP, Pasquet E (2007) Synchronous intercontinental splits between assemblages of woodpeckers suggested by molecular data. Zool Scr 36:11–25

Gavryushkina A, Heath TA, Ksepka DT, Stadler T, Welch D, Drummond AJ (2017) Bayesian total evidence dating reveals the recent crown radiation of penguins. Syst Biol 66:57–73

Gibb GC, England R, Hartig G, McLenachan PA, Taylor Smith BL, McComish BJ, Cooper A, Penny D (2015) New Zealand passerines help clarify the diversification of major songbird lineages during the Oligocene. Genome Biol Evol 7:2983–2995

Goldberg EE, Lancaster LT, Ree RH (2011) Phylogenetic inference of reciprocal effects between geographic range evolution and diversification. Syst Biol 60:451–465

Goswami A, Upchurch P (2010) The dating game: a reply to Heads (2010). Zool Scr 39:406–409

Graur D, Martin W (2004) Reading the entrails of chickens: molecular timescales of evolution and the illusion of precision. Trends Genet 20:80–86

Haines WP, Schmitz P, Rubinoff D (2014) Ancient diversification of *Hyposmocoma* moths in Hawaii. Nat Commun 5:3502

Heads M (2010) Evolution and biogeography of primates: a new model based on molecular phylogenetics, vicariance and plate tectonics. Zool Scr 39:107–127

Heads M (2012) Bayesian transmogrification of clade divergence dates: a critique. J Biogeogr 39:1749–1756

Heath TA, Huelsenbeck JP, Stadler T (2014) The fossilized birth–death process for coherent calibration of divergence-time estimates. Proc Natl Acad Sci USA 111:E2957–E2966

Ho SYW, Tong KJ, Foster CSP, Ritchie AM, Lo N, Crisp MD (2015) Biogeographic calibrations for the molecular clock. Biol Lett 11:20150194

Höhna S, Landis MJ, Heath TA, Boussau B, Lartillot N, Moore BR, Huelsenbeck JP, Ronquist F (2016) RevBayes: Bayesian phylogenetic inference using graphical models and interactive model-specification language. Syst Biol 65:726–736

Hugall AF, Foster R, Lee MSY (2007) Calibration choice, rate smoothing, and the pattern of tetrapod diversification according to the long nuclear gene RAG-1. Syst Biol 56:543–563

Jaanusson V (1976) Faunal dynamics in the middle Ordovician (Viruan) of Balto-Scandia. In: Bassett MG (ed) The Ordovician system: Proceedings of a Palaeontological Association symposium. University of Wales Press, Cardiff, pp 301–326

Jordan S, Simon C, Polhemus D (2003) Molecular systematics and adaptive radiation of Hawaii's endemic damselfly genus *Megalagrion* (Odonata: Coenagrionidae). Syst Biol 52:89–109

Joyce WG (2007) Phylogenetic relationships of Mesozoic turtles. B Peabody Mus Nat Hist 48:3–102

Joyce WG, Parham JF, Lyson TR, Warnock RCM, Donoghue PCJ (2013) A divergence dating analysis of turtles using fossil calibrations: an example of best practices. J Paleontol 87:612–634

Joyce WG, Rabi M, Clark JM, Xu X (2016) A toothed turtle from the late Jurassic of China and the global biogeographic history of turtles. BMC Evol Biol 16:236

Kodandaramaiah U (2011) Tectonic calibrations in molecular dating. Curr Zool 57:116–124

Landis MJ (2017) Biogeographic dating of speciation times using paleogeographically informed processes. Syst Biol 66:128–144

Landis MJ, Matzke NJ, Moore BR, Huelsenbeck JP (2013) Bayesian analysis of biogeography when the number of areas is large. Syst Biol 62:789–804

Landis MJ, Freyman WA, Baldwin BG (2018) Retracing the Hawaiian silversword radiation despite phylogenetic, biogeographic, and paleogeographic uncertainty. Evolution 72:2343–2359

Landis MJ, Edwards EJ, Donoghue MJ (2021) Modeling phylogenetic biome shifts on a planet with a past. Syst Biol (in press)

Lemey P, Rambaut A, Drummond AJ, Suchard MA (2009) Bayesian phylogeography finds its roots. PLOS Comput Biol 5:e1000520

Lemmon AA, Lemmon EM (2008) A likelihood framework for estimating phylogeographic history on a continuous landscape. Syst Biol 57:544–561

Lv Y, He K, Klaus S, Brown RM, Li J (2018) A comprehensive phylogeny of the genus *Kurixalus* (Rhacophoridae, Anura) sheds light on the geographical range evolution of frilled swamp treefrogs. Mol Phylogenet Evol 121:224–232

Mansion G, Zeltner L (2004) Phylogenetic relationships within the New World endemic *Zeltnera* (Gentianaceae-Chironiinae) inferred from molecular and karyological data. Am J Bot 91:2069–2086

Marshall CR (1990) The fossil record and estimating divergence times between lineages: maximum divergence times and the importance of reliable phylogenies. J Mol Evol 30:400–408

Marshall CR (2008) A simple method for bracketing absolute divergence times on molecular phylogenies using multiple fossil calibration points. Am Nat 171:726–742

Matzke NJ (2014) Model selection in historical biogeography reveals that founder-event speciation is a crucial process in island clades. Syst Biol 63:951–970

McCulloch GA, Wallis GP, Waters JM (2016) A time-calibrated phylogeny of southern hemisphere stoneflies: testing for Gondwanan origins. Mol Phylogenet Evol 96:150–160

Meseguer AS, Lobo JM, Ree R, Beerling DJ, Sanmartín I (2015) Integrating fossils, phylogenies, and niche models into biogeography to reveal ancient evolutionary history: the case of *Hypericum* (Hypericaceae). Syst Biol 64:215–232

Mildenhall DC, Mortimer N, Bassett KN, Kennedy EM (2014) Oligocene paleogeography of New Zealand: maximum marine transgression. New Zeal J Geol Geop 57:107–109

Moyle RG (2004) Calibration of molecular clocks and the biogeographic history of Crypteroniaceae. Evolution 58:1871–1873

Near TJ, Meylan PA, Shaffer HB (2004) Assessing concordance of fossil calibration points in molecular clock studies: an example using turtles. Am Nat 165:137–146

Nee S, May RM, Harvey PH (1994) The reconstructed evolutionary process. Philos Trans R Soc B 344:305–311

Ogilvie HA, Vaughan TG, Matzke NJ, Slater GJ, Stadler T, Welch D, Drummond AJ (2018) Inferring species trees using integrative models of species evolution. bioRxiv. https://doi.org/10.1101/242875

Paleobiology Database (2018) The Paleobiology Database. https://paleobiodb.org

Parham JF, Irmis RB (2007) Caveats on the use of fossil calibrations for molecular dating: a comment on Near et al. Am Nat 171:132–136

Parham JF, Donoghue PCJ, Bell CJ, Calway TD, Head JJ, Holroyd PA, Inoue JG, Irmis RB, Joyce WG, Ksepka DT, Patané JSL, Smith ND, Tarver JE, van Tuinen M, Yang Z, Angielczyk KD, Greenwood JM, Hipsley CA, Jacobs L, Makovicky PJ, Müller J, Smith KT, Theodor JM, Warnock RCM (2012) Best practices for justifying fossil calibrations. Syst Biol 61:346–359

Plana V, Gascoigne A, Forrest LL, Harris D, Pennington RT (2004) Pleistocene and pre-Pleistocene *Begonia* speciation in Africa. Mol Phylogenet Evol 31:449–461

Pyron RA (2011) Divergence time estimation using fossils as terminal taxa and the origins of Lissamphibia. Syst Biol 60:466–481

Quintero I, Landis MJ (2020) Interdependent phenotypic and biogeographic evolution driven by biotic interactions. Syst Biol 69:739–755

Quintero I, Keil P, Jetz W, Crawford FW (2015) Historical biogeography using species geographical ranges. Syst Biol 64:1059–1073

Ree RH, Sanmartín I (2009) Prospects and challenges for parametric models in historical biogeographical inference. J Biogeogr 36:1211–1220

Ree RH, Smith SA (2008) Maximum likelihood inference of geographic range evolution by dispersal, local extinction, and cladogenesis. Syst Biol 57:4–14

Ree RH, Moore BR, Webb CO, Donoghue MJ (2005) A likelihood framework for inferring the evolution of geographic range on phylogenetic trees. Evolution 59:2299–2311

Renner SS (2005) Relaxed molecular clocks for dating historical plant dispersal events. Trends Plant Sci 10:550–558

Román-Palacios C, Tavera J, Castañeda M (2018) When did anoles diverge? An analysis of multiple dating strategies. Mol Phylogenet Evol 127:655–668

Ronquist F, Teslenko M, van der Mark P, Ayres DL, Darling A, Höhna S, Larget B, Liu L, Suchard MA, Huelsenbeck JP (2012) MrBayes 3.2: efficient Bayesian phylogenetic inference and model choice across a large model space. Syst Biol 61:539–542

Rutschmann F, Eriksson T, Schönenberger J, Conti E (2004) Did Crypteroniaceae really disperse out of India? Molecular dating evidence from *rbc*L, *ndh*F, and *rpl*16 intron sequences. Int J Plant Sci 165:S69–S83

Sanmartín I, Mark PVD, Ronquist F (2008) Inferring dispersal: a Bayesian approach to phylogeny-based island biogeography, with special reference to the Canary Islands. J Biogeogr 35:428–449

Stadler T, Gavryushkina A, Warnock RCM, Drummond AJ, Heath TA (2018) The fossilized birth-death model for the analysis of stratigraphic range data under different speciation modes. J Theor Biol 447:41–55

Strauss D, Sadler PM (1989) Classical confidence intervals and Bayesian probability estimates for ends of local taxon ranges. Math Geol 21:411–427

Sukumaran J, Economo EP, Knowles LL (2016) Machine learning biogeographic processes from biotic patterns: a new trait-dependent dispersal and diversification model with model choice by simulation-trained discriminant analysis. Syst Biol 65:525–545

Swart BL, von der Heyden S, Bester-van der Merwe A, Roodt-Wilding R (2015) Molecular systematics and biogeography of the circumglobally distributed genus *Seriola* (Pisces: Carangidae). Mol Phylogenet Evol 93:274–280

Swenson U, Nylinder S, Munzinger J (2014) Sapotaceae biogeography supports New Caledonia being an old Darwinian island. J Biogeogr 41:797–809

Tagliacollo VA, Duke-Sylvester SM, Matamoros WA, Chakrabarty P, Albert JS (2015) Coordinated dispersal and pre-Isthmian assembly of the Central American ichthyofauna. Syst Biol 66:183–196

Thacker CE (2017) Patterns of divergence in fish species separated by the Isthmus of Panama. BMC Evol Biol 17:111

Thorne J, Kishino H, Painter IS (1998) Estimating the rate of evolution of the rate of molecular evolution. Mol Biol Evol 15:1647–1657

Turtle Taxonomy Working Group (2017) Turtles of the world: Annotated checklist and atlas of taxonomy, synonymy, distribution, and conservation status. In: Rhodin AGJ, Iverson JB, van Dijk PP, Saumure RA, Buhlmann KA, Pritchard PCH, Mittermeier RA (eds) Conservation biology of freshwater turtles and tortoises: a compilation project of the IUCN/SSC Tortoise and Freshwater Turtle Specialist Group. Chelonian Res Monogr 7:1–292

Wallace AR (1855) On the law which has regulated the introduction of new species. Ann Mag Nat Hist 16:184–196

Wallace AR (1876) The geographical distribution of animals. Macmillan, London

Warnock RCM, Parham JF, Joyce WG, Lyson TR, Donoghue PCJ (2015) Calibration uncertainty in molecular dating analyses: there is no substitute for the prior evaluation of time priors. Proc R Soc B 282:20141013

Webb CO, Ree RH (2012) Historical biogeography inference in Malesia. In: Gower D, Johnson K, Richardson J, Rosen B, Ruber L, Williams S (eds) Biotic evolution and environmental change in Southeast Asia. Cambridge University Press, Cambridge, UK, pp 191–215

Yang Z, Rannala B (2006) Bayesian estimation of species divergence times under a molecular clock using multiple fossil calibrations with soft bounds. Mol Biol Evol 23:212–226

Zuckerkandl E, Pauling L (1962) Molecular disease, evolution, and genic heterogeneity. In: Kasha M, Pullman B (eds) Horizons in biochemistry. Academic, New York, pp 189–225

Estimating Evolutionary Rates and Timescales from Time-Stamped Data

10

Sebastian Duchêne and David A. Duchêne

Abstract

Methods of molecular dating are playing increasingly valuable roles in evolutionary biology. These methods require independent information to calibrate the molecular clock and obtain meaningful estimates of evolutionary rates and times. One source of such information is the age of the molecular samples, such that the data are said to be time-stamped. In this chapter, we present an outline of current practice and the latest advances in methods for molecular dating using time-stamped data. In addition, there is a broad range of approaches for identifying whether time-stamped data contain sufficient information for estimating evolutionary rates and timescales. We describe a fully Bayesian approach for this purpose and illustrate its performance in analyses of sequence data from H1N1 influenza virus and from *Mycobacterium tuberculosis*. The approaches outlined here provide the foundations for the analysis of time-stamped data in the era of high-throughput sequencing and high-performance computing.

Keywords

Tip-dating · Measurably evolving populations · Tests of temporal structure · Bayesian inference · Microbial evolution · Ancient DNA

10.1 Introduction

Molecular clock models in phylogenetics are widely used for estimating evolutionary rates and timescales. In addition to information about genetic divergence, molecular clocks often use information about the timing of evolutionary events, also known as a time calibration. Such calibrations provide the raw material for estimating absolute evolutionary rates and times from sequence data. A popular source of calibrating information for molecular clock analyses is the timing of sample collection (Rambaut 2000). Data sets that contain samples collected at different points in time are described as time-stamped or heterochronous. In contrast, isochronous data sets contain samples of similar or identical ages and their evolution is most appropriately represented using an ultrametric time-tree (phylogenetic tree with branch lengths in time units and where ultrametricity means that the distance from the root to each of the tips is the

S. Duchêne
Department of Biochemistry and Molecular Biology, Bio21 Molecular Sciences and Biotechnology Institute, University of Melbourne, Melbourne, VIC, Australia

D. A. Duchêne (✉)
Centre for Evolutionary Hologenomics, University of Copenhagen, Copenhagen, Denmark
e-mail: david.duchene@sund.ku.dk

S. Y. W. Ho (ed.), *The Molecular Evolutionary Clock*, https://doi.org/10.1007/978-3-030-60181-2_10

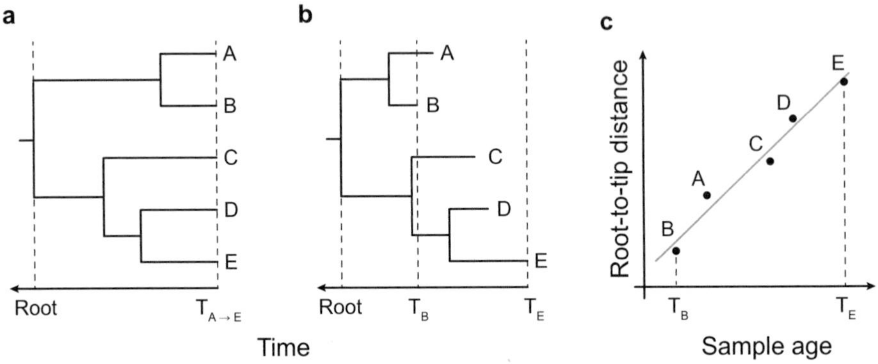

Fig. 10.1 Examples of (**a**) isochronous (i.e., ultrametric) and (**b**) heterochronous trees. Data sets with time-stamped sequences are expected to produce heterochronous trees where an appreciable amount of evolutionary change has occurred over the sampling window. The limited sampling window in **a** is insufficient to measure evolutionary change, while that in **b** between T_B and T_E is a candidate for being measurably evolving. (**c**) The root-to-tip distances plotted as a function of their sampling times. The evolutionary rate is the slope of the regression line and is intuitively equivalent to the difference in root-to-tip distances between any pair of tips (such as tips A and B) divided by the difference in their sampling times (T_A and T_B). The *x*-intercept corresponds to the time to the most recent common ancestor and the degree to which the points deviate from the line (R^2) reflects the extent to which the data have departed from strictly clocklike evolution

same). For the sampling times to be useful for calibration of the molecular clock, the period of collection, or sampling window, must have been sufficient for molecular evolution to have left a signature (Fig. 10.1). Samples that have been collected over sufficiently broad periods of time to accumulate evolutionary change are said to come from 'measurably evolving populations' (Drummond et al. 2002, 2003).

Many of the principles of phylogenetic inference using molecular clocks in isochronous data also hold for time-stamped data, such as the requirement of using substitution and molecular clock models. The central difference between analyses of these two types of data is methods of fitting a molecular clock and the statistical tests used to confirm that the sampling times span a sufficiently long time, known as assessment of temporal structure (Rieux and Balloux 2016).

Time-stamped data have most frequently been used for the study of evolutionary events involving individuals sampled from a single population or species, as opposed to divergence events among species or higher taxonomic groups. This means that the principles of microevolution and population genetics often play an important role in analyses of time-stamped data (Arbogast et al. 2002). Combining the methods used in phylogenetics and population genetics largely relies on genealogy-based inference using the principles outlined in coalescent theory (Kingman 1982; Griffiths and Tavaré 1994). By drawing from population genetics theory, analyses of time-stamped data can lead to a range of insights about demography and epidemiology. The power of these approaches is exemplified by the thriving field of phylodynamics (Grenfell et al. 2004).

The growth of efficient, low-cost sequencing has had a substantial impact on the analysis of time-stamped data. Before the major advances in sequencing and computational technologies, studies of pathogen populations using time-stamped data were restricted to RNA viruses (Drummond et al. 2003). This was due to the high rates of evolution of RNA viruses compared with those of other microbes, which allowed for the accumulation of sufficient molecular change over calendar time, even for short genomic regions. Nowadays, high-throughput sequencing allows the extraction of vast amounts of data, including complete genomes, from more slowly evolving microbes. Similarly, novel sequencing technologies have revolutionized the extraction of target DNA from highly degraded samples, making way for whole-genome analysis of ancient DNA from plants and animals (Millar et al. 2008; Der Sarkissian et al. 2015). As a consequence, methods of analysing these data have also seen appreciable progress, for instance by lifting their former restriction to population-level processes (e.g., Stadler et al. 2013; Grealy et al. 2017) or to rapidly evolving microbes (Biek et al. 2015; Menardo et al. 2019).

Time-stamped data have the advantage that sampling times alone can be used to calibrate the molecular clock, often without the need for other forms of calibration using divergence times or rates. Incorporating sample ages as time calibrations is done in the same way as for node calibrations, either by treating each sample age as a fixed time point or by specifying a probability distribution that accounts for age uncertainty (Rieux and Balloux 2016). In this chapter, we discuss the methods of data collection and analysis of two commonly used types of time-stamped data: those coming from populations of pathogens and those sampled from subfossil material as extracted using ancient DNA techniques.

10.2 Measurably Evolving Populations

10.2.1 Microbial Evolution over Calendar Time

Early phylogenetic analyses of RNA viruses revealed that their substitution rates were sufficiently high that the viruses were able to accumulate an appreciable number of substitutions over weeks or months (Holmes 2009). For example, influenza viruses have been found to evolve at rates of up to 10^{-2} substitutions per site per year; with a genome size of around 13.5 kb, they can accumulate several substitutions per day (Duffy and Holmes 2009). Human immunodeficiency virus (HIV) also undergoes very rapid evolutionary change, with a rate of about 10^{-3} substitutions per site per year (Duchêne et al. 2014a), and can accumulate at least one substitution per week (assuming a genome size of about 9.5 kb). In a seminal study, Korber et al. (2000) took advantage of the rapid evolution of HIV to calibrate the molecular clock to date its origin in human populations, which revealed that some strains of HIV probably originated in the early twentieth century.

Whole-genome sequencing has revolutionized studies of more slowly evolving microbes, notably bacteria. The evolutionary rates of bacteria are much lower than those of viruses, implying that they would need a much wider sampling window than viruses for their evolutionary rates to be estimated reliably. However, bacteria also have much larger genomes than viruses, such that even with lower rates it is sometimes possible to treat them as measurably evolving. As a case in point, estimates of the evolutionary rate of *Salmonella enterica* range from about 10^{-7} to about 10^{-6} substitutions per site per year (Duchêne et al. 2016b), and hence are at least three orders of magnitude lower than those of some RNA viruses. If only a small portion of its ~5.3 Mb genome is sequenced, for example 10 kb, it would take about 10 years to observe a single substitution. In contrast, when the complete genome is sequenced, up to four substitutions might be observed per month and the samples can be treated as measurably evolving (Zhou et al. 2018). As a result, the growing prevalence of whole-genome sequencing means that many bacteria can now be analysed as measurably evolving populations (Biek et al. 2015).

Although whole-genome sequencing has expanded the range of analyses that are possible in microbes, it has also revealed biological patterns that are not correctly modelled by standard techniques. The most notable problem is homologous recombination, which is very common in some bacterial groups (Yahara et al. 2016). The most obvious limitation of phylogenetic analyses of data sets with substantial recombination is that the whole genome cannot be assumed to follow a single phylogenetic tree topology and that estimates of branch lengths will be incorrect (Hedge and Wilson 2014). While some methods explicitly attempt to model recombination events (Didelot and Wilson 2015;

Vaughan et al. 2017), the most popular approach is to remove recombining regions and to conduct phylogenetic analysis on a 'core genome' that includes only sites that have been inherited vertically (Croucher et al. 2014). Failing to account for recombination can give the impression of an erratic molecular clock, and removing such regions can improve the extent to which the data can be treated as measurably evolving (Schultz et al. 2016). It is important to note that downstream analyses based on estimates of evolutionary timescales, such as skyline plots (Pybus et al. 2000), can produce biased inferences when recombinant regions are removed (Lapierre et al. 2016). Accordingly, it is preferable to model recombination explicitly, although this is usually computationally intensive.

10.2.2 Ancient DNA for Temporal Calibration

DNA taken from subfossils of plants and animals usually comes from highly degraded material and requires specialized extraction techniques (Gamba et al. 2016). Until recently, ancient DNA was primarily retrieved from mitochondrial genomes, which are more abundant and have a lower rate of degradation than nuclear genomes (Allentoft et al. 2012). The mitochondrial DNA molecule usually has a highly stable circular structure and has additional protection from decay due to the double membrane of the organelle. In most animals, the rate of evolution of mitochondrial DNA is much higher than that of most nuclear DNA. These characteristics make mitochondrial DNA particularly useful for inferring population-level dynamics over short geological timescales (de Bruyn et al. 2011; Ho and Shapiro 2011). Fast-evolving ancient DNA has been instrumental for inferring population-size fluctuations in a great range of taxa, including the woolly mammoth (Palkopoulou et al. 2013), steppe bison (Shapiro et al. 2004), musk ox (Campos et al. 2010), collared lemming (Brace et al. 2012), and hominids (Posth et al. 2017), among many others (e.g., Lorenzen et al. 2011).

The advent of genome-scale sequencing technologies has greatly facilitated the recovery of ancient DNA data. High-throughput sequencing methods can target highly fragmented DNA molecules, which enables vast amounts of nuclear DNA to be retrieved. This has allowed whole-genome sequences to be recovered from ancient remains (Prüfer et al. 2014). Similarly, it is now commonplace to recover sequence data from materials with trace amounts of the target DNA (Grealy et al. 2017). As a result of novel extraction and sequencing technologies, older samples can now be included in genetic studies.

Some ancient tissue samples used for ancient DNA sequencing have known ages, for instance as documented dates of collection, but others are too old for their ages to be known exactly. Therefore, the ages of samples in time-stamped data sets can have a degree of uncertainty that should not be ignored in phylogenetic analysis. One common source of age uncertainty in ancient DNA samples that are less than around 55,000 years old is that arising from radiocarbon dating (Guilderson et al. 2005). An additional complication is that radiocarbon dates are different from absolute ages, due to fluctuations in atmospheric ^{14}C content. To allow the radiocarbon date estimates to be compared with the timing of other events, such as those of climatic changes, radiocarbon ages need to be converted to calendar time. This conversion can be done by using estimates of atmospheric ^{14}C content in the past, which are becoming increasingly accurate (Stuiver and Reimer 1993; Reimer et al. 2013).

The distribution of uncertainty that emerges from radiocarbon dating can be multimodal, so using a point summary such as the mean or median is a poor description of sample ages (Molak et al. 2015). To solve this problem, some phylogenetics software allow the implementation of parametric distributions to account for uncertainty in sample ages (Shapiro et al. 2011). There are also applications that allow the use of nonparametric distributions to model the uncertainty in radiocarbon dates (Molak et al. 2015). Nonetheless, using the point mean or median estimates of sample ages of time-stamped data strikingly often leads to reasonable estimates of uncertainty in times and rates (Molak et al. 2013).

An ancient sample can also be dated using indirect methods. The age estimate of the archaeological or stratigraphic location of a sample, or ages estimated from nearby samples, can be used for calibration. However, dating based on the boundaries of stratigraphic layers is often associated with much greater uncertainty than direct estimates. Dates estimated using this method can also be highly inaccurate if the deposit has been reburied or mixed.

10.3 Popular Approaches for Molecular Dating Using Time-Stamped Data

Since the early 2000s, a range of methods have been developed for calibrating the molecular clock using sampling times: root-to-tip regression, likelihood or optimality methods, and Bayesian inference. The intuition behind using sampling times for calibration is that the evolutionary rate is approximately the difference in evolutionary distance between a pair of tips divided by the difference in their sampling times. In the phylogenetic tree in Fig. 10.1b, the rate of evolution can be calculated as the difference in the root-to-tip distance between tips B and E divided by their difference in sampling times ($T_E - T_B$). To obtain a time-tree, the branch lengths of the phylogenetic tree (in units of substitutions per site) can be divided by the evolutionary rate estimate (substitutions per site per year). Clearly, the inclusion of a larger number of time-stamped tips gives more opportunities to calculate the evolutionary rate, thereby improving its accuracy. A fundamental consideration with all methods that use time-stamped data is that the estimates depend on the position of the root, which can be selected or estimated in a number of ways.

10.3.1 Root-to-Tip Regression

One of the earliest molecular clock approaches to time-stamped data was implemented by Korber et al. (2000) to infer the age of the most recent common ancestor of HIV pandemic strains. The data consisted of molecular sequences of the *gag* and *env* genes, with the samples collected over about 10 years. Their method consisted of inferring a phylogenetic tree using maximum likelihood and assuming a constant evolutionary rate (i.e., a strict molecular clock). They conducted a regression of the distance from the root of the tree to each of the tips as a function of their sampling times. The expectation is that samples that are collected later should have undergone more molecular evolutionary change than those closer to the root of the tree.

In such root-to-tip regression, the slope of the line corresponds to the evolutionary rate and the *x*-intercept is the age of the most recent common ancestor (Fig. 10.1c). The optimal position of the root is that which maximizes clocklike behaviour, which is typically quantified with the R^2 of the regression, although a range of regression statistics can be used. Alternatively, the position of the root can be specified by including an outgroup taxon. The root-to-tip regression method is implemented in the program TempEst (Rambaut et al. 2016).

Benefits of the root-to-tip regression include that it only requires a phylogenetic tree with branch lengths in units of evolutionary distance that can be inferred using different methods (distance-based, maximum-likelihood, or Bayesian approaches), it is computationally very efficient, and it gives a measure of the clocklike behaviour of the data. Empirical studies suggest that it can produce estimates of evolutionary rates that are comparable to those of more sophisticated methods (Duchêne et al. 2016a; Tong et al. 2018). However, the root-to-tip regression has some important limitations. Measuring the root-to-tip distance for every tip means that there is substantial pseudoreplication because the path from the root to each of the tips will go through the internal branches multiple times and it does not report uncertainty in a meaningful way. In turn, using a p-value to determine the significance of the association of evolutionary distance and time is statistically invalid (Rambaut et al. 2016). Although the phylogenetic tree is used to measure evolutionary distance, the branching order is not taken into account in the regression so that this potentially useful information is discarded. Finally, modelling rate variation among lineages is not straightforward. For these reasons, the root-to-tip regression is mostly used for visual inspection of the data, rather than as a rigorous molecular clock method (see Sect. 10.4).

10.3.2 Optimality Methods

Approaches based on optimizing a function to fit a molecular clock fall in the category of *optimality methods* and include those based on maximum likelihood, least squares, and genetic distance. Rambaut (2000) devised a likelihood function where branch lengths in the tree are treated as the product of evolutionary rates and times. Given a phylogenetic tree and sampling times, it is possible to estimate the evolutionary rate that maximizes this likelihood. This can be performed under the assumption that there is a strict molecular clock, or by allowing rates among branches to be governed by a probability distribution (Seo et al. 2002; Volz and Frost 2017; Sagulenko et al. 2018). Nonparametric methods also optimize a likelihood (or penalized likelihood) function to fit a molecular clock with different degrees of rate variation among lineages (Sanderson 2003; Fourment and Holmes 2014; Chap. 12). There exist several software programs to fit molecular clocks to time-stamped data using likelihood, including TreeDater (Volz and Frost 2017), TreeTime (Sagulenko et al. 2018), TipDate (Rambaut 2000), Physher (Fourment and Holmes 2014), and r8s (Sanderson 2003).

In the program LSD, To et al. (2016) implemented a least-squares dating method that is similar in principle to the Langley–Fitch model (Langley and Fitch 1974), which assumes a strict molecular clock. The new method differs in that errors in evolutionary rates are assumed to follow a Gaussian, rather than a Poisson, distribution. The objective function depends on the evolutionary rate and the branch lengths. The optimization is conducted via weighted least squares, where the weights are the uncertainty of the Gaussian distribution that governs rates (To et al. 2016). This method assumes a strict molecular clock and aims to minimize evolutionary rate variation among lineages. To obtain uncertainty in the estimates of node times and evolutionary rates, LSD conducts a parametric bootstrap of branches. The position of the root can be optimized in the program, or specified using an outgroup or a particular branch. A useful feature of this method is that it is possible to estimate the ages of samples with unknown collection times.

The optimality methods described here are computationally very efficient, which makes them amenable to very large data sets. For example, LSD has been used to infer the evolutionary rate and timescale of over 1000 strains of influenza within a few minutes on a standard laptop (To et al. 2016). Such computational efficiency is due to the fact that these methods require an estimated phylogenetic tree as an input, instead of inferring the tree directly from the sequence data as is the case with most Bayesian methods. The most obvious limitation is that any uncertainty in LSD estimates typically reflects evolutionary rate variation but not uncertainty in the tree topology or branch lengths. However, these sources of uncertainty can be incorporated using indirect methods, such as repeating the analyses on a set of bootstrap trees.

10.3.3 Bayesian Methods

Most Bayesian molecular clock methods naturally incorporate uncertainty in the estimates of the tree topology, branch lengths, and evolutionary rates via the posterior distribution (see Chap. 6). They can also implement sophisticated models to describe complex patterns of evolutionary rate variation and demographic dynamics. It is of particular relevance to ancient DNA studies that Bayesian methods allow the researcher to assign a prior distribution for the ages of tips, for example to reflect the uncertainty in ^{14}C dating, and their posterior distribution will be estimated as for other parameters (Shapiro et al. 2011). The most widely used programs that incorporate a full Bayesian model include BEAST 1 (Suchard et al. 2018) and BEAST 2 (Bouckaert et al. 2019), MrBayes (Ronquist et al. 2012b), and RevBayes (Höhna et al. 2016).

In its simplest form, the full Bayesian model consists of a time-tree prior (du Plessis and Stadler 2015) to describe the branching process, a molecular clock model to describe the prior on branch rates, and a substitution model. The phylogenetic likelihood of the sequence data given the tree and the substitution model is calculated by treating branch lengths as the product of times (from the time-tree prior) and rates (from the clock model) (Heath and Moore 2014; Bromham et al. 2018). The position of the root of the tree is informed by the tree prior, instead of being optimized independently as in optimality methods and the root-to-tip regression. The range of clock models that can be used is the same as that for isochronous data, but only some of the available tree priors are valid for heterochronous data.

The most common tree priors posit that branching events are described by either a coalescent or a birth–death process (Drummond and Stadler 2015). Coalescent models are backwards-in-time processes that are conditioned on the ages and number of samples. The rate at which lineages coalesce back in time is determined by a mathematical function of population size over time (Rosenberg and Nordborg 2002). For example, an exponential function can be used to estimate the growth rate of a pathogen population based on the temporal distribution of nodes (Volz et al. 2009). An array of flexible skyline-plot methods can also use the coalescent to infer more complex population dynamics using nonparametric and semiparametric approaches (Ho and Shapiro 2011). Because coalescent models do not explicitly describe the sampling process, they only require a few modifications to make them applicable to heterochronous data (Rodrigo and Felsenstein 1999; Drummond et al. 2002).

Birth–death models are forwards-in-time processes and they have an expectation of the number of samples and of their ages. The simplest model is known as the Yule process and it assumes constant diversification and no extinction, or death, of lineages (Yule 1924). The result of the Yule process is always an isochronous time-tree, so it cannot be used for analyses of heterochronous data. A birth–death process with explicit sampling assumes that lineages can go extinct and can be sampled with some probability (Stadler 2010), and hence can be applied to heterochronous data. A key consideration relating to the birth–death model is that the sampling parameter should reflect the process under which the data were sampled; the constant birth–death assumes constant sampling effort over time and lineages, whereas the birth–death skyline allows the user to specify periods of time with variable sampling effort (Stadler et al. 2013). There also exist multiple birth–death models that allow certain lineages to be sampled with a higher probability (Stadler and Bonhoeffer 2013). Recent studies have suggested that the choice of sampling scheme can have a considerable effect on the birth–death tree prior, producing time priors for internal nodes that are more informative than those under the coalescent (Boskova et al. 2018). As with all Bayesian analyses, it is important to choose a prior distribution that is reasonable for the data at hand.

Some Bayesian methods do not implement a full Bayesian model. Instead of relying on sequence data, they assume an estimate of the phylogenetic tree with branch lengths (Thorne et al. 1998; Yang 2007; Didelot et al. 2018). These approaches are usually more computationally efficient than those that use the full Bayesian model. However, they currently have a limited range of tree priors available and their computational efficiency comes at the expense of ignoring phylogenetic uncertainty.

10.4 Verifying Temporal Structure

Estimating evolutionary rates and times using time-stamped data requires sufficient molecular evolution between sampling times (Duchêne et al. 2015b; Murray et al. 2015). If this requisite is met, the data are said to have temporal structure. If the molecular data have evolved too slowly relative to the timeframe covered by the samples, then they might not have temporal structure and can produce spurious inferences of evolutionary rates and times (Rambaut 2000). Failing a test of temporal structure generally means that either a more rapidly evolving molecular marker must be sampled, or the sampling window must be widened by the inclusion of new samples from times outside the existing window. Below we outline the methods that have been proposed to test whether time-stamped data have temporal structure.

10.4.1 Root-to-Tip Regression

A fast and popular approach to test temporal structure is to employ a root-to-tip regression under the assumption that the data follow a molecular clock, as described above. The test only requires estimation of the root-to-tip distances, which are the summed lengths of branches from the root of the tree to each tip. This method tests for a linear relationship between the molecular substitutions accumulated and the ages of the samples (Fitch et al. 1991; Fig. 10.1c). The slope must be positive because it is a crude estimate of the evolutionary rate, and a high R^2 coefficient of determination indicates clocklike evolution (Korber et al. 2000).

The root-to-tip regression has several known shortcomings. In many time-structured data sets, the samples come from only a small number of time points; this means that the results could be based on only a small number of data points, leading to low statistical power. In addition, many data sets violate the assumption of the molecular clock, such that a poor root-to-tip regression can lead to falsely taking the data as lacking temporal structure (Firth et al. 2010; Duchêne et al. 2020). More critically, the root-to-tip measurements used in this method are not statistically independent, as explained in Sect. 10.3.1. Nonetheless, root-to-tip regression is extremely fast and it is commonly used as an exploratory diagnostic of the reliability of rate estimates (Duchêne et al. 2016a; Rambaut et al. 2016; Tong et al. 2018).

10.4.2 Date-Randomization Test

A more robust test of temporal structure known as the date-randomization test involves permuting the dates of samples (Ramsden et al. 2008). The goal of permuting the sample ages is to create data sets where the association between sample age and molecular evolution is broken. A large collection of data sets with randomized tips can be taken to represent a null distribution of rate estimates. Temporal structure is said to be lacking if the rate estimates obtained with the correct sampling times resemble those estimated from the date-randomized replicates (Fig. 10.2a).

The date-randomization test can be used to evaluate temporal structure using Bayesian and optimality methods, and two test criteria have been proposed. The first criterion (CR1) assesses whether the mean rate estimated from the empirical data falls within the 95% credible interval of the rate estimates from the date-randomized replicates (Fig. 10.2a). The second criterion (CR2) assesses whether the 95% credible interval of the rate estimates with correct sampling times overlaps with the range of those from the date-randomized replicates (Duffy and Holmes 2009; Ramsden et al. 2009). CR2 provides a more conservative assessment and is recommended, with minimal chances of failing to reject a data set with no temporal structure (Duchêne et al. 2015b). However, this criterion also brings a moderate chance of incorrectly rejecting data sets as lacking structure, equivalent to a high Type II error rate. An implementation of this test in LSD, or any optimality method, is computationally less expensive and it is feasible to conduct a large number of

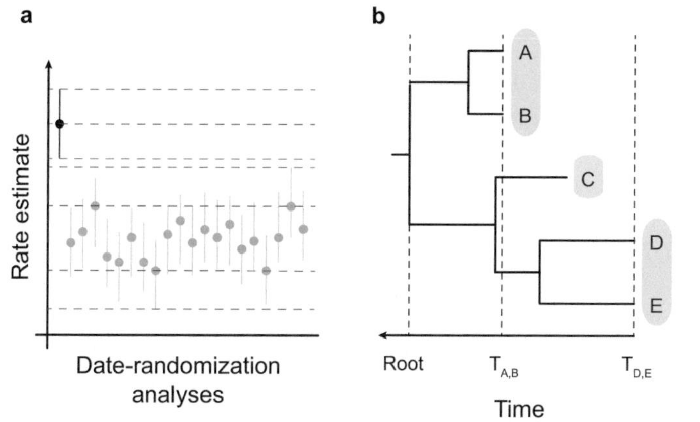

Fig. 10.2 (**a**) Example of results from a date-randomization test indicating strong temporal structure. Evolutionary rate estimates correspond to the estimate with the correct sampling times (black) and those from 20 date-randomized replicates (grey). Solid circles are the mean rate estimates and the error bars are the 95% credible intervals. The blue dashed lines denote the mean rate values with the correct sampling times and the range in mean rates coming from the randomizations, as used in the CR1 method of testing temporal structure. Similarly, the red dashed lines denote the 95% credible intervals from the data with correct sampling times and the randomizations, and can be used as a stringent criterion for assessing temporal structure CR2. In CR2, the data are considered to have strong temporal structure if the credible interval for the estimate with correct sampling times does not overlap with those from any of the date-randomized replicates. The tree in **b** presents an example of phylogenetic and temporal clustering, where samples A and B have similar sampling times to each other, as do samples D and E. In this case, a cluster-based date-randomization test is more appropriate

randomizations. In this case, one can compute where the estimate with correct sampling times falls with respect to those from the randomizations, providing the equivalent of a frequentist *p*-value (Duchêne et al. 2018). Although a large number of randomizations is desirable, several studies have used 20 with reasonable results (e.g., Kerr et al. 2012; Duchêne et al. 2015b).

A critical consideration when performing the date-randomization test comes about when data have a nonuniform temporal sampling. In many time-structured data sets, dates are grouped in such a way that close relatives have similar sampling ages, a pattern known as phylogenetic and temporal clustering (Fig. 10.2b).

In cases of nonuniform temporal sampling, the temporal and phylogenetic information is confounded and this can lead to severe overestimation of molecular rates. A possible reason for the poor rate estimates is that such data sets provide few independent instances of comparison between molecular and temporal data, and therefore less information about molecular rates (Murray et al. 2015). Interestingly, data sets that yield highly phylogenetically imbalanced trees (those that look pectinate or comb-like) also tend to yield overestimates of molecular rates (Duchêne et al. 2015a), which might in part be explained by the common confounding of temporal and phylogenetic data observed in imbalanced trees.

Most tests of temporal structure fail to reject data sets when the temporal and phylogenetic information are confounded. This means that data will be falsely identified as containing temporal structure. One solution to this problem is to use the clustered date-randomization test (Duchêne et al. 2015b), in which sampling times are permuted among samples but not among those that share the same age. Such a clustered approach to date randomization leads to a reliable test of temporal structure (Duchêne et al. 2015b; Murray et al. 2015).

10.4.3 Bayesian Test of Model Fit

The statistical fit of models with different sample dates can provide an alternative test of temporal structure. For example, treating a heterochronous data set as isochronous is expected to lead to a poorer statistical fit than if samples are assigned their true dates. In Bayesian molecular dating, testing for temporal structure using model fit is done by estimating the marginal likelihood of two different models: one using the empirical sampling times, and one where all the samples are assigned the most recent date (Baele et al. 2012; Murray et al. 2015). If the data contain temporal structure, the marginal likelihood for the model with the original sample dates is expected to be superior. Approximate methods for computing marginal likelihoods are often regarded as computationally expensive and sometimes unreliable, but some are likely to be sufficient (e.g., path sampling, stepping-stone sampling; Xie et al. 2011; Baele et al. 2013). Marginal likelihoods can be readily estimated using popular software such as BEAST.

Analyses of empirical data have shown that this method can be misleading if a poor marginal-likelihood estimator is used, with a tendency to support the presence of temporal structure even in analyses that yield incorrect estimates of evolutionary rates and times (Murray et al. 2015). However, recent work has demonstrated that a highly accurate estimator, generalized stepping-stone sampling (Fan et al. 2011; Baele et al. 2016), can effectively detect temporal structure in simulations and empirical data (Duchêne et al. 2020).

10.4.4 Comparing the Prior and Posterior to Assess Temporal Structure

The broad uptake of the date-randomization test is largely due to the possibility of implementing it in popular Bayesian frameworks, such as BEAST. However, the interpretation of its result is not strictly Bayesian, instead bearing some resemblance to frequentist methods; the goal is to test a hypothesis (whether the data have temporal structure or not) with some confidence level (similar to p-value testing using a significance value, α). In contrast, a fully Bayesian approach should assess statistical support for including sampling times (Baele et al. 2012; Murray et al. 2015; Duchêne et al. 2020) or assessing the extent to which the sequence data and sampling times are informative about the inferences. The former method has been previously assessed (see Sect. 10.4.3), but the latter has received limited attention.

In general, sequence data are considered informative if the posterior distribution is considerably different from the prior (with the notable exception of internal-node calibrations; Heath and Moore 2014). The expectation is that with informative sequence data, the posterior should be more precise and closer to the true value than the prior, a behaviour also known as statistical consistency. However, even sequence data with very low information content can drive, and sometimes mislead, estimates of some parameters in the full phylogenetic model. As a case in point, Möller et al. (2018) found that uninformative sequence data can produce precise, but incorrect, estimates of tree length and of the evolutionary rate. This probably occurs because sequence data that are uninformative for estimating evolutionary rates and timescales can still contain sufficient information to resolve the topology. Under these circumstances, a limited set of trees will be sampled and lead to a posterior that is much more precise than the prior. Other parameters, including the age of the root node, do not appear to suffer from this problem.

Here we describe an approach that involves comparing the prior and posterior distributions of different parameters to assess information content in molecular sequence data and their association with their sampling times. Our method of assessing temporal structure consists of quantifying information content in the posterior relative to that of the prior for the age of the root node. A simple measure is to take the 95% quantile width divided by the mean, an analogue to the coefficient of variation and referred to here as CV. We calculate this for the prior, CV_{prior}, and for the posterior, $CV_{posterior}$, and take the ratio $CV_{ratio} = CV_{prior} / CV_{posterior}$. A CV_{ratio} of 1 means that the prior and posterior are equally informative, whereas a CV_{ratio} of more than 1 means that the posterior is more informative than the prior.

We expect that data with temporal structure should have a higher CV_{ratio} than those with no temporal structure. However, this can depend on the parameter in question and its corresponding prior. For example, if the evolutionary rate has a very broad prior, even sequence data with no temporal structure can produce a posterior that is much more informative than the prior, with a potentially large but misleading CV_{ratio}. This is expected because the rate will be a function of the number of variable sites and the prior on the age of the root node. In contrast, the age of the root node will require data with strong temporal structure to obtain an informative posterior and high CV_{ratio}.

To determine the behaviour of this approach, we simulated the evolution of DNA sequences using parameters inferred for H1N1 influenza virus, which typically has clocklike behaviour and strong temporal structure (Hedge et al. 2013). We used the HKY+Γ substitution model, a strict molecular clock with an evolutionary rate of 3.66×10^{-3} substitutions per site per year, and an exponential coalescent process for the branching times. One hundred data sets were simulated to have temporal structure, with sampling times that span 7 months and which match those of some data sets of the 2009 influenza pandemic in North America (Hedge et al. 2013), while another 100 were generated on ultrametric trees and with no temporal structure. All data sets contained 50 samples, sequence lengths of 13,156 nt, and about 350 variable sites to match typical genome data sets from influenza virus.

We analysed the data in BEAST 2.5 (Bouckaert et al. 2014, 2019) using a substitution model and tree prior that matched those used to generate the data, and a Markov chain Monte Carlo simulation with 5×10^7 steps, sampling every 5000 steps. For the data with temporal structure, we used the correct sampling times for calibration, but, for the data with no temporal structure, we set sampling times from a typical influenza outbreak (Hedge et al. 2013). We used a relaxed-clock model with a lognormal distribution. This model has good performance even for data that follow a strict clock (Drummond et al. 2006), and it can accommodate apparent rate variation among lineages that might arise when specifying sampling times for the data with no temporal structure. The priors were all proper, such that each integrates to 1, and were selected according to previous analyses of these data (Duchêne et al. 2019). Ideally, one could compare the prior selected for each parameter with its marginal posterior distribution. However, such user-specified priors often differ from the marginal prior, particularly those for ages of nodes (including that of the root node) which can interact with the topology and other priors (Duchêne et al. 2014b). To obtain the marginal prior one can run the analyses without sequence data (equivalent to selecting the option 'sample from prior' in BEAST 2).

The simulations demonstrate that analyses of data with temporal structure result in a posterior that is much more informative than the prior (Fig. 10.3a, b), with a CV_{ratio} between 3 and 11 for the evolutionary rate and between 4 and 13 for the age of the root node. In 97 out of 100 simulations the posterior 95% credible interval of the evolutionary rate included the value used to generate the data. Analyses of the data with no temporal structure yielded rate estimates that never included the true evolutionary rate, but the posterior for this parameter was nonetheless always more informative than the prior (Fig. 10.3c), with a CV_{ratio} between 1 and 12. The CV_{ratio} values of the evolutionary rate are similar for both sets of simulations, despite those with no temporal structure always yielding incorrect rate estimates. This illustrates the point that comparing the prior and posterior of the rate can provide a misleading assessment of temporal structure (Fig. 10.4a). This probably occurs because sequence data are informative about the topology and the total amount of sequence divergence even in the absence of temporal structure.

The CV_{ratio} of the age of the root node is a more useful diagnostic to assess temporal structure than that of the evolutionary rate (e.g.,

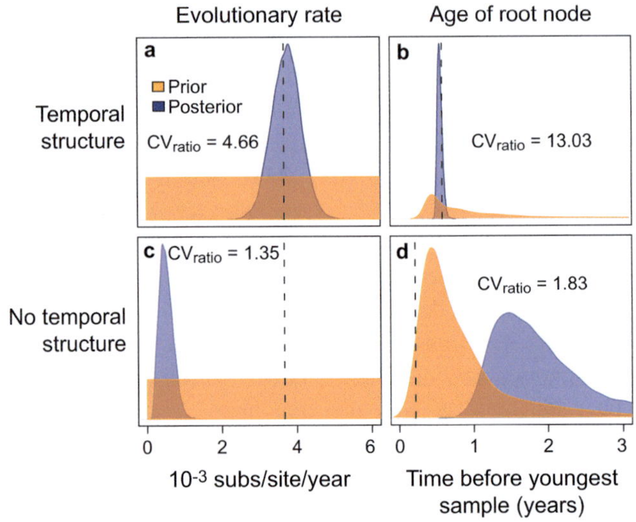

Fig. 10.3 Prior and posterior densities of the mean evolutionary rate and the age of the root node for a simulated data set with temporal structure (**a**, **b**) and without temporal structure (**c**, **d**). CV_{ratio} is a measure of information content. For the prior and the posterior, we calculate the 95% interval width divided by the mean, and the ratio of this quantity of the prior and the posterior is the CV_{ratio}. A value of 1 indicates that the prior and posterior are similarly informative, with higher values suggesting a more informative posterior. The dashed line corresponds to the 'true' value used to generate the data

Fig. 10.3b, d). Its value ranged between 4.8 and 14 for the simulated data with temporal structure and between 1 and 2.5 for those with no temporal structure (Fig. 10.4). According to these results, a posterior for the age of the root node that is about fivefold more informative than the prior

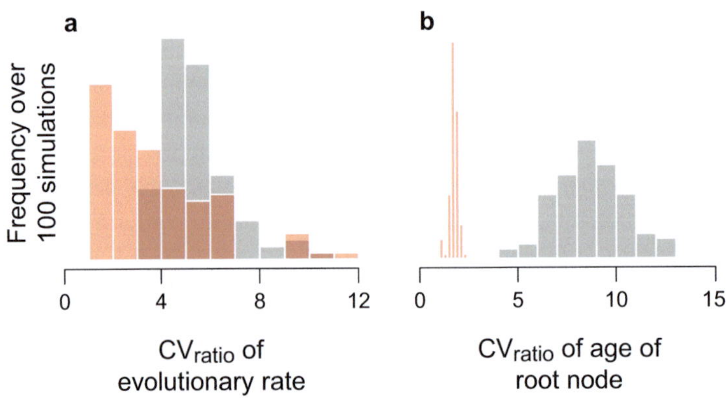

Fig. 10.4 Histograms of the CV_{ratio} for the evolutionary rate and age of the root node for 100 simulations with temporal structure (grey) and without temporal structure (red). Higher values indicate a more informative posterior distribution relative to the prior. (**a**) CV_{ratio} of the evolutionary rate is similar whether the data have temporal structure or not, but, in the data with no temporal structure, this parameter was never estimated correctly, meaning that this statistic is misleading for assessing temporal structure. (**b**) In contrast, CV_{ratio} of the age of the root node is much higher for the simulated data with temporal structure. The distribution of CV_{ratio} with no temporal structure (red) is lower and does not overlap with that from the simulated data with temporal structure (grey). As such, CV_{ratio} of the age of the root node is an informative statistic for assessing temporal structure

($CV_{ratio} > 5$) can be used as evidence of temporal structure.

Although this test of temporal structure appears to be effective, it requires careful consideration of the priors, especially those on node times as imposed by the tree prior. Here we have used a coalescent tree prior that is conditioned on the sampling times. In contrast, birth–death tree priors can provide information about sampling times and, if provided, they can be treated as part of the data to inform diversification parameters and the age of the root node (Boskova et al. 2018). For this reason, it is important to sample from the prior to determine whether it is reasonable (Heled and Drummond 2012). An obvious problem is when the prior on the age of the root node is over-informative, such that the posterior is very similar regardless of whether the data have strong temporal structure or not. In such circumstances, inferences of evolutionary rates and times are driven by the tree prior and temporal structure might only have a small impact. Although this is sometimes desirable, for example when internal-node calibrations are used in combination with sampling times, it should be explicitly acknowledged when interpreting the estimates.

10.5 Heterochronous Data Analysis in Practice

To provide an illustration of how temporal structure can be evaluated using Bayesian methods, we present here an analysis of two empirical data sets. Although there has been extensive use and validation of the date-randomization test and the root-to-tip regression, less attention has been given to comparing prior and posterior distributions to assess temporal structure. We analysed two previously published data sets of 2009 H1N1 influenza virus (Hedge et al. 2013) and of an outbreak of the bacterium *Mycobacterium tuberculosis* in the Swiss city of Bern (Kühnert et al. 2018) to show how the results from the simulations in Sect. 10.4.4 can be applied to empirical data. The influenza data set consists of 100 whole genomes collected between February and August 2009 in North America, while the *M. tuberculosis* data set consists of 68 samples collected in Bern over a 10-year period. Our analyses are similar to those described in Sect. 10.4.4, with the same tree prior, substitution model, and Markov chain Monte Carlo settings.

The evolutionary rate estimates from both data sets were similar to those of the original studies, at 0.22 SNPs per genome per year for *M. tuberculosis*, and 3.66×10^{-3} substitutions per site per year for H1N1 influenza, although slightly lower for *M. tuberculosis*, reported at about 0.5 SNPs per genome per year by Kühnert et al. (2018). The estimate of the age of the root node of influenza is around the start of 2009, which is consistent with the expected origin of the 2009 influenza outbreak in the Northern Hemisphere (Fig. 10.5). According to the simulations in Sect. 10.4.4, a CV_{ratio} for the age of the root node of at least 5 would indicate evidence for temporal structure. As such, there appears to be strong evidence of temporal structure for the influenza data set, with a CV_{ratio} of 8.91, whereas that for the *M. tuberculosis* data is only 1.32. The low CV_{ratio} of the *M. tuberculosis* data is consistent with a low R^2 (0.05) from a root-to-tip regression in the original study (Kühnert et al. 2018). Comparing prior and posterior distributions of the age of the root node appears to be effective for analyses of empirical data. It has the key benefits of an intuitive interpretation and ease of use.

10.6 Conclusions and Future Directions

Calibrating the molecular clock using heterochronous data has been valuable for estimating evolutionary rates and timescales in rapidly evolving organisms and in ancient DNA studies. There has been dramatic progress since the proposal of the early root-to-tip regression and strict-clock methods (Korber et al. 2000; Seo et al. 2002), towards incorporating more sophisticated models of rate variation (Ho and Duchêne 2014; Bromham et al. 2018), modelling

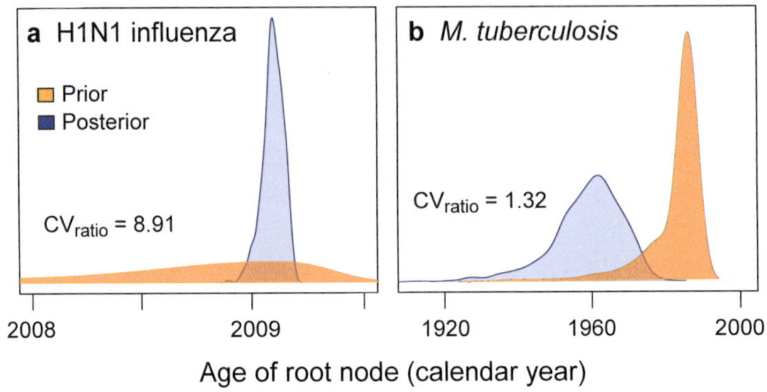

Fig. 10.5 Posterior density of the age of the root node for two empirical data sets of (**a**) H1N1 influenza virus and (**b**) *Mycobacterium tuberculosis*. CV_{ratio} of the root height is reported in each case, and it is much higher for the H1N1 influenza data set and within the range of values from the simulated data with temporal structure, indicating that it has strong temporal structure. In contrast, the *M. tuberculosis* data set has a much lower CV_{ratio} that lies within the range of values from the simulated data with no temporal structure

uncertainty in sampling times (Shapiro et al. 2011; Molak et al. 2013), and handling very large data sets (To et al. 2016).

With most current methods it is important to verify temporal structure to avoid misleading inferences. Several methods to do this have been described in this chapter, but regardless of the choice of method for assessing temporal structure, the results should be carefully considered before any of the data are discarded. For example, many bacterial data sets have strong temporal structure, but it is often obscured by recombination (Schultz et al. 2016), so correctly accounting for recombination is an important development (Vaughan et al. 2017; Didelot et al. 2018). In some viruses, notably Hepatitis B virus, even data sets that include samples from about 500 years ago still show little temporal structure, a pattern that has been attributed to mutational saturation (Patterson Ross et al. 2018). Accordingly, developing more realistic substitution and molecular clock models is likely to improve the resulting inferences. In cases when the best available methods still detect no temporal structure in the data, it might be necessary to resort to adding calibrating information via internal-node calibration or previous rate estimates, to widen the sampling window, or to sequence more informative genomic regions.

Bayesian approaches have been particularly popular because they allow simultaneous estimation of a multitude of parameters of interest, such as migration rates or epidemiological spread (Lemey et al. 2009; Kühnert et al. 2011), and because they can combine different sources of information for calibration (Ronquist et al. 2012a; Zhang et al. 2015). Recent developments, mostly in the Bayesian framework, include models that allow ancient samples to be placed as direct ancestors to modern samples (Gavryushkina et al. 2014), and those that can treat fossil taxa as tips in the phylogenetic tree instead of using them indirectly for internal-node calibrations (Heath et al. 2014). The flexibility of many Bayesian software programs, such as BEAST 2 and RevBayes (Höhna et al. 2016; Bouckaert et al. 2019), presents a key opportunity to develop more realistic approaches for including heterochronous data in complex evolutionary scenarios.

References

Allentoft ME, Collins M, Harker D, Haile J, Oskam CL, Hale ML, Campos PF, Samaniego JA, Gilbert MTP, Willerslev E, Zhang G, Scofield RP, Holdaway RN, Bunce M (2012) The half-life of DNA in bone: measuring decay kinetics in 158 dated fossils. Proc R Soc B 279:4724–4733

Arbogast BS, Edwards SV, Wakeley J, Beerli P, Slowinski JB (2002) Estimating divergence times from molecular data on phylogenetic and population genetic timescales. Annu Rev Ecol Syst 33:707–740

Baele G, Lemey P, Bedford T, Rambaut A, Suchard MA, Alekseyenko AV (2012) Improving the accuracy of demographic and molecular clock model comparison while accommodating phylogenetic uncertainty. Mol Biol Evol 29:2157–2167

Baele G, Li WLS, Drummond AJ, Suchard MA, Lemey P (2013) Accurate model selection of relaxed molecular clocks in Bayesian phylogenetics. Mol Biol Evol 30:239–243

Baele G, Lemey P, Suchard MA (2016) Genealogical working distributions for Bayesian model testing with phylogenetic uncertainty. Syst Biol 65:250–264

Biek R, Pybus OG, Lloyd-Smith JO, Didelot X (2015) Measurably evolving pathogens in the genomic era. Trends Ecol Evol 30:306–313

Boskova V, Stadler T, Magnus C (2018) The influence of phylodynamic model specifications on parameter estimates of the Zika virus epidemic. Virus Evol 4: vex044

Bouckaert R, Heled J, Kühnert D, Vaughan T, Wu C-H, Xie D, Suchard MA, Rambaut A, Drummond AJ (2014) BEAST 2: a software platform for Bayesian evolutionary analysis. PLOS Comput Biol 10: e1003537

Bouckaert R, Vaughan TG, Barido-Sottani J, Duchêne S, Fourment M, Gavryushkina A, Heled J, Jones G, Kühnert D, De Maio N, Matschiner M, Mendes FK, Müller NF, Ogilvie HA, du Plessis L, Popinga A, Rambaut A, Rasmussen D, Siveroni I, Suchard MA, Wu C-H, Xie D, Zhang C, Stadler T, Drummond AJ (2019) BEAST 2.5: an advanced software platform for Bayesian evolutionary analysis. PLOS Comput Biol 15:e1006650

Brace S, Palkopoulou E, Dalén L, Lister AM, Miller R, Otte M, Germonpré M, Blockley SPE, Stewart JR, Barnes I (2012) Serial population extinctions in a small mammal indicate Late Pleistocene ecosystem instability. Proc Natl Acad Sci USA 109:20532–20536

Bromham L, Duchêne S, Hua X, Ritchie AM, Duchêne DA, Ho SYW (2018) Bayesian molecular dating: opening up the black box. Biol Rev 93:1165–1191

Campos PF, Willerslev E, Sher A, Orlando L, Axelsson E, Tikhonov A, Aaris-Sørensen K, Greenwood AD, Kahlke R-D, Kosintsev P, Krakhmalnaya T, Kuznetsova T, Lemey P, MacPhee R, Norris CA, Shepherd K, Suchard MA, Zazula GD, Shapiro B, Gilbert MTP (2010) Ancient DNA analyses exclude humans as the driving force behind late Pleistocene musk ox (*Ovibos moschatus*) population dynamics. Proc Natl Acad Sci USA 107:5675–5680

Croucher NJ, Page AJ, Connor TR, Delaney AJ, Keane JA, Bentley SD, Parkhill J, Harris SR (2014) Rapid phylogenetic analysis of large samples of recombinant bacterial whole genome sequences using Gubbins. Nucleic Acids Res 43:e15

de Bruyn M, Hoelzel AR, Carvalho GR, Hofreiter M (2011) Faunal histories from Holocene ancient DNA. Trends Ecol Evol 26:405–413

Der Sarkissian C, Allentoft ME, Ávila-Arcos MC, Barnett R, Campos PF, Cappellini E, Ermini L, Fernández R, da Fonseca R, Ginolhac A, Hansen AJ, Jónsson H, Korneliussen T, Margaryan A, Martin MD, Moreno-Mayar JV, Raghavan M, Rasmussen M, Velasco MS, Schroeder H, Schubert M, Seguin-Orlando A, Wales N, Gilbert MTP, Willerslev E, Orlando L (2015) Ancient genomics. Philos Trans R Soc B 370:20130387

Didelot X, Wilson DJ (2015) ClonalFrameML: efficient inference of recombination in whole bacterial genomes. PLOS Comput Biol 11:e1004041

Didelot X, Croucher NJ, Bentley SD, Harris SR, Wilson DJ (2018) Bayesian inference of ancestral dates on bacterial phylogenetic trees. Nucleic Acids Res 46: e134

Drummond AJ, Stadler T (2015) Evolutionary trees. In: Drummond AJ, Bouckaert R (eds) Bayesian evolutionary analysis with BEAST. Cambridge University Press, Cambridge, UK, pp 21–43

Drummond AJ, Nicholls GK, Rodrigo AG, Solomon W (2002) Estimating mutation parameters, population history and genealogy simultaneously from temporally spaced sequence data. Genetics 161:1307–1320

Drummond AJ, Pybus OG, Rambaut A, Forsberg R, Rodrigo AG (2003) Measurably evolving populations. Trends Ecol Evol 18:481–488

Drummond AJ, Ho SYW, Phillips MJ, Rambaut A (2006) Relaxed phylogenetics and dating with confidence. PLOS Biol 4:e88

Duchêne S, Holmes EC, Ho SYW (2014a) Analyses of evolutionary dynamics in viruses are hindered by a time-dependent bias in rate estimates. Proc R Soc B 281:20140732

Duchêne S, Lanfear R, Ho SYW (2014b) The impact of calibration and clock-model choice on molecular estimates of divergence times. Mol Phylogenet Evol 78:277–289

Duchêne D, Duchêne S, Ho SYW (2015a) Tree imbalance causes a bias in phylogenetic estimation of evolutionary timescales using heterochronous sequences. Mol Ecol Resour 15:785–794

Duchêne S, Duchêne D, Holmes EC, Ho SYW (2015b) The performance of the date-randomization test in phylogenetic analyses of time-structured virus data. Mol Biol Evol 32:1895–1906

Duchêne S, Geoghegan JL, Holmes EC, Ho SYW (2016a) Estimating evolutionary rates using time-structured data: a general comparison of phylogenetic methods. Bioinformatics 32:3375–3379

Duchêne S, Holt KE, Weill F-X, Le Hello S, Hawkey J, Edwards DJ, Fourment M, Holmes EC (2016b) Genome-scale rates of evolutionary change in bacteria. Microb Genom 2:e000094

Duchêne S, Duchêne DA, Geoghegan JL, Dyson ZA, Hawkey J, Holt KE (2018) Inferring demographic parameters in bacterial genomic data using Bayesian and hybrid phylogenetic methods. BMC Evol Biol 18:95

Duchêne S, Bouckaert R, Duchene DA, Stadler T, Drummond AJ (2019) Phylodynamic model adequacy using posterior predictive simulations. Syst Biol 68:358–364

Duchêne S, Stadler T, Ho SYW, Duchêne DA, Dhanasekaran V, Baele G (2020) Bayesian evaluation of temporal signal in measurably evolving populations. Mol Biol Evol 37:3363–3379

Duffy S, Holmes EC (2009) Validation of high rates of nucleotide substitution in geminiviruses: phylogenetic evidence from East African cassava mosaic viruses. J Gen Virol 90:1539–1547

du Plessis L, Stadler T (2015) Getting to the root of epidemic spread with phylodynamic analysis of genomic data. Trends Microbiol 23:383–386

Fan Y, Wu R, Chen M-H, Kuo L, Lewis PO (2011) Choosing among partition models in Bayesian phylogenetics. Mol Biol Evol 28:523–532

Firth C, Kitchen A, Shapiro B, Suchard MA, Holmes EC, Rambaut A (2010) Using time-structured data to estimate evolutionary rates of double-stranded DNA viruses. Mol Biol Evol 27:2038–2051

Fitch WM, Leiter JME, Li X, Palese P (1991) Positive Darwinian evolution in human influenza A viruses. Proc Natl Acad Sci USA 88:4270–4274

Fourment M, Holmes EC (2014) Novel non-parametric models to estimate evolutionary rates and divergence times from heterochronous sequence data. BMC Evol Biol 14:163

Gamba C, Hanghøj K, Gaunitz C, Alfarhan AH, Alquraishi SA, Al-Rasheid KAS, Bradley DG, Orlando L (2016) Comparing the performance of three ancient DNA extraction methods for high-throughput sequencing. Mol Ecol Resour 16:459–469

Gavryushkina A, Welch D, Stadler T, Drummond AJ (2014) Bayesian inference of sampled ancestor trees for epidemiology and fossil calibration. PLOS Comput Biol 10:e1003919

Grealy A, Phillips M, Miller G, Gilbert MTP, Rouillard J-M, Lambert D, Bunce M, Haile J (2017) Eggshell palaeogenomics: Palaeognath evolutionary history revealed through ancient nuclear and mitochondrial DNA from Madagascan elephant bird (*Aepyornis* sp.) eggshell. Mol Phylogenet Evol 109:151–163

Grenfell BT, Pybus OG, Gog JR, Wood JLN, Daly JM, Mumford JA, Holmes EC (2004) Unifying the epidemiological and evolutionary dynamics of pathogens. Science 303:327–332

Griffiths RC, Tavaré S (1994) Sampling theory for neutral alleles in a varying environment. Philos Trans R Soc B 344:403–410

Guilderson TP, Reimer PJ, Brown TA (2005) The boon and bane of radiocarbon dating. Science 307:362–364

Heath TA, Moore BR (2014) Bayesian inference of species divergence times. In: Chen M-H, Kuo L, Lewis PO (eds) Bayesian phylogenetics: methods, algorithms, and applications. CRC Press, Boca Raton, FL, pp 277–318

Heath TA, Huelsenbeck JP, Stadler T (2014) The fossilized birth–death process for coherent calibration of divergence-time estimates. Proc Natl Acad Sci USA 111:E2957–E2966

Hedge J, Wilson DJ (2014) Bacterial phylogenetic reconstruction from whole genomes is robust to recombination but demographic inference is not. mBio 5:e02158–e02114

Hedge J, Lycett SJ, Rambaut A (2013) Real-time characterization of the molecular epidemiology of an influenza pandemic. Biol Lett 9:20130331

Heled J, Drummond AJ (2012) Calibrated tree priors for relaxed phylogenetics and divergence time estimation. Syst Biol 61:138–149

Ho SYW, Duchêne S (2014) Molecular-clock methods for estimating evolutionary rates and time scales. Mol Ecol 23:5947–5975

Ho SYW, Shapiro B (2011) Skyline-plot methods for estimating demographic history from nucleotide sequences. Mol Ecol Resour 11:423–434

Höhna S, Landis MJ, Heath TA, Boussau B, Lartillot N, Moore BR, Huelsenbeck JP, Ronquist F (2016) RevBayes: Bayesian phylogenetic inference using graphical models and an interactive model-specification language. Syst Biol 65:726–736

Holmes EC (2009) The evolution and emergence of RNA viruses. Oxford University Press, Oxford, UK

Kerr PJ, Ghedin E, DePasse JV, Fitch A, Cattadori IM, Hudson PJ, Tscharke DC, Read AF, Holmes EC (2012) Evolutionary history and attenuation of myxoma virus on two continents. PLOS Pathog 8: e1002950

Kingman JFC (1982) The coalescent. Stoch Proc Appl 13:235–248

Korber B, Muldoon M, Theiler J, Gao F, Gupta R, Lapedes A, Hahn BH, Wolinsky S, Bhattacharya T (2000) Timing the ancestor of the HIV-1 pandemic strains. Science 288:1789–1796

Kühnert D, Wu C-H, Drummond AJ (2011) Phylogenetic and epidemic modeling of rapidly evolving infectious diseases. Infect Genet Evol 11:1825–1841

Kühnert D, Coscolla M, Brites D, Stucki D, Metcalfe J, Fenner L, Gagneux S, Stadler T (2018) Tuberculosis outbreak investigation using phylodynamic analysis. Epidemics 25:47–53

Langley CH, Fitch WM (1974) An examination of the constancy of the rate of molecular evolution. J Mol Evol 3:161–177

Lapierre M, Blin C, Lambert A, Achaz G, Rocha EPC (2016) The impact of selection, gene conversion, and biased sampling on the assessment of microbial demography. Mol Biol Evol 33:1711–1725

Lemey P, Rambaut A, Drummond AJ, Suchard MA (2009) Bayesian phylogeography finds its roots. PLOS Comput Biol 5:e1000520

Lorenzen ED, Nogués-Bravo D, Orlando L, Weinstock J, Binladen J, Marske KA, Ugan A, Borregaard MK,

Gilbert MTP, Nielsen R, Ho SYW, Goebel T, Graf KE, Byers D, Stenderup JT, Rasmussen M, Campos PF, Leonard JA, Koepfli K-P, Froese D, Zazula G, Stafford TW, Aaris-Sørensen K, Batra P, Haywood AM, Singarayer JS, Valdes PJ, Boeskorov G, Burns JA, Davydov SP, Haile J, Jenkins DL, Kosintsev P, Kuznetsova T, Lai X, Martin LD, McDonald HG, Mol D, Meldgaard M, Munch K, Stephan E, Sablin M, Sommer RS, Sipko T, Scott E, Suchard MA, Tikhonov A, Willerslev R, Wayne RK, Cooper A, Hofreiter M, Sher A, Shapiro B, Rahbek C, Willerslev E (2011) Species-specific responses of Late Quaternary megafauna to climate and humans. Nature 479:359–364

Menardo F, Duchêne S, Brites D, Gagneux S (2019) The molecular clock of *Mycobacterium tuberculosis*. PLOS Pathog 15:e1008067

Millar CD, Huynen L, Subramanian S, Mohandesan E, Lambert DM (2008) New developments in ancient genomics. Trends Ecol Evol 23:386–393

Molak M, Lorenzen ED, Shapiro B, Ho SYW (2013) Phylogenetic estimation of timescales using ancient DNA: the effects of temporal sampling scheme and uncertainty in sample ages. Mol Biol Evol 30:253–262

Molak M, Suchard MA, Ho SYW, Beilman DW, Shapiro B (2015) Empirical calibrated radiocarbon sampler: a tool for incorporating radiocarbon-date and calibration error into Bayesian phylogenetic analyses of ancient DNA. Mol Ecol Resour 15:81–86

Möller S, du Plessis L, Stadler T (2018) Impact of the tree prior on estimating clock rates during epidemic outbreaks. Proc Natl Acad Sci USA 115:4200–4205

Murray GGR, Wang F, Harrison EM, Paterson GK, Mather AE, Harris SR, Holmes MA, Rambaut A, Welch JJ (2015) The effect of genetic structure on molecular dating and tests for temporal signal. Methods Ecol Evol 7:80–89

Palkopoulou E, Dalén L, Lister AM, Vartanyan S, Sablin M, Sher A, Edmark VN, Brandström MD, Germonpré M, Barnes I, Thomas JA (2013) Holarctic genetic structure and range dynamics in the woolly mammoth. Proc R Soc B 280:20131910

Patterson Ross Z, Klunk J, Fornaciari G, Giuffra V, Duchêne S, Duggan AT, Poinar D, Douglas MW, Eden J-S, Holmes EC (2018) The paradox of HBV evolution as revealed from a 16th century mummy. PLOS Pathog 14:e1006750

Posth C, Wißing C, Kitagawa K, Pagani L, van Holstein L, Racimo F, Wehrberger K, Conard NJ, Kind CJ, Bocherens H, Krause J (2017) Deeply divergent archaic mitochondrial genome provides lower time boundary for African gene flow into Neanderthals. Nat Commun 8:16046

Prüfer K, Racimo F, Patterson N, Jay F, Sankararaman S, Sawyer S, Heinze A, Renaud G, Sudmant PH, de Filippo C, Li H, Mallick S, Dannemann M, Fu Q, Kircher M, Kuhlwilm M, Lachmann M, Meyer M, Ongyerth M, Siebauer M, Theunert C, Tandon A, Moorjani P, Pickrell J, Mullikin JC, Vohr SH, Green RE, Hellmann I, Johnson PLF, Blanche H, Cann H, Kitzman JO, Shendure J, Eichler EE, Lein ES, Bakken TE, Golovanova LV, Doronichev VB, Shunkov MV, Derevianko AP, Viola B, Slatkin M, Reich D, Kelso J, Pääbo S (2014) The complete genome sequence of a Neanderthal from the Altai Mountains. Nature 505:43–49

Pybus OG, Rambaut A, Harvey PH (2000) An integrated framework for the inference of viral population history from reconstructed genealogies. Genetics 155:1429–1437

Rambaut A (2000) Estimating the rate of molecular evolution: incorporating non-contemporaneous sequences into maximum likelihood phylogenies. Bioinformatics 16:395–399

Rambaut A, Lam TT, Carvalho LM, Pybus OG (2016) Exploring the temporal structure of heterochronous sequences using TempEst (formerly Path-O-Gen). Virus Evol 2:vew007

Ramsden C, Melo FL, Figueiredo LM, Holmes EC, Zanotto PMA, VGDN Consortium (2008) High rates of molecular evolution in hantaviruses. Mol Biol Evol 25:1488–1492

Ramsden C, Holmes EC, Charleston MA (2009) Hantavirus evolution in relation to its rodent and insectivore hosts: no evidence for codivergence. Mol Biol Evol 26:143–153

Reimer PJ, Bard E, Bayliss A, Beck JW, Blackwell PG, Ramsey CB, Buck CE, Cheng H, Edwards RL, Friedrich M, Grootes PM, Guilderson TP, Haflidason H, Hajdas I, Hatté C, Heaton TJ, Hoffmann DL, Hogg AG, Hughen KA, Kaiser KF, Kromer B, Manning SW, Niu M, Reimer RW, Richards DA, Scott EM, Southon JR, Staff RA, Turney CSM, van der Plicht J (2013) IntCal13 and Marine13 radiocarbon age calibration curves 0–50,000 years cal BP. Radiocarbon 55:1869–1887

Rieux A, Balloux F (2016) Inferences from tip-calibrated phylogenies: a review and a practical guide. Mol Ecol 25:1911–1924

Rodrigo AG, Felsenstein J (1999) Coalescent approaches to HIV population genetics. In: Crandall KA (ed) The evolution of HIV. Johns Hopkins University Press, Baltimore, MD, pp 233–272

Ronquist F, Klopfstein S, Vilhelmsen L, Schulmeister S, Murray DL, Rasnitsyn AP (2012a) A total-evidence approach to dating with fossils, applied to the early radiation of the Hymenoptera. Syst Biol 61:973–999

Ronquist F, Teslenko M, Van Der Mark P, Ayres DL, Darling A, Höhna S, Larget B, Liu L, Suchard MA, Huelsenbeck JP (2012b) MrBayes 3.2: efficient Bayesian phylogenetic inference and model choice across a large model space. Syst Biol 61:539–542

Rosenberg NA, Nordborg M (2002) Genealogical trees, coalescent theory and the analysis of genetic polymorphisms. Nat Rev Genet 3:380–390

Sagulenko P, Puller V, Neher RA (2018) TreeTime: maximum-likelihood phylodynamic analysis. Virus Evol 4:vex042

Sanderson MJ (2003) r8s: inferring absolute rates of molecular evolution and divergence times in the absence of a molecular clock. Bioinformatics 19:301–302

Schultz MB, Duy PT, Nhu TDH, Wick RR, Ingle DJ, Hawkey J, Edwards DJ, Kenyon JJ, Nguyen PHL, Campbell JI, Thwaites G, Nguyen TKN, Hall RM, Fournier-Level A, Baker S, Holt KE (2016) Repeated local emergence of carbapenem-resistant *Acinetobacter baumannii* in a single hospital ward. Microb Genom 2:e000050

Seo TK, Thorne JL, Hasegawa M, Kishino H (2002) A viral sampling design for testing the molecular clock and for estimating evolutionary rates and divergence times. Bioinformatics 18:115–123

Shapiro B, Drummond AJ, Rambaut A, Wilson MC, Matheus PE, Sher AV, Pybus OG, Gilbert MTP, Barnes I, Binladen J, Willerslev E, Hansen AJ, Baryshnikov GF, Burns JA, Davydov S, Driver JC, Froese DG, Harington CR, Keddie G, Kosintsev P, Kunz ML, Martin LD, Stephenson RO, Storer J, Tedford R, Zimov S, Cooper A (2004) Rise and fall of the Beringian steppe bison. Science 306:1561–1565

Shapiro B, Ho SYW, Drummond AJ, Suchard MA, Pybus OG, Rambaut A (2011) A Bayesian phylogenetic method to estimate unknown sequence ages. Mol Biol Evol 28:879–887

Stadler T (2010) Sampling-through-time in birth-death trees. J Theor Biol 167:696–404

Stadler T, Bonhoeffer S (2013) Uncovering epidemiological dynamics in heterogeneous host populations using phylogenetic methods. Philos Trans R Soc B 368:20120198

Stadler T, Kühnert D, Bonhoeffer S, Drummond AJ (2013) Birth–death skyline plot reveals temporal changes of epidemic spread in HIV and hepatitis C virus (HCV). Proc Natl Acad Sci USA 110:228–233

Stuiver M, Reimer PJ (1993) Extended ^{14}C data base and revised CALIB 3.0 ^{14}C age calibration program. Radiocarbon 35:215–230

Suchard MA, Lemey P, Baele G, Ayres DL, Drummond AJ, Rambaut A (2018) Bayesian phylogenetic and phylodynamic data integration using BEAST 1.10. Virus Evol 4:vey016

Thorne JL, Kishino H, Painter IS (1998) Estimating the rate of evolution of the rate of molecular evolution. Mol Biol Evol 15:1647–1657

To T-H, Jung M, Lycett S, Gascuel O (2016) Fast dating using least-squares criteria and algorithms. Syst Biol 65:82–97

Tong KJ, Duchêne DA, Duchêne S, Geoghegan JL, Ho SYW (2018) A comparison of methods for estimating substitution rates from ancient DNA sequence data. BMC Evol Biol 18:70

Vaughan TG, Welch D, Drummond AJ, Biggs PJ, George T, French NP (2017) Inferring ancestral recombination graphs from bacterial genomic data. Genetics 205:857–870

Volz EM, Frost SDW (2017) Scalable relaxed clock phylogenetic dating. Virus Evol 3:vex025

Volz EM, Kosakovsky Pond SL, Ward MJ, Brown AJL, Frost SDW (2009) Phylodynamics of infectious disease epidemics. Genetics 183:1421–1430

Xie W, Lewis PO, Fan Y, Kuo L, Chen MH (2011) Improving marginal likelihood estimation for Bayesian phylogenetic model selection. Syst Biol 60:150–160

Yahara K, Didelot X, Jolley KA, Kobayashi I, Maiden MCJ, Sheppard SK, Falush D (2016) The landscape of realized homologous recombination in pathogenic bacteria. Mol Biol Evol 33:456–471

Yang Z (2007) PAML 4: phylogenetic analysis by maximum likelihood. Mol Biol Evol 24:1586–1591

Yule GU (1924) A mathematical theory of evolution, based on the conclusions of Dr J. C. Willis, F.R.S. Philos Trans R Soc B 213:21–87

Zhang C, Stadler T, Klopfstein S, Heath TA, Ronquist F (2015) Total-evidence dating under the fossilized birth–death process. Syst Biol 65:228–249

Zhou Z, Lundstrøm I, Tran-Dien A, Duchêne S, Alikhan N-F, Sergeant MJ, Langridge G, Fotakis AK, Nair S, Stenøien HK (2018) Pan-genome analysis of ancient and modern *Salmonella enterica* demonstrates genomic stability of the invasive Para C lineage for millennia. Curr Biol 28:2420–2428

11 Total-Evidence Dating and the Fossilized Birth–Death Model

Alexandra Gavryushkina and Chi Zhang

Abstract

Molecular clock dating has been widely used to study the evolutionary history of species. Total-evidence dating is an approach where morphological and temporal data from fossils, together with morphological and molecular sequence data from extant species, are analysed jointly to infer dated phylogenetic trees and evolutionary parameters. The data used for such an analysis are generated by biological and geological processes and collected by researchers according to their interests. There is a large amount of stochasticity involved in these processes and the data are often incomplete: the species have undergone speciation and extinction over time; fossils are only discovered at random points in time; and only a fraction of species (out of many more unobserved or unsampled species) are included in an analysis following a chosen sampling strategy. Therefore, in a Bayesian model-based framework, it is very important to account for the processes that generated the data. The fossilized birth–death model describes the stochastic processes of speciation and extinction, the distribution of available fossils over time, and the sampling of fossil and extant species. This model has been used productively in total-evidence dating analyses, and several variations of the model have been developed. In this chapter, we introduce these models and give examples of their applications in joint or total-evidence dating analyses. We also introduce the Lewis Mk morphological substitution model and its several extensions. We highlight specifics of the implementations of the fossilized birth–death and Lewis Mk models in the Bayesian software packages BEAST 2 (version 2.6.1) and MrBayes (version 3.2.7). The chapter ends with a comparison of the total-evidence and node-dating approaches.

Keywords

Total-evidence dating · Fossilized birth–death model · Bayesian phylogenetics · Speciation · Fossilization · Morphological data

11.1 Introduction

Molecular clock dating is an approach to estimating species divergence times and evolutionary rates (dos Reis et al. 2016). Data from

A. Gavryushkina (✉)
Department of Biochemistry, University of Otago, Dunedin, New Zealand
e-mail: sasha.gavryushkina@otago.ac.nz

C. Zhang
Key Laboratory of Vertebrate Evolution and Human Origins, and Center for Excellence in Life and Paleoenvironment, Institute of Vertebrate Paleontology and Paleoanthropology, Chinese Academy of Sciences, Beijing, China
e-mail: zhangchi@ivpp.ac.cn

S. Y. W. Ho (ed.), *The Molecular Evolutionary Clock*, https://doi.org/10.1007/978-3-030-60181-2_11

the fossils are used to inform the divergence times, while the molecular clock assumption is essential for the evolutionary rates. Early efforts focused on node-dating approaches, while total-evidence dating has become a promising alternative in recent years (Pyron 2011; Ronquist et al. 2012, 2016; Zhang et al. 2016; Gavryushkina et al. 2017). In molecular clock dating, the Bayesian framework is typically employed for its flexibility and ability to account for various sources of information and uncertainty in the analysis (see Chap. 6).

Modelling of the tree-generating process is a very important part of Bayesian total-evidence dating. Initial applications of total-evidence dating did not pay much attention to the choice of the model for this process. However, it has been shown that it has a large effect on the estimated dates (Zhang et al. 2016; Gavryushkina et al. 2017). Here we describe the fossilized birth–death (FBD) model (Heath et al. 2014), which was recently developed to account for the diversification and sampling processes. The two main advantages of this model are that it directly accounts for the stochasticity of the fossilization and fossil-sampling processes and that the probability density function has a closed-form expression which makes it inexpensive for Bayesian inference. To account for different sampling strategies and the variability of the speciation and sampling processes, several extensions of this model have been developed and are introduced in this chapter.

Another important component of total-evidence dating analysis is the model of morphological character evolution (see also Chap. 7). The Lewis Mk model (Lewis 2001) is the only model of discrete character evolution that has been used for total-evidence dating. We introduce this model and its several extensions and discuss potential directions for improved modelling of morphological character evolution.

We advocate for the statistically rigorous total-evidence dating approach as an alternative to the node-dating approach. At the end of this chapter, we compare the two approaches and discuss the advantages and shortcomings of the total-evidence dating approach using the FBD model.

11.2 The Concept of Joint Inference (Tip-Dating) and Total-Evidence Dating

The *joint inference* approach is a method that uses both comparative (molecular and/or morphological) data and temporal fossil data in a joint analysis that acknowledges that the two sources of data come from processes that are not independent. This approach has also been termed tip-dating in the literature, but we avoid this term because it draws an analogy with the *node-dating* or calibration approach, which is different from the joint inference approach in principle. Another reason is that, under the model that we introduce here, fossils are not necessarily tips in phylogenies, which makes the tip-dating term misleading (see also Gavryushkina 2017).

Bayesian joint inference takes the comparative data of fossil and extant species and temporal data of fossil species as inputs. The comparative data (D) are either morphological traits of fossil and extant species, or morphological traits of fossil and extant species and molecular sequences of extant species. In the latter case, the joint inference approach has been termed *total-evidence* dating (Ronquist et al. 2012; Chap. 7). In both cases, it is important that we have enough morphological data for both fossil and extant species.

The temporal data are usually presented as stratigraphic ranges of fossil species based on the geological information. A stratigraphic range represents the age difference between the first and last occurrences of the fossil species, and also reflects the uncertainty in the estimated ages of the fossil specimens (see Chap. 8). Thus, for a given fossil species i with a true fossilization time x_i, we only know an interval $[\tau_{i,1}, \tau_{i,2}]$ that contains x_i. Let τ denote a collection of such intervals (stratigraphic ranges) for all of the fossil species. Note that, in this approach, we assume that every fossil species is sampled only once, that is, we only discover a single fossil specimen for each species. This is a very strong assumption, and we discuss how it can be avoided in Sect. 11.3.4.

Then, in the Bayesian analysis, we estimate the posterior distribution of the dated phylogenetic tree (T), diversification and sampling parameters (η), and parameters of the models of morphological and molecular evolution (θ) after observing the data: morphological traits and molecular sequences (D) and stratigraphic ranges (τ). The posterior probability can be written as follows:

$$P(T, \eta, \theta | D, \tau) = \frac{P(D, \tau | T, \eta, \theta) P(T, \eta, \theta)}{P(D, \tau)}.$$

Note that morphological and molecular data D and stratigraphic intervals τ are not directly dependent. The sequence data do not directly depend on the diversification and sampling parameters but only on the parameters of the evolutionary model and the phylogeny. Similarly, the assigned stratigraphic intervals only depend on the true sampling times (and possibly other parameters that we do not model here):

$$P(D, \tau | T, \eta, \theta) = P(D | T, \theta) P(\tau | T).$$

We assume that the diversification-sampling process is independent of the processes of molecular and morphological evolution:

$$P(T, \eta, \theta) = P(T, \eta) P(\theta) = P(T | \eta) P(\eta) P(\theta).$$

As mentioned above, we do not model the process in which stratigraphic ranges are assigned to fossils here, but assume that the probability of assigning range $[\tau_1, \tau_2]$ to a fossil with a true preservation time t is constant; that is, $P(\tau | T)$ is constant. We obtain:

$$P(T, \eta, \theta | D, \tau) \propto P(D | T, \theta) P(T | \eta) P(\eta) P(\theta). \quad (11.1)$$

A graphical representation of the model is shown in Fig. 11.1.

In the following section, we will focus on the models that are used in the joint inference to describe the process that generates the phylogenetic tree with observed fossil and extant samples.

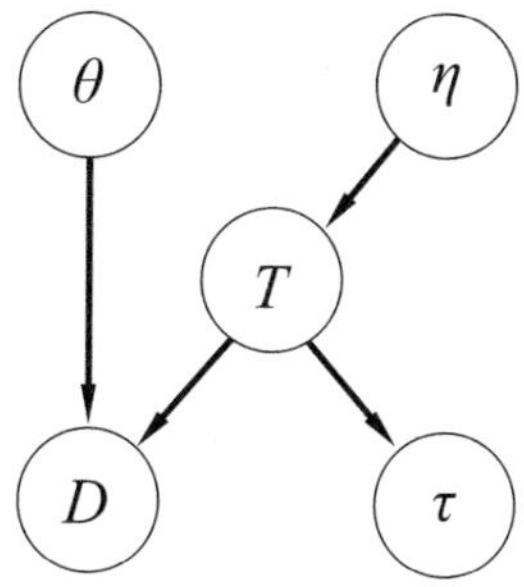

Fig. 11.1 Graphical representation of the Bayesian network for the joint inference approach. θ denotes the parameters of the evolutionary models of molecular and morphological data, η the diversification and sampling model parameters, T the dated phylogeny, D the molecular and morphological data, and τ the stratigraphic ranges

Such a model will define the term $P(T|\eta)$ in Eq. (11.1). In Sect. 11.4, we will discuss the models of morphological evolution that define a part of the term $P(D|T, \theta)$.

11.3 The Tree-Branching, Fossil-Sampling Process

An important part of the inference is the model that describes how the phylogenetic tree with fossil and extant samples arises. The total-evidence approach has been used with only a few models that describe the tree-branching, fossil-sampling process. The models that were used in initial applications did not model the sampling of extinct and extant lineages but either assumed that all lineages are sampled (Pyron 2011) or ignored the sampling process completely (Ronquist et al. 2012). Stadler (2010) and Didier et al. (2012) extended the classical model of speciation, the birth–death model (Kendall 1948), to account for the sampling process. We describe this model in the following section.

11.3.1 Fossilized Birth–Death Process

The fossilized birth–death (FBD) process (Stadler 2010; Didier et al. 2012) is a continuous-time Markov process (Grimmett and Stirzaker 2001)

that starts with one lineage at some point in time. As time goes on the lineage can split (bifurcate) or terminate (go extinct). At each split, a new lineage arises that starts a new FBD process. Thus, the process can generate many lineages that exist at the same time. If a lineage terminates, the process stops for that lineage. At the same time, every lineage arising in the process can be instantly sampled. This process is modelled as a combination of three independent Poisson processes (Grimmett and Stirzaker 2001): bifurcation (with rate λ), termination (with rate μ), and sampling (with rate ψ). Thus, the waiting time until a bifurcation event is exponentially distributed with rate λ. Similarly, the waiting times until an extinction event and a sampling event are exponentially distributed with rates μ and ψ, respectively.

The FBD process assumes that at any point in time several coexisting lineages develop independently of each other; however, the rates are equal for all lineages. The process can finish because of the termination of all lineages or can run for infinitely long. To restrict the space of the outcomes, we usually use some type of stopping condition. For example, we might only allow the process to run for a certain period of time or until a certain number of lineages is reached. Here we only consider the first case, that is, the process stops after a certain period of time (for the other case, see Stadler 2010). The FBD process describes events in the past, so the time is usually expressed in time units before the present. That is, the process starts at a positive time, the time of origin t_{or}, and finishes at time zero (the present time). The lineages that survived until the present time are additionally sampled with the same constant probability ρ.

We use the described theoretical process to model the biological process of speciation, the geological process of fossilization, and the process by which species are chosen for analysis. In this way, the splits correspond to speciation events and terminations to extinction events. For the two sampling efforts, ψ-sampling corresponds to a series of events: fossilization, preservation, discovery, identification, and inclusion in an analysis. ρ-sampling corresponds to the choice of extant species for an analysis.

The process described here generates bifurcating phylogenetic trees in which lineages are marked at various points of time with fossil occurrences and at time zero with extant samples (Fig. 11.2). However, it is not possible to estimate the *complete tree* because we only observe (sample) a part of this tree and, therefore, we can only estimate this observed part of the tree called a *reconstructed tree* or *sampled tree*. In the complete tree, all fossil samples are points inside (along) branches, that is, a sampling point never coincides with a bifurcation or a lineage

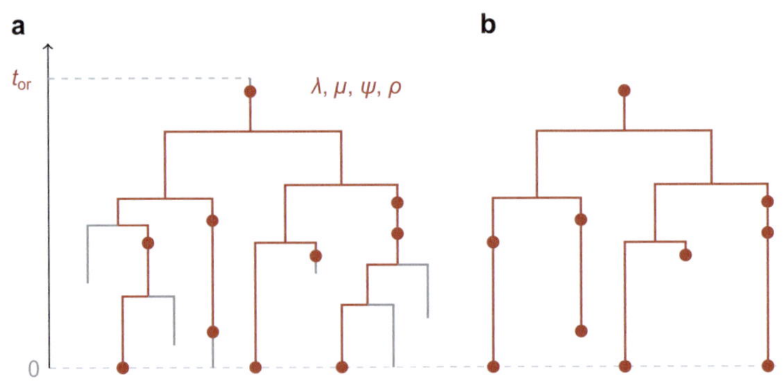

Fig. 11.2 (**a**) A *complete tree* generated by the fossilized birth–death process with parameters (t_{or}, λ, μ, ψ, ρ). Filled circles are sampling points of fossils at nonzero times (in the past) and of extant samples at time zero (present time). The observed part of the phylogeny is highlighted in red on the complete tree. (**b**) The corresponding *sampled tree*. Some fossil samples become tips and some remain inside the branches (sampled ancestors) in the sampled tree

termination. In a sampled tree, however, some of the sampling points remain inside the branches and some become tips if the lineage or its descendent lineages were not sampled again (either via ψ-sampling in the past or via ρ-sampling at the present). Until recently, the former type of sampling points were not allowed in the reconstructed trees. Gavryushkina et al. (2014) suggested inferring trees in which samples can lie inside the branches and called such sampling points *sampled ancestors*. In species phylogenies, sampled ancestors represent fossil species that are direct ancestors of other sampled species or clades. This explains why the 'tip-dating' term is misleading (Gavryushkina 2017).

The FBD model has been implemented in the SA package (Gavryushkina et al. 2014) for BEAST 2 (Bouckaert et al. 2014) and in MrBayes (Ronquist et al. 2012; Zhang et al. 2016), with some differences in the available options that we describe below. It is also available in RevBayes (Höhna et al. 2016). Thus, a total-evidence analysis using the FBD model can be performed in several major software packages for Bayesian phylogenetics.

There are several variations in sampling conditions that are assumed by the process. First, we can consider all sampled trees generated by an FBD process that stops after a fixed time t_{or} with fixed parameters λ, μ, ψ, and ρ. There exists a closed-form expression for the probability density function over the space of these sampled trees (Eq. (3) in Stadler (2010), up to a constant for labelled trees). Second, one can consider only the trees that survived until the present and in which at least one lineage was sampled at the present time. This condition is usually called *conditioning on sampling of at least one extant individual.* The probability density function over such trees also has a closed-form expression [Stadler 2010; Eq. (2) in Gavryushkina et al. (2014)]. Third, we can *condition on sampling of at least one individual* (not necessarily extant) because only non-empty samples are observed. All of these options are available in BEAST 2.

The motivation for conditioning on sampling of at least one lineage comes from simulation studies where one simulates trees and then uses these trees (or sequence data evolved along these trees) to re-estimate the parameters of the model. In such studies only non-empty trees are used and, therefore, conditioning on sampling of at least one lineage makes the estimates of the parameters from simulated data more accurate. It is not clear, however, whether to condition on sampling when analysing extant clades. On one hand, we only analyse non-empty data sets. On the other hand, the fact that the clade did not go extinct (or unobserved) might be due to the diversification and fossilization parameters being higher; conditioning on sampling would result in underestimation of these parameters.

For some applications, it is more convenient to describe the process as starting with a bifurcation event (Heath et al. 2014). In this case, the process starts at some point in time with two lineages (the root of the complete tree). This assumption is often referred to as *conditioning on the root* rather than *conditioning on the origin.* The probability density function for sampled trees generated by the process conditioning on the root and conditioning on sampling of at least one extant individual on each side of the root was derived by Stadler (2010) and defined by Eq. (3) in Gavryushkina et al. (2014). In MrBayes, the process is always assumed to be conditioned on the root. Moreover, it is assumed that there is at least one sampled lineage (not necessarily extant) on each side of the root. There is an equivalent option in BEAST 2.

The BEAST 2 implementation allows two different parameterizations, λ, μ, ψ, ρ, and d, v, s, ρ, where:

$$d = \lambda - \mu > 0,$$

$$v = \frac{\mu}{\lambda},$$

$$s = \frac{\psi}{\psi + \mu},$$

while MrBayes only allows the d, v, s parameterization. Note that the alternative parameterization implies that speciation always exceeds extinction ($d = \lambda - \mu > 0$). This might not be a plausible assumption and the original parameterization should be used if one wishes to allow a negative net diversification rate (d).

The FBD model can also be used to analyse purely extinct taxa. In this case, we would also assume that the process started at the time of origin (or the root) and finished at the present. Our knowledge that the clade has gone extinct can be interpreted as sampling at present with probability 1 even though no extant samples exist. That is, we intended to include *every* extant species belonging to this clade by comparing the morphology of the fossils with that of all the extant species, but none of the existing species could be assigned to this clade. Thus, an analysis of an extinct clade using the FBD model would require fixing the ρ parameter to 1.0 and not conditioning on the sampling of at least one individual at the present. However, this possibility is not available in BEAST 2. One could still analyse extinct clades by not specifying the ρ parameter (ρ is then fixed to zero by the program). This would not be equivalent to modelling the extinct clades as described above, but rather would treat the most recent fossil as the last ψ sample of the process and assume that sampling of extant species was not attempted. It means that such an analysis does not have the information that the clade has gone extinct, but assumes instead that there could still be unsampled extant lineages.

MrBayes does not support analysing purely extinct taxa in an appropriate way (as described above). A possible solution is to treat the most recent fossils as 'extant' species, by adjusting the stratigraphic ranges of the older fossils relative to the most recent fossil (time zero), and setting ρ to represent the putative sampling proportion of the most recent fossils.

As described in the previous section, the ages of the fossils are rarely known with certainty. One might fix the fossil age to the midpoint or a random point drawn from within the stratigraphic range, but it has been shown that ignoring fossil age uncertainties can lead to biased divergence-time estimates (Barido-Sottani et al. 2019). The implementations in BEAST 2 and MrBayes allow a time range $[\tau_1, \tau_2]$ to be specified for each of the fossil species; the unknown age of the fossil is averaged over all possible values within the specified range in the inference.

The total-evidence approach using the FBD model was initially applied to a penguin data set consisting of 35 fossils and 19 extant species (Gavryushkina et al. 2017). The analysis of morphological data from extant and fossil species, molecular data from extant species, and fossil stratigraphic intervals yielded a surprising result. The age of the crown penguin radiation was estimated to be substantially younger than in previous studies: 12.7 Myr ago compared with 20.4 Myr ago (the youngest estimate by Subramanian et al. 2013). This was attributed to both the improved modelling of the tree-branching and fossil-sampling process and the inclusion of all available stem fossils. This analysis brought another interesting insight into penguin fossil ancestry: a fossil species *Spheniscus muizoni* was inferred as the direct ancestor of the extant *Spheniscus* clade in 61% of the posterior trees.

11.3.2 Varying Diversification and Sampling Rates

The FBD model assumes that the diversification and fossil-sampling rates are constant through time. The extension of the FBD process that accounts for rate variation over time is called the *skyline* FBD model (Stadler et al. 2013; Gavryushkina et al. 2014), where the variation in rates is modelled in a piecewise manner. The time is divided into k intervals and the diversification and sampling rates are constant within each interval but can vary between intervals. Let $t_0 = t_{or}$ be the time of origin, $t_1 > \ldots > t_{k-1}$ are some times in the past, and $t_k = 0$ is the present time, then each interval $[t_i, t_{i-1}]$ for $i \in \{1, \ldots, k\}$ has rates λ_i, μ_i, and ψ_i and the last interval $[t_k, t_{k-1}]$ also has the sampling-at-present probability ρ. The probability density of a sampled tree (Fig. 11.3) generated by the skyline

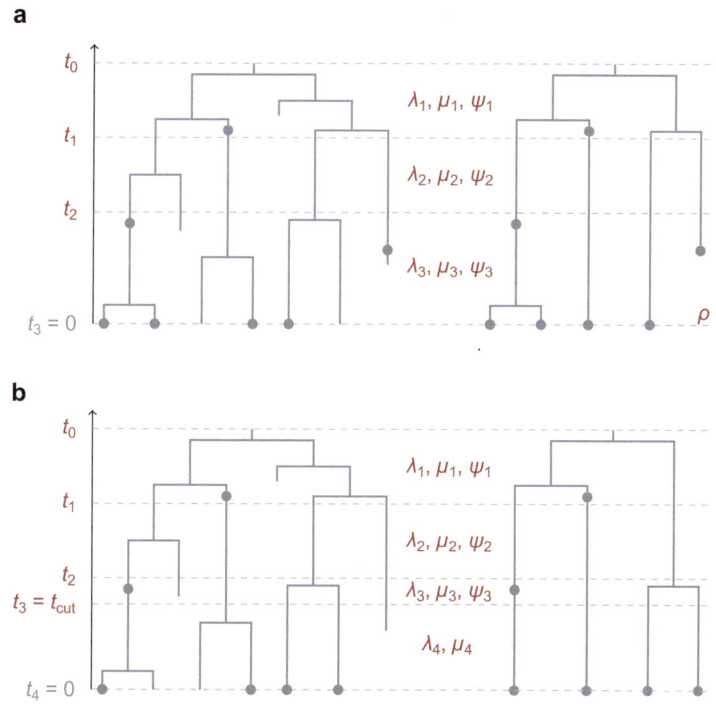

Fig. 11.3 Skyline fossilized birth–death (FBD) process with uniform and diversified extant sampling. (**a**) A complete tree generated under the skyline FBD process with uniform sampling (left) and a corresponding sampled tree (right). Each of the intervals has distinct rates λ, μ, and ψ. Uniform ρ sampling is applied at time zero. (**b**) The same complete tree generated under the skyline FBD process but with diversified sampling (left) and the corresponding sampled tree (right). No fossils are sampled after the cut-off time t_{cut}. Five lineages existed at time t_{cut}, four of which gave rise to surviving lineages or clades. Exactly one extant species from each of the four clades is sampled

FBD process, conditioning on sampling of at least one individual at the present time, is defined by Eq. (7) in Gavryushkina et al. (2014). If the process starts with a bifurcation event (at the root), and at least one individual is sampled on each side of the root, the probability density function is defined by Eq. (1) in Zhang et al. (2016) (setting ρ_i and N_i to zero for $i \in 1, \ldots, l-1$ and $\rho_l = \rho$ and multiplying by a constant for labelled trees).

The skyline version of the FBD model is implemented in the BDSKY package for BEAST 2. This implementation permits the user to fix the times of the rate shifts (t_i) or to specify the number of intervals, k. In the latter case, the period between the time of origin and the present time is divided into k equal intervals. This implies that the time shifts will be relative to the time of origin and will not be fixed within an analysis if the time of origin is not fixed. One can condition either on the root or on the origin, either condition or not condition on the sampling of at least one individual, and either condition or not condition on sampling of at least one extant individual. MrBayes implemented the skyline FBD model with several variations. First, the process starts from the root with two lineages instead of one from the origin. Second, it conditions on sampling at least one individual on each side of the root. Third, it can only specify absolute and fixed rate-shifting times.

11.3.3 Diversified Sampling of Extant Species

The FBD models described above assume a uniform sampling of extant species, meaning that we do not make any preferences (e.g., sampling one clade more densely than another) when choosing which extant species to be included in the analysis. For large clades and higher-level taxa, however, it is common to include the most diverse sample of the extant representatives of the clade (Höhna et al. 2011). Zhang et al. (2016) developed a variant of the FBD process where this sampling bias is taken into account and called it the *diversified FBD process*.

Consider a skyline FBD process with two modifications. First, the sampling rate ψ is fixed to zero within the most recent interval. The beginning of this interval is called the *cut-off time* ($t_{\mathrm{cut}} = t_{k-1}$) and we assume that no fossils are sampled after the cut-off time. The second modification is in the sampling scheme for extant species. Consider all lineages that existed at the cut-off time and survived until the present. Each of these lineages gives rise to a group of extant species (a clade). The model assumes that exactly one species from each of these groups is sampled at the present time. Such an assumption is more appropriate for higher taxa, such as in previous studies of hymenopterans and mammals (Zhang et al. 2016; Ronquist et al. 2016).

Thus, the model parameters are (t_0, t_1, ..., $t_{k-1} = t_{\mathrm{cut}}$, $t_k = 0$), (λ_i, μ_i, ψ_i) for $i \in \{1, \ldots, k-1\}$, and λ_k, μ_k. The probability density of a sampled tree generated by this process is defined by Eq. (7) in Zhang et al. (2016) (again setting ρ_i and N_i to zero for $i \in 1, \ldots, l-1$ and $\rho_l = 1$ and multiplying by a constant for labelled trees).

The skyline FBD model has been applied to a data set from Hymenoptera (Ronquist et al. 2012) under both random and diversified sampling of extant species (Zhang et al. 2016). The data set included 60 extant and 45 fossil hymenopterans and consisted of one morphological and seven molecular partitions. The geological time was divided into three intervals, with the rates shifting at 252 Myr ago and 66 Myr ago. All fossils were included in the middle interval. The analyses also explored two relaxed-clock models, the autocorrelated lognormal (TK02, Thorne and Kishino 2002) and the independent gamma rates (IGR, Lepage et al. 2007), shared (linked) across the partitions. The age estimates, particularly for the initial radiation of Hymenoptera, were quite different among these model assumptions. The age estimates for Hymenoptera were oldest when assuming random extant sampling and the TK02 clock model (365.3 Myr), and youngest when assuming diversified extant sampling and the IGR clock model (251.7 Myr). This disparity suggested that violation of the sampling assumptions of the FBD process can have a strong influence on the age estimates. The authors reasoned that diversified sampling and the IGR clock model provided the best fit for the higher-level taxon sampling and dramatic evolutionary rate variation across lineages, and the ages inferred were the most in line with the fossil record. Nevertheless, it would be important to examine the different assumptions in the FBD model and the influence of violating these assumptions on the posterior estimates in practice.

11.3.4 Accounting for Multiple Fossils of the Same Species

In Sect. 11.2, we made an assumption that every fossil species is sampled only once. However, palaeontological data sets often include several fossil specimens, from different localities and different stratigraphic layers, assigned to the same species. Incorporating these data into the inference would restrict the space of possible phylogenies, because fossil samples of the same species cannot be separated by a split, and would enable a more accurate inference through improved estimates of sampling rates.

In this section, we introduce a model that relaxes the ‘single fossil per species’ assumption. To achieve this, one first needs to model how lineages are divided into species through time. Until this point, the FBD model has only assumed that coexisting lineages belong to different species but has assumed no knowledge as to whether the same lineage represents the same species at different points in time. This also implied non-oriented trees, that is, there is no distinction between the lineage that continues the original lineage and the other that starts a new FBD process at each split. Stadler et al. (2018) introduced a mixed speciation model that assigns species through time on semi-oriented trees. Following Foote (1996), they assumed three types of speciation: *asymmetric speciation* where, after a split, one lineage continues the ancestral species and the other lineage starts a new species; *symmetric speciation* where the ancestral species goes extinct at a branching event that gives rise to two new species; and *anagenetic speciation*, where speciation occurs without branching because one species goes extinct and gives rise to another species that continues the lineage.

The mixed speciation process extends the fossilized birth–death process by assuming that every bifurcation event is an instance of a symmetric speciation event with probability β and an asymmetric speciation event with probability $1 - \beta$. Additionally, anagenetic speciation events occur with rate λ_a. Then ψ- and ρ-sampling occurs as before. Thus, the FBD process is a special case of the mixed speciation process when $\beta = \lambda_a = 0$. Such an extension reflects the biological process of speciation and explains the disparity in the estimates of the speciation and extinction rates from two sources of data: phylogenies inferred from molecular sequences and fossil stratigraphic ranges (Silvestro et al. 2018).

If we keep track of the types of speciation events, the directionality of asymmetric speciation events, and the times and locations of anagenetic speciation events, then we always know the time boundaries of species in a complete tree generated by this process. Then we can group all ψ- and ρ-sampled nodes belonging to the same species. The oldest and youngest samples of these groups will define observed segments of lineages fully belonging to the same species. We call such segments *stratigraphic ranges* of species.

The aim is to use this process to infer phylogenies from the data composed of groups of specimens with estimated ages assigned to species, that is, from stratigraphic ranges. The sampled tree is then defined as the observed part of the complete tree together with the knowledge of which fossils form stratigraphic ranges. In a sampled tree, we no longer know the time boundaries of species because only some of the branching events and none of the anagenetic events are observed. However, we still know which fossils belong to the same species. We call the process that generates such sampled trees on stratigraphic ranges a *stratigraphic-range FBD process*.

Stadler et al. (2018) have shown that the probability density function for sampled trees generated by the stratigraphic-range FBD process has a closed-form expression. Therefore, there is no computational barrier to using this model in a Bayesian joint or total-evidence inference. The behaviour of this model and its advantages for joint inference are yet to be investigated once the implementation of a full Bayesian inference becomes available.

11.4 The Role of Morphological Data

Morphological data are crucial for a total-evidence or joint inference analysis. The topological placement of fossils and the lengths of the branches in the estimated tree are both influenced by the morphological data. Although one can use the FBD model in a dating analysis with only molecular data (Heath et al. 2014), fossils in such an analysis must be pre-assigned to clades in the molecular phylogeny. This again involves comparing morphology between extinct and extant species. Without (or with too little) morphological data, that is, with uncertain topological placement of fossils, the posterior distribution of node ages will reflect the prior distribution of the parameters of the FBD model informed by only the number and ages of fossil and extant samples. Furthermore, inappropriate morphological data or models can have negative impacts on the estimated dates.

As introduced in Sect. 11.2, the input data D for a Bayesian joint or total-evidence inference analysis consists of morphological data or a combination of morphological and molecular data. Let $D = (M, S)$, where M is morphological data from fossil and extant species and S is molecular data from extant species (if present). The morphological data matrix M consists of a collection of discrete characters or traits for each species, each with a number of possible states. For example, position 21 of the morphological character matrix used for the analysis of penguins describes the trait 'iris colour' (Gavryushkina et al. 2017). It has six possible states: dark, reddish-brown, claret red, yellow, white, and silvery grey. These are coded with numbers 0–5. Then 5 at position 21 for Eudyptula minor (little blue penguin) reflects the grey colour of the eyes in these penguins.

We assume that these sets of characters evolve along the tree similarly to molecular sequences, that is, the state of a character can change from one to another along the branches of the tree. Given the amount of neutral molecular evolution and the complex nature of phenotype–genotype relationships, we further assume here that morphological and molecular data evolve independently. However, this assumption can be partly relaxed by linking molecular and morphological clock models (Ronquist et al. 2012; Zhang et al. 2016). Then the term $P(D|T, \theta)$ in Eq. (11.1) can be written as:

$$P(D|T,\theta) = P(M|T,\theta_M)P(S|T,\theta_S), \quad (11.2)$$

where θ_M and θ_S are the parameters of the models of morphological and molecular evolution, respectively. The term $P(M|T, \theta_M)$ defines how the morphological characters change over time and involves two components: a clock model, which describes how the average number of changes per character per unit of time varies among the branches; and a character substitution model, which describes the relative frequency of particular state-to-state changes. The term 'substitution' here is borrowed from models of molecular evolution. In the following section, we introduce the Lewis Mk morphological character substitution model and describe several generalizations (see also Chap. 7). Then we discuss several problems concerning the availability and quality of morphological data.

11.4.1 Lewis Mk and Mkv Model

The Lewis Mk model (Lewis 2001) is widely used for morphological character evolution. It is a generalization of the Jukes–Cantor model (Jukes and Cantor 1969) for four-state nucleotide substitution. The model assumes that every state has the same instantaneous rate of changing into every other state (Fig. 11.4a). For k states ($k \geq 2$), the (instantaneous) substitution rate matrix is:

$$Q = a\begin{bmatrix} 1-k & 1 & \cdots & 1 \\ 1 & 1-k & \cdots & 1 \\ \vdots & \vdots & \ddots & \vdots \\ 1 & 1 & \cdots & 1-k \end{bmatrix},$$

where a is the instantaneous rate of change between states. Thus, the transition probability matrix $P(t)$ has elements:

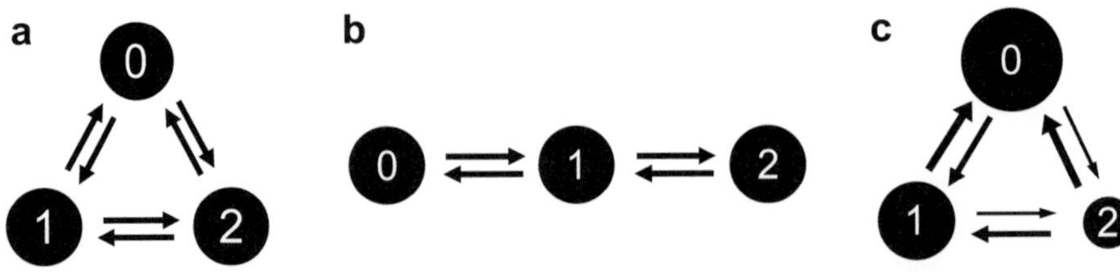

Fig. 11.4 (**a**) Lewis Mk model, (**b**) ordered states with equal stationary frequencies, and (**c**) a more general model with unequal stationary frequencies. The size of each circle represents the stationary frequency of the corresponding state. The thickness of each arrow represents the instantaneous rate of change

$$P_{ii}(t) = \frac{1}{k} + \frac{k-1}{k} e^{-kat},$$

$$P_{ij}(t) = \frac{1}{k} - \frac{1}{k} e^{-kat}.$$

The stationary distribution of the k states is $(1/k, \ldots, 1/k)$. For the likelihood calculation where branch lengths are typically measured by the expected number of substitutions per character, $a = 1/(k - 1)$.

Palaeontologists typically code variable characters only, that is, the characters that have at least some variation in states among the species of interest. Applying the Mk model directly in this case will lead to overestimation of the branch lengths. To account for such an *acquisition bias*, one needs to divide the likelihood for each character c by the probability of c being variable, that is, 1 minus the probability of c being constant. This correction has been named the Mkv model (Lewis 2001).

11.4.2 Modelling Substitution Rate Heterogeneity

The assumption of an equal rate of change between states is considered a poor fit for certain morphological characters. For example, some characters might be specified as 'ordered', that is, instantaneous change is only allowed between adjacent states. In this case, only the elements adjacent to the diagonal have rate 1, while the rest of the rates are 0 in the Q matrix ($k \geq 3$).

$$Q = a\begin{bmatrix} -1 & 1 & 0 & \cdots & 0 \\ 1 & -2 & 1 & \cdots & 0 \\ 0 & 1 & -2 & \cdots & 0 \\ \vdots & \vdots & \vdots & \ddots & \vdots \\ 0 & 0 & 0 & \cdots & -1 \end{bmatrix}.$$

This model has the same equal stationary frequencies $(1/k, \ldots, 1/k)$ as the Mk model but takes a longer time to move between nonadjacent states.

Another model, described by Wright et al. (2016), extends the Mk model and has unequal stationary frequencies $(\pi_0, \ldots, \pi_{k-1})$. It is a generalization of the F81 model (Felsenstein 1981) for k states ($k \geq 2$).

$$Q = a\begin{bmatrix} \pi_0 - 1 & \pi_1 & \cdots & \pi_{k-1} \\ \pi_0 & \pi_1 - 1 & \cdots & \pi_{k-1} \\ \vdots & \vdots & \ddots & \vdots \\ \pi_0 & \pi_1 & \cdots & \pi_{k-1} - 1 \end{bmatrix}.$$

This model becomes the Mk model when the π_i s are all equal.

More generally, to account for both the ordering of states and unequal stationary frequencies, the substitution rate matrix ($k \geq 3$) can be written as:

$$Q = a\begin{bmatrix} . & \pi_1 & b\pi_2 & \cdots & b\pi_{k-1} \\ \pi_0 & . & \pi_2 & \cdots & b\pi_{k-1} \\ b\pi_0 & \pi_1 & . & \cdots & b\pi_{k-1} \\ \vdots & \vdots & \vdots & \ddots & \vdots \\ b\pi_0 & b\pi_1 & b\pi_2 & \cdots & . \end{bmatrix}.$$

The parameter b controls the bias towards ordered characters, where $b = 0$ models completely ordered characters. For $0 < b < 1$, instantaneous change tends to be more likely between adjacent states than nonadjacent ones.

Since the substitution process is completely determined by the Q matrix, one might even extend the model to be analogous to the GTR model (Tavaré 1986) by adding more free parameters. However, it appears less meaningful to estimate all such parameters from morphological characters. One concern, as mentioned by Wright et al. (2016), is that morphological character states do not carry common meaning across characters because each describes a distinct trait (i.e., '0' for one character might have a completely different meaning from '0' for another).

In practice, the unequal stationary frequencies can be averaged over through the use of priors in the Bayesian analysis. This is straightforward for two-state characters, because there is a single parameter π_0 in the model ($\pi_1 = 1 - \pi_0$). One can assign a beta(α, α) distribution to π_0 with mean 0.5, discretize the density into several categories with equal probabilities, and average over the values of π_0 in the likelihood calculation (Wright et al. 2016), a strategy similar to discretizing the gamma distribution for rate variation across sites (Yang 1994). The posterior estimate of α indicates the degree of asymmetry of character state substitutions, with larger α corresponding to more similar rates of change and thus similar state frequencies on average for the characters analysed. The limiting condition of $\alpha = \infty$ corresponds to the Mk model.

For characters with more than two states, one can assign a symmetric Dirichlet distribution with a single parameter α for the stationary frequencies $\{\pi_i\}$. However, applying this strategy is not straightforward, because there is no simple way to discretize a Dirichlet distribution. In practice, the stationary distribution of $\{\pi_i\}$ needs to be coestimated. For the purpose of interpretation, it is better to consistently code ancient traits as smaller numbers (e.g., '0') than for derived traits, to avoid arbitrary labelling.

Lewis Mk and Mkv models are implemented in the MM package for BEAST 2 and in MrBayes. Both software packages allow the user to specify unequal frequencies and to assign a Dirichlet prior for the frequencies. BEAST 2 will estimate the overall frequencies of all of the characters with the same number of states (if the morphological matrix is partitioned into groups of characters with the same number of states). MrBayes will estimate the frequencies of each character with more than two states and average over the frequencies for two-state characters using a discrete beta prior described above.

11.4.3 Other Models and Morphological Data

The specifics of the morphological data should be reflected in the statistical models that we use for the inference. The total-evidence approach is a recent method and the models of morphological evolution used in it have not been tested thoroughly. Better models can potentially be developed to improve the inferences.

Several other variants of the Lewis Mk model have been introduced. Klopfstein et al. (2015) applied a nonstationary Markov model to the Hymenoptera data set, which resulted in improved precision in the divergence-time estimates. Hoyal Cuthill (2015) considered Lewis models with different state spaces: finite, inertial (locally finite), and infinite. She simulated patterns of homoplasy from these different models and compared them with the patterns of several empirical data sets, most of which were compatible with the inertial model and some with the finite model.

No specific clock models, the second component of the term $P(M|T, \theta_M)$ in Eq. (11.2), have been developed for morphological data. They are simply adapted from clock models that were designed for molecular evolution. The most common models that have been used for morphological data are the strict clock (Zuckerkandl and Pauling 1962) and uncorrelated relaxed-clock (Drummond et al. 2006; Lepage et al. 2007). Only a few studies have attempted to compare and test different clock models for morphological data (Zhang et al. 2016; King et al. 2017). When morphological data are limited, the temporal data might drive the estimated divergence times and the application of the relaxed-clock model might result in unexpected patterns of rate variation among branches.

We usually assume that different characters evolve independently of each other and share the same rates of evolution. However, some characters might be correlated and evolve together (Lee and Palci 2015; dos Reis et al. 2016). For example, several characters that represent different measurements (e.g., in different dimensions) of the same phenotypic feature will most likely evolve in a similar way. Lee (2016) partially addressed this issue by partitioning a large morphological data set of mammals based on the similarity of the relative branch lengths reconstructed from different proposed partitions, then applying a distinct relaxed-clock model to each of the partitions in the joint inference. This resulted in an improved recovery of the topological relationships. Thus, better statistical models accounting for character correlation are needed.

Goloboff et al. (2019) showed that morphological characters have distinct patterns of rate variation among lineages. They showed in simulation studies that this had a slight effect on the undated phylogeny estimated from morphological data using Bayesian inference. More research is needed to assess how violating the Lewis Mk model assumption affects the divergence times from joint inference and whether better modelling is possible.

The availability of morphological data might be an issue for joint inference. First of all, there need to be enough morphological data to inform the fossil placements in the tree and to date the phylogeny reliably. The majority of extant species have not been coded for morphological characters (Lee and Palci 2015), making them unavailable to be included in a joint inference with the related fossil species for which morphological data are available. Guillerme and Cooper (2016) assessed the impact of missing morphological data on the inferred topology in a total-evidence analysis. They showed that the ability to recover a tree topology decreases with decreasing morphological data. Decreasing the number of living taxa with morphological characters and decreasing the total number of morphological characters had the worst effect, whereas missing characters in the fossil data affected the inference to a lesser degree.

Incomplete morphological data are not always due to the natural reasons of poor preservation or limited access to geological sites. For example, the available morphological data have often been collected with different inference approaches in mind. In some cases, this issue might be able to be addressed by improved modelling. Typically, morphological character matrices constructed for parsimony analyses do not contain so-called parsimony-uninformative characters: constant characters and autapomorphies. The latter are characters for which only one state can be present in more than one species. For example, a data column '000100200' is an autapomorphy. Matzke and Irmis (2018) showed that the exclusion of autapomorphies led to the morphological clock rate increasing by an order of magnitude in a joint inference analysis of early eureptiles. In simulation studies, they also showed that when the morphological data are clocklike, the exclusion of autapomorphies leads to biased branch-length estimates and, in particular, underestimation of the terminal branch lengths. A simple solution by Lewis (2001), where the likelihood is conditioned on nonrecording for constant characters (Mkv model), cannot be easily transformed to ascertaining for autapomorphies, for computational reasons. Hoyal Cuthill (2015) also noted that it is not possible to distinguish between the homoplasy patterns from finite and inertial models from data sets where parsimony-uninformative characters have been excluded. Studies are now filling in this gap by updating morphological data sets with autapomorphies (Cau 2017; King et al. 2017).

To date, most studies applying the joint inference approach have been using species-level fossil data. That is, each fossil species is represented by a single fossil sample, with morphological codings merged from many available fossil specimens of this species. While including only a single fossil sample per species violates the assumptions of the FBD model and might lead to biases in the diversification and sampling parameters, merging morphological codings from different specimens also misleads the models of morphological evolution. First of all, such data are not accurate because the characters merged from specimens of distinct ages could have evolved over the time elapsed between the fossilization events. On the other hand, characters with identical states in several specimens from different stratigraphic layers indicate the period of time during which no changes occurred in these characters, improving the estimates of the morphological evolutionary rates.

Another approach is to include each fossil specimen with its own morphological codings and treat it as a distinct sample. The first and only study that has applied joint inference with the FBD model to specimen-level data is by Cau (2017), who analysed tooth plates from individual fossils of Mesozoic dipnoans to test for chronostratigraphic relationships among these fossil specimens (that is, whether different specimens represent the same species at different points in time). However, specimen-level data are rarely available and can be impractical to collect and/or analyse because of their size. The model introduced in Sect. 11.3.4 could overcome the issue of violating the FBD model assumption when the size of the specimen-level data set becomes prohibitive, but it does not solve the problems arising from merging morphological codings. A better modelling of morphological evolution might be able to address this problem.

The last issue that we discuss here is discretizing continuous characters. In other words, if a trait represents a continuous characteristic (e.g., length) then its values will be discretized into several categories (e.g., short, medium, and long). Many traits are presented by continuous measurements and morphometrics, but the models (Mk model in particular) that have been used in joint inference analyses do not account for this. In phylogenetics, the drift of continuous characters has been modelled by a Brownian motion (Felsenstein 1973) or Ornstein–Uhlenbeck process (Felsenstein 1988). The former has no bound on the possible value of the trait in the long run, because the variance grows in proportion to time. The latter has a long-term mean and the process will fluctuate around this mean in the long run. The models of continuous character evolution have great potential to be integrated into a joint or total-evidence dating analysis to take advantage of different types of morphological data. Álvarez-Carretero et al. (2019) recently attempted joint inference with a combination of continuous traits and molecular sequences. They implemented the Brownian motion process (Felsenstein 1973) and accounted for correlations among the traits. In their approach, the topology of the tree needs to be fixed.

11.5 Calibration Approach Versus Joint Inference

In dating analyses based on the calibration approach, calibration densities are used as prior distributions for the ages of particular nodes, which we refer to as *calibration nodes* (Tavaré et al. 2002; Yang and Rannala 2006; Ho and Phillips 2009). The topology of extant species is often estimated prior to the dating analysis and calibration nodes are fixed in the tree. If the topology and the divergence dates are coestimated in the dating analysis, a calibration node is defined as the most recent common ancestor of a group of extant species. In both cases, dating with calibrations is performed in two steps. First, temporal fossil data are transformed into calibration densities and assigned to calibration nodes. Second, a separate dating analysis of extant species is performed. This analysis uses molecular sequences of extant species as data and calibration densities as prior distributions on the ages of calibration nodes to infer dated phylogenies (Chap. 8). Here we first describe the limitations of the calibration approach and where joint inference can overcome them. Then we discuss the limitations of the joint inference approach and the directions for further improvement.

11.5.1 Total-Evidence Dating Overcomes Limitations of the Calibration Approach

To calibrate a phylogeny with a fossil, one needs to first assign the fossil to a clade and then transform the fossil age into a calibration. Often fossils are assigned to clades based on apomorphies (traits that are unique for the clade) without any phylogenetic analysis or, where phylogenetic analyses do take place, usually using non-statistical parsimony methods. In either case, incorrectly assigned fossils will calibrate the wrong nodes. In a joint inference analysis, the positions of the fossils are estimated from morphological and temporal data based on statistical models that describe how the diversification and sampling occurs (FBD model) and how the morphology evolves (Lewis Mk and clock models).

Even when a careful phylogenetic analysis is performed, some fossils cannot be unambiguously assigned to any clade and, following best practice (Parham et al. 2012), these fossils will be discarded from a calibration analysis. Often only a few exemplars from the candidate fossils are used in the analysis, typically the oldest fossil of a clade and possibly others that were used indirectly to identify the tails of the calibration density. It is also recommended to only use well-preserved fossils with good age estimates, in order to minimize errors (Benton and Donoghue 2007). In contrast, a joint inference can analyse all available fossils (up to a reasonable number). In the analysis of the penguin data set by Gavryushkina et al. (2017), including a large collection of stem fossils (the majority of which were not used in previous dating analyses) resulted in a dramatic decrease in the estimated age of the crown penguin radiation.

Even if a fossil is correctly identified, phylogenetically placed, and dated, the way in which the fossil is transformed to a calibration density might greatly deviate from an accurate representation of any uncertainty. The calibration densities are specified ad hoc without a statistical analysis of the temporal data. Moreover, even if the minimum bound on the age of the calibration node can be reliably chosen, maximum bounds are much more difficult to justify. Statistical methods have been developed to define the calibration densities (Bapst 2013; Matschiner et al. 2017), but these methods only directly use the age of the oldest fossils (the ages of other fossils might be used indirectly, that is, to infer sampling and extinction rates in a prior analysis).

After the calibrations have been specified, they need to be correctly incorporated into the analysis. In a Bayesian framework, the calibration densities are seen as prior information about particular divergence times. These densities interact with each other and with the rest of the prior model in a couple of ways. The first type of interaction is because there is an internal dependence of the timing events on a phylogeny: deeper divergences must occur earlier than more shallow divergences. Thus, if there are two clades with one nested inside the other and each has a probability density calibrating its age, then the actual prior probability densities will differ from the ones specified by experts (Ho and Phillips 2009; Warnock et al. 2012). Further, there is a tree-wide prior distribution (the tree-generating model in case of the joint inference) that describes the distribution of all branching times in a phylogeny, which should be conditioned on the given calibration densities (conditional construction). In the case of a fixed topology, such a model is feasible (Yang and Rannala 2006). However, when the topology is coestimated with the divergence times, the exact modelling becomes computationally intensive; theoretically inaccurate multiplicative construction has been used instead (Inoue et al. 2010; Heled and Drummond 2012). The joint inference does not require specifying calibration densities and, therefore, does not suffer from these problems.

The final and the most important problem with calibrations is that they are done sequentially. First of all, the multiple steps create more opportunities for errors. For example, there can be errors in fossil age estimates, topological placement of fossils, transforming fossil ages to calibration densities, and incorporating the densities into an analysis. These errors can accumulate and propagate to the final estimates of divergence times. Secondly, and most importantly, the sequential nature of this procedure assumes independence of the processes that generate the data for different stages of the analysis. However, fossil samples come from the same underlying phylogeny on which molecular and morphological data evolved. Thus, if there is a conflicting phylogenetic signal in the fossil, molecular, and morphological data, a joint analysis of these data can resolve the conflict or average over the phylogenetic inputs from the different data. Even where there is no conflict in the different data, incorrect assumptions of independence might lead to an inaccurate posterior distribution. This problem is solved by hierarchical modelling and analysing the data together in the joint inference.

11.5.2 Limitations of the Joint Inference Approach

There are several aspects that can make the joint inference approach impossible or inaccurate. We have already discussed the problems connected to the availability and modelling of morphological data in Sect. 11.4.3. Another major problem is the impact of the tree-generating model on the estimated ages and other model parameters. Finally, the full Bayesian total-evidence inference might not scale well with the large amounts of molecular data that are now available.

It is very important to account for the sampling process of fossil and extant species in a joint inference. Matzke and Wright (2016) showed that estimates of divergence times in the mammalian family Canidae shifted by as much as ∼30 Myr when they used models that did or did not account for the fossil sampling process. However, violation of the assumptions in the FBD process might also lead to large differences in the posterior estimates.

For extant species, either random (uniform) or diversified sampling is assumed, both of which are extreme cases. In practice, sampling is likely to be uniform for some clades but diversified for others, or might be completely different from either. We have seen in Sect. 11.3.3 that different assumptions about sampling of extant species changed the estimated time of the origin of Hymenoptera from ∼330 Myr ago to ∼250 Myr ago.

For fossil sampling, the simple FBD model assumes that the lineages on the complete tree are uniformly sampled through time. The skyline FBD model relaxes this assumption by allowing variation in the fossil sampling rate through time. However, a nonuniform fossil sampling among clades has not been considered in an FBD model. Some parts of the phylogeny might be sampled more intensively, owing to higher fossilization and/or preservation or better access to geological sites. The biases in fossil sampling scheme are probably more difficult to take into account. Similarly, the diversification rates are assumed to be equal in different parts of the phylogeny (throughout the tree in the constant-rate model or between rate-shifting times in the skyline model). The impacts of violating these assumptions (and probably others) on the estimated ages in the joint inference have not yet been investigated comprehensively, and the FBD model needs to be appropriately extended where possible.

Including outgroup species might also violate the assumption of uniform sampling if the sampling strategy for the outgroup species was different from that for the ingroup species. For example, all extant representatives of the ingroup clade but only a few extant species from the outgroup clades are typically included in an analysis. Similarly, all available ingroup fossils but only a few fossils (out of many more available) from the outgroup clades might be used. The effect of such a biased sampling on the posterior estimates has not been investigated. The joint inference analysis using the FBD model does not require outgroup species and we generally do not recommend including them if the sampling of outgroup species is different from that of ingroup species.

Another major problem for a joint inference, and for dating analyses in general, is the availability of the data. As mentioned in Sect. 11.4.3, limited morphological data might be an issue. However, limited temporal data is also a problem. When there are only a few fossils in the clade of interest or only little morphological data, the result is usually slow convergence in the Bayesian analysis and very broad credible intervals for the estimated ages. In some cases, this can be overcome by including a sister clade that has a richer fossil record; however, this needs to be done cautiously when including outgroup species, and the same fossil and extant species-sampling strategies should be used for the clade of interest and the sister clade. Another solution is to use informative prior distributions on the parameters of the FBD model. Information about the diversification or fossil sampling rates can possibly be obtained from other studies, for example, that analysed larger clades in which the clade of interest is nested.

Note that the problem of having a limited number of fossils cannot be overcome by the calibration approach either, because too few calibration points would also imply very large uncertainties. However, if the calibration densities come from other sources of information such as biogeographic events (see Chap. 9), then the calibration approach can be used in combination with the total-evidence approach (O'Reilly and Donoghue 2016). Nevertheless, the problem of how to correctly incorporate such calibration densities remains open.

Finally, although the total-evidence approach is a rigorous statistical method, it comes with large computational costs that are sometimes prohibitive for the increasingly large data sets that are available today. Much of the computational burden is in coestimating divergence dates and topological relationships. Kumar and Hedges (2016) discussed what they call fourth-generation, non-model-based methods that estimate divergence times on fixed phylogenies from large data sets (Tamura et al. 2012; To et al. 2015). The total-evidence approach is limited to data sets that are relatively small, such as particular clades in the tree of life, and is the most effective for clades with good fossil records. For supertrees (Matschiner et al. 2017) and larger data sets, these non-model-based methods might be more practical (Chap. 12). The methods that use a pre-estimated phylogeny, such as calibration approaches or FBD with fixed phylogeny (Heath et al. 2014), might be more feasible options. Another option is to use approximate maximum-likelihood approaches (Sagulenko et al. 2018). There is also ongoing research into improving the speed of algorithms used in Bayesian phylogenetic inference (Aberer et al. 2014; Zhang et al. 2020). Implementing such algorithms in combination with the FBD model would enable a full Bayesian total-evidence analysis of larger data sets.

11.6 Concluding Remarks

The joint inference or total-evidence approach with the FBD model is an advanced statistical method that objectively transforms the fossil record into absolute time constraints on a phylogeny. This method should be used to infer dated phylogenies, especially for data sets where the sampling of fossil and extant species is either unbiased or the sampling scheme can be directly modelled, where morphological and stratigraphic data from fossil and extant species are available and appropriate, and where the volume of the data is moderate.

The models used in the total-evidence dating approach have the flexibility to accommodate various evolutionary processes. For example, the skyline FBD process can account for variation in rates of diversification and fossil sampling over time, as well as different sampling schemes for extant taxa. The Mk model can be extended to incorporate heterogeneity in substitution rates. The Bayesian framework also makes it feasible to take into account fossil age and topological uncertainties. In the meantime, the approach is still in its early stages and is under active development. There is still abundant work that needs to be done to improve the models and implementations.

References

Aberer AJ, Kobert K, Stamatakis A (2014) ExaBayes: massively parallel Bayesian tree inference for the whole-genome era. Mol Biol Evol 31:2553–2556

Álvarez-Carretero S, Goswami A, Yang Z, dos Reis M (2019) Bayesian estimation of species divergence times using correlated quantitative characters. Syst Biol 4:393

Bapst DW (2013) A stochastic rate-calibrated method for time-scaling phylogenies of fossil taxa. Methods Ecol Evol 4:724–733

Barido-Sottani J, Aguirre-Fernández G, Hopkins MJ, Stadler T, Warnock R (2019) Ignoring stratigraphic age uncertainty leads to erroneous estimates of species divergence times under the fossilized birth-death process. Proc R Soc B 286:20190685

Benton MJ, Donoghue PCJ (2007) Paleontological evidence to date the tree of life. Mol Biol Evol 24:26–53

Bouckaert R, Heled J, Kühnert D, Vaughan T, Wu C-H, Xie D, Suchard MA, Rambaut A, Drummond AJ (2014) BEAST 2: a software platform for Bayesian evolutionary analysis. PLOS Comput Biol 10: e1003537

Cau A (2017) Specimen-level phylogenetics in paleontology using the fossilized birth-death model with sampled ancestors. PeerJ 5:e3055
Didier G, Royer-Carenzi M, Laurin M (2012) The reconstructed evolutionary process with the fossil record. J Theor Biol 315:26–37
dos Reis M, Donoghue PCJ, Yang Z (2016) Bayesian molecular clock dating of species divergences in the genomics era. Nat Rev Genet 17:71–80
Drummond AJ, Ho SYW, Phillips MJ, Rambaut A (2006) Relaxed phylogenetics and dating with confidence. PLOS Biol 4:e88
Felsenstein J (1973) Maximum-likelihood estimation of evolutionary trees from continuous characters. Am J Hum Genet 25:471–492
Felsenstein J (1981) Evolutionary trees from DNA sequences: a maximum likelihood approach. J Mol Evol 17:368–376
Felsenstein J (1988) Phylogenies and quantitative characters. Annu Rev Ecol Syst 19:445–471
Foote M (1996) On the probability of ancestors in the fossil record. Paleobiology 22:141–151
Gavryushkina A (2017) Sampled ancestors and dating in Bayesian phylogenetics. PhD thesis, University of Auckland, Auckland
Gavryushkina A, Welch D, Stadler T, Drummond AJ (2014) Bayesian inference of sampled ancestor trees for epidemiology and fossil calibration. PLOS Comput Biol 10:e1003919
Gavryushkina A, Heath TA, Ksepka DT, Stadler T, Welch D, Drummond AJ (2017) Bayesian total-evidence dating reveals the recent crown radiation of penguins. Syst Biol 66:57–73
Goloboff PA, Pittman M, Pol D, Xu X (2019) Morphological data sets fit a common mechanism much more poorly than DNA sequences and call into question the Mkv model. Syst Biol 68:494–504
Grimmett G, Stirzaker D (2001) Probability and random processes. Oxford University Press, Oxford, UK
Guillerme T, Cooper N (2016) Effects of missing data on topological inference using a total evidence approach. Mol Phylogenet Evol 94:146–158
Heath TA, Huelsenbeck JP, Stadler T (2014) The fossilized birth-death process for coherent calibration of divergence-time estimates. Proc Natl Acad Sci USA 111:E2957–E2966
Heled J, Drummond AJ (2012) Calibrated tree priors for relaxed phylogenetics and divergence time estimation. Syst Biol 61:138–149
Ho SYW, Phillips MJ (2009) Accounting for calibration uncertainty in phylogenetic estimation of evolutionary divergence times. Syst Biol 58:367–380
Höhna S, Stadler T, Ronquist F, Britton T (2011) Inferring speciation and extinction rates under different sampling schemes. Mol Biol Evol 28:2577–2589
Höhna S, Landis MJ, Heath TA, Boussau B, Lartillot N, Moore BR, Huelsenbeck JP, Ronquist F (2016) RevBayes: Bayesian phylogenetic inference using graphical models and an interactive model-specification language. Syst Biol 65:726–736
Hoyal Cuthill JF (2015) The morphological state space revisited: what do phylogenetic patterns in homoplasy tell us about the number of possible character states? Interface Focus 5:20150049
Inoue J, Donoghue PCJ, Yang Z (2010) The impact of the representation of fossil calibrations on Bayesian estimation of species divergence times. Syst Biol 59:74–89
Jukes TH, Cantor CR (1969) Evolution of protein molecules. In: Munro HN (ed) Mammalian protein metabolism. Academic, New York, pp 21–123
Kendall DG (1948) On the generalized "birth-and-death" process. Ann Math Stat 19:1–15
King B, Qiao T, Lee MSY, Zhu M, Long JA (2017) Bayesian morphological clock methods resurrect placoderm monophyly and reveal rapid early evolution in jawed vertebrates. Syst Biol 66:499–516
Klopfstein S, Vilhelmsen L, Ronquist F (2015) A nonstationary Markov model detects directional evolution in hymenopteran morphology. Syst Biol 64:1089–1103
Kumar S, Hedges SB (2016) Advances in time estimation methods for molecular data. Mol Biol Evol 33:863–869
Lee MSY (2016) Multiple morphological clocks and total-evidence tip-dating in mammals. Biol Lett 12:20160033
Lee MSY, Palci A (2015) Morphological phylogenetics in the genomic age. Curr Biol 25:R922–R929
Lepage T, Bryant D, Philippe H, Lartillot N (2007) A general comparison of relaxed molecular clock models. Mol Biol Evol 24:2669–2680
Lewis PO (2001) A likelihood approach to estimating phylogeny from discrete morphological character data. Syst Biol 50:913–925
Matschiner M, Musilová Z, Barth JM, Starostová Z, Salzburger W, Steel M, Bouckaert R (2017) Bayesian phylogenetic estimation of clade ages supports trans-Atlantic dispersal of cichlid fishes. Syst Biol 66:3–22
Matzke NJ, Irmis RB (2018) Including autapomorphies is important for paleontological tip-dating with clocklike data, but not with non-clock data. PeerJ 6:e4553
Matzke NJ, Wright A (2016) Inferring node dates from tip dates in fossil Canidae: the importance of tree priors. Biol Lett 12:20160328
O'Reilly JE, Donoghue PCJ (2016) Tips and nodes are complementary not competing approaches to the calibration of molecular clocks. Biol Lett 12:20150975
Parham JF, Donoghue PCJ, Bell CJ, Calway TD, Head JJ, Holroyd PA, Inoue JG, Irmis RB, Joyce WG, Ksepka DT, Patané JSL, Smith ND, Tarver JE, van Tuinen M, Yang Z, Angielczyk KD, Greenwood JM, Hipsley CA, Jacobs L, Makovicky PJ, Müller J, Smith KT, Theodor JM, Warnock RCM, Benton MJ (2012) Best practices for justifying fossil calibrations. Syst Biol 61:346–359
Pyron RA (2011) Divergence time estimation using fossils as terminal taxa and the origins of Lissamphibia. Syst Biol 60:466–481

Ronquist F, Klopfstein S, Vilhelmsen L, Schulmeister S, Murray DL, Rasnitsyn AP (2012) A total-evidence approach to dating with fossils, applied to the early radiation of the Hymenoptera. Syst Biol 61:973–999

Ronquist F, Lartillot N, Phillips MJ (2016) Closing the gap between rocks and clocks using total-evidence dating. Philos Trans R Soc B 371:20150136

Sagulenko P, Puller V, Neher RA (2018) TreeTime: maximum-likelihood phylodynamic analysis. Virus Evol 4:vex042

Silvestro D, Warnock RCM, Gavryushkina A, Stadler T (2018) Closing the gap between palaeontological and neontological speciation and extinction rate estimates. Nat Commun 9:5237

Stadler T (2010) Sampling-through-time in birth-death trees. J Theor Biol 267:396–404

Stadler T, Kühnert D, Bonhoeffer S, Drummond AJ (2013) Birth–death skyline plot reveals temporal changes of epidemic spread in HIV and hepatitis C virus (HCV). Proc Natl Acad Sci USA 110:228–233

Stadler T, Gavryushkina A, Warnock RC, Drummond AJ, Heath TA (2018) The fossilized birth-death model for the analysis of stratigraphic range data under different speciation modes. J Theor Biol 447:41–55

Subramanian S, Beans-Picón G, Swaminathan SK, Millar CD, Lambert DM (2013) Evidence for a recent origin of penguins. Biol Lett 9:20130748

Tamura K, Battistuzzi FU, Billing-Ross P, Murillo O, Filipski A, Kumar S (2012) Estimating divergence times in large molecular phylogenies. Proc Natl Acad Sci USA 109:19333–19338

Tavaré S (1986) Some probabilistic and statistical problems on the analysis of DNA sequences. Lect Math Life Sci 17:57–86

Tavaré S, Marshall CR, Will O, Soligo C, Martin RD (2002) Using the fossil record to estimate the age of the last common ancestor of extant primates. Nature 416:726–729

Thorne JL, Kishino H (2002) Divergence time and evolutionary rate estimation with multilocus data. Syst Biol 51:689–702

To T-H, Jung M, Lycett S, Gascuel O (2015) Fast dating using least-squares criteria and algorithms. Syst Biol 65:82–97

Warnock RC, Yang Z, Donoghue PCJ (2012) Exploring uncertainty in the calibration of the molecular clock. Biol Lett 8:156–159

Wright AM, Lloyd GT, Hillis DM (2016) Modeling character change heterogeneity in phylogenetic analyses of morphology through the use of priors. Syst Biol 65:602–611

Yang Z (1994) Maximum likelihood phylogenetic estimation from DNA sequences with variable rates over sites: approximate methods. J Mol Evol 39:306–314

Yang Z, Rannala B (2006) Bayesian estimation of species divergence times under a molecular clock using multiple fossil calibrations with soft bounds. Mol Biol Evol 23:212–226

Zhang C, Stadler T, Klopfstein S, Heath TA, Ronquist F (2016) Total-evidence dating under the fossilized birth-death process. Syst Biol 65:228–249

Zhang C, Huelsenbeck JP, Ronquist F (2020) Using parsimony-guided tree proposals to accelerate convergence in Bayesian phylogenetic inference. Syst Biol 69:1016–1032

Zuckerkandl E, Pauling L (1962) Molecular disease, evolution, and genic heterogeneity. In: Pullman M, Kasha B (eds) Horizons in biochemistry. Academic, New York, pp 189–225

Part IV

Phylogenomics

Efficient Methods for Dating Evolutionary Divergences

12

Qiqing Tao, Koichiro Tamura, and Sudhir Kumar

Abstract

Reliable estimates of divergence times are crucial for biological studies to decipher temporal patterns of macro- and microevolution of genes and organisms. Molecular sequences have become the primary source of data for estimating divergence times. The sizes of molecular data sets have grown quickly due to the development of inexpensive sequencing technology. To deal with the increasing volumes of molecular data, many efficient dating methods are being developed. These methods not only relax the molecular clock and offer flexibility to use multiple clock calibrations, but also complete calculations much more quickly than Bayesian approaches. Here, we discuss the theoretical and practical aspects of these non-Bayesian approaches and present a guide to using these methods effectively. We suggest that the computational speed and reliability of non-Bayesian relaxed-clock methods offer opportunities for enhancing scientific rigour and reproducibility in biological research for large and small data sets.

Keywords

Molecular dating · Strict clock · Local clock · Calibration · Maximum likelihood · RelTime

12.1 Introduction

Computational methods to estimate divergence times of genes and species from molecular data have enjoyed a long history of development, spanning more than 50 years (dos Reis et al. 2016; Kumar and Hedges 2016). Divergence times derived by using these methods and molecular data have illuminated the role of geological history in shaping the emergence of species (Hedges et al. 1996; Hedges and Kumar 2009), tempo and mode of speciation (Hedges et al. 2015; Marin et al. 2017), dynamics of genome evolution through gene duplication

Q. Tao
Institute for Genomics and Evolutionary Medicine, Temple University, Philadelphia, PA, USA

Department of Biology, Temple University, Philadelphia, PA, USA
e-mail: qiqing.tao@temple.edu

K. Tamura
Center for Genomics and Bioinformatics, Tokyo Metropolitan University, Tokyo, Japan

Department of Biological Sciences, Tokyo Metropolitan University, Tokyo, Japan
e-mail: ktamura@tmu.ac.jp

S. Kumar (✉)
Institute for Genomics and Evolutionary Medicine, Temple University, Philadelphia, PA, USA

Department of Biology, Temple University, Philadelphia, PA, USA

Center for Excellence in Genome Medicine and Research, King Abdulaziz University, Jeddah, Saudi Arabia
e-mail: s.kumar@temple.edu

S. Y. W. Ho (ed.), *The Molecular Evolutionary Clock*, https://doi.org/10.1007/978-3-030-60181-2_12

(Huerta-Cepas and Gabaldón 2011; Jiao et al. 2011; Yu et al. 2017), and evolution of pathogens (Faria et al. 2014; Worobey et al. 2014; Biek et al. 2015; Metsky et al. 2017). Every year, hundreds of studies report estimates of species divergence times, enabling the assembly of the grand time-tree of life and revealing the fundamental biological processes underlying species diversity (Hedges et al. 2015).

Early statistical methodologies of molecular clock dating (Zuckerkandl and Pauling 1962) were based on the assumption of a constant rate of evolution over time and across lineages (strict clock) and used fossil-age calibrations as point values (Kumar 2005). Over the last two decades, molecular dating methods have become increasingly sophisticated and embrace greater biological realism. They now relax the strict-clock assumption and have the ability to estimate divergence times even when molecules have evolved with vastly different evolutionary rates across loci and lineages (Ho 2014; Ho and Duchêne 2014; Kumar and Hedges 2016). Many modern approaches are also available to incorporate detailed information from the fossil record to generate time-calibrated phylogenies (time-trees).

Statistical development of molecular dating methods remains vibrant even after six decades of development. It is at the centre of systematics, biodiversity, and genome evolution research owing to the ease with which large sequence data sets can now be assembled (Kumar and Hedges 2016). Chronologies of molecular dating methods and their statistical properties have been presented in recent years (Kumar 2005; Ho 2014; Ho and Duchêne 2014; Kumar and Hedges 2016; dos Reis et al. 2016). Therefore, here we focus on a more pragmatic account of molecular dating methods, aimed at assisting researchers to select and utilize available methods.

12.1.1 Non-Bayesian Versus Bayesian Methods

Increased sophistication of molecular dating methods has often been accompanied by increased demand for computational time and memory. There exists a clear dichotomy of molecular dating methods based on their computational resource requirements for large data sets. Bayesian methods are computationally demanding because of their need for extensive sampling from the posterior distribution using the Markov chain Monte Carlo (MCMC) approach (Bromham et al. 2018). The computational burden is usually very high for large data sets and grows with the number of sequences (Crosby and Williams 2017; Tamura et al. 2018). In addition, problems in MCMC mixing can increase the computational time further (Bhatnagar et al. 2011). Sometimes, there is a need to run multiple Bayesian analyses to test different prior assumptions and calibration settings, which might result in the requirement for high-performance computing infrastructure.

In contrast, many non-Bayesian methods tend to have much smaller computational needs, while still allowing rates to vary throughout the tree. For example, both penalized likelihood (Sanderson 2002) and RelTime (Tamura et al. 2012, 2018) are very fast and known to be accurate. Although their computational requirements increase linearly with the number of sequences and sites, they still take orders of magnitude less time than the Bayesian methods (Fig. 12.1). Computational time demands of these non-Bayesian methods are essentially the same as the time taken to estimate branch lengths of a phylogeny, for example by using the maximum-likelihood method. Non-Bayesian methods can also be applied directly to a phylogeny with branch lengths (phylogram), which decreases the computational times further for very large data sets.

In this chapter, our focus is on providing practitioners with a guide to effectively using non-Bayesian methods for molecular dating. We also discuss the advantages and disadvantages of these methods, because the best approach depends on the size of the available data, degree of rate variation among species and loci, nature of clock calibrations, and the availability of computing resources. Table 12.1 shows a summary of different non-Bayesian methods, their statistical properties, and the software packages in which they are implemented.

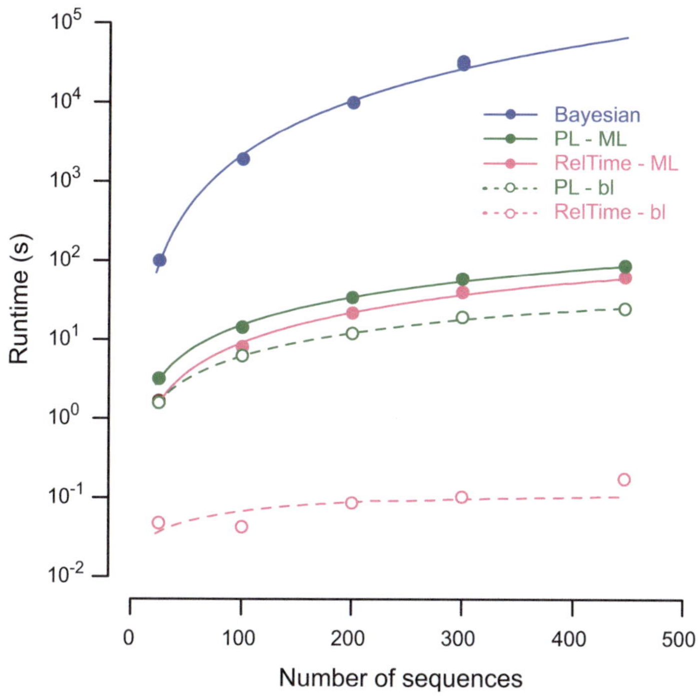

Fig. 12.1 Computational times required by Bayesian, penalized likelihood (PL), and RelTime methods to estimate divergence times for data sets containing an increasing number of sequences (*n*). Bayesian (blue solid line) is the computational time of the Bayesian method using molecular sequences as input. PL-bl (green dashed line) and RelTime-bl (pink dashed line) are the computational times of PL and RelTime methods using phylogenetic trees with branch lengths as input. PL-ML (green solid line) and RelTime-ML (pink solid line) are the total computational times required by PL and RelTime methods using molecular sequences as input, which are the sum of the computational time of maximum likelihood (ML) inferences of branch lengths and the computational time of PL-bl and RelTime-bl. ML inferences of branch lengths were conducted in MEGA X (Kumar et al. 2018). Bayesian, PL, and RelTime analyses were conducted in MCMCTree (Yang 2007), treePL (Smith and O'Meara 2012), and MEGA X, respectively. All times were estimated on a single-core computer by using an alignment of 4493 sites that was simulated with extensive rate variation (RR50 from Tamura et al. 2012). For this data set, the best-fit exponential equation is $0.06 \times n^{2.28}$, $0.08 \times n^{1.16}$, $0.07 \times n^{0.97}$, $0.03 \times n^{1.27}$, and $0.01 \times n^{0.44}$ for Bayesian, PL-ML, PL-bl, RelTime-ML, and RelTime-bl, respectively

12.2 A Practical Guide to Selecting Non-Bayesian Methods

12.2.1 Using Strict- and Local-Clock Methods

The simplest scenario for molecular dating is when the evolutionary rates are the same (or very similar) across different evolutionary lineages. In this case, methods that assume a strict clock will usually be reliable and produce the most precise time estimates. This assumption was commonly employed in the earliest molecular dating studies that produced many fundamental biological insights, including the finding that humans shared a most recent common ancestor with chimpanzees only five million years (Myr) ago, rather than 15–20 Myr ago based on the classification of *Homo* as a sister group to apes in the early 1960s (Sarich and Wilson 1967, 1973).

Interestingly, methods based on the strict clock continue to be developed and used today. For example, the mean path length (MPL) method (Britton et al. 2002), implemented in the software PATHd8 (Britton et al. 2007), has been used in many recent studies (e.g., Louca et al. 2018; Lu

Table 12.1 A summary of available efficient non-Bayesian dating methods

Software	Statistical basis[a]	Clock type[b]	Calibration type[c]	Confidence interval	References
Lintre	Regression	SC	F	Bootstrap	Takezaki et al. (1995)
PATHd8	MPL	SC	F, B	Bootstrap	Britton et al. (2007)
DAMBE	LS	SC, LC, RC	F, B, S	Bootstrap	Xia and Yang (2011), Xia (2018a)
r8s	LF, NPRS, PL	SC, LC, DC, RC	F, B, S	Bootstrap	Sanderson (1997, 2002, 2003)
treePL	PL	SC, RC	F, B	Likelihood	Smith and O'Meara (2012)
Ape—chronos & chronoMPL	PL, MPL	SC, DC, RC	F, B	Bootstrap	Paradis (2013)
MEGA X—RelTime, RTDT	RRF	SC, RC	F, B, D, R, S	Analytical	Kumar et al. (2018), Tamura et al. (2018), Tao et al. (2019), Miura et al. (2020)
TipDate	Regression	SC	S	Likelihood	Rambaut (2000)
TREBLE	UPGMA	SC	S	Bootstrap	Yang et al. (2007)
Physher	ML	SC, LC, DC	S	Bootstrap	Fourment and Holmes (2014)
LSD	LS	SC, RC	S	Bootstrap	To et al. (2016)
treedater	LS, ML	SC, RC	S	Bootstrap	Volz and Frost (2017)
TreeTime	ML	SC, RC	S	Likelihood	Sagulenko et al. (2018)

[a]*MPL* mean path length, *LS* least squares, *LF* Langley–Fitch method (Langley and Fitch 1974), *NPRS* nonparametric rate-smoothing, *PL* penalized likelihood, *ML* maximum likelihood, *RRF* relative-rate framework
[b]*SC* strict clock, *LC* local multi-rate clock, *DC* discrete multi-rate clock, *RC* relaxed clock
[c]*F* fixed node calibration, *B* node calibration boundary, *D* node calibration density, *R* substitution rate, *S* sampling tip date

et al. 2018). This method assumes that the ratio of ages between two nodes in a phylogeny is proportional to the ratio of their average node-to-tip distances. Therefore, it is only suitable for data sets in which the substitution rates are strictly or nearly constant among lineages throughout the phylogeny (Britton et al. 2002).

The problem of the equal-rates assumption is illustrated in the analysis of a phylogeny consisting of two clades (X and Y) with an outgroup (Fig. 12.2a). Each clade contains two orthologous DNA sequences of zinc-finger genes *zfx* and *zfy*; this is a simple gene-family tree with two genes that arose from a gene duplication prior to the divergence of human and mouse. Molecular dating methods should produce the same values for t_X and t_Y because they refer to the same evolutionary event: the divergence between human and mouse. Therefore, the expected ratio of t_X and t_Y is 1, which is what a molecular dating method should produce despite the fact that mouse *zfx* gene has evolved four times more quickly than the human *zfx*, and the mouse *zfy* gene has evolved seven times more quickly than the human *zfy*.

Analysis of this data set by the MPL approach in the PATHd8 software produced a $t_X/t_Y = 0.43$, which is much smaller than 1. It estimated that the divergence between human and mouse in clade Y (t_Y) happened much earlier than the same event in clade X (t_X) (Fig. 12.2b). This result is clearly inconsistent with the phylogenetic tree in Fig. 12.2a and shows that strict-clock methods produce biologically incorrect results if they are used for data sets in which evolutionary rates vary extensively among lineages. Smith and O'Meara (2012) have also reported that PATHd8 was not so reliable in analyses of empirical data sets and simulated data sets when evolutionary rates varied. Another least-squares method (Xia and Yang 2011) minimizes the residual sum of squares of patristic distance and distance computed by the rate and time under the global clock. This method, implemented in the DAMBE software (Xia 2018a), also produced an incorrect date ratio of 0.37 (Fig. 12.2c).

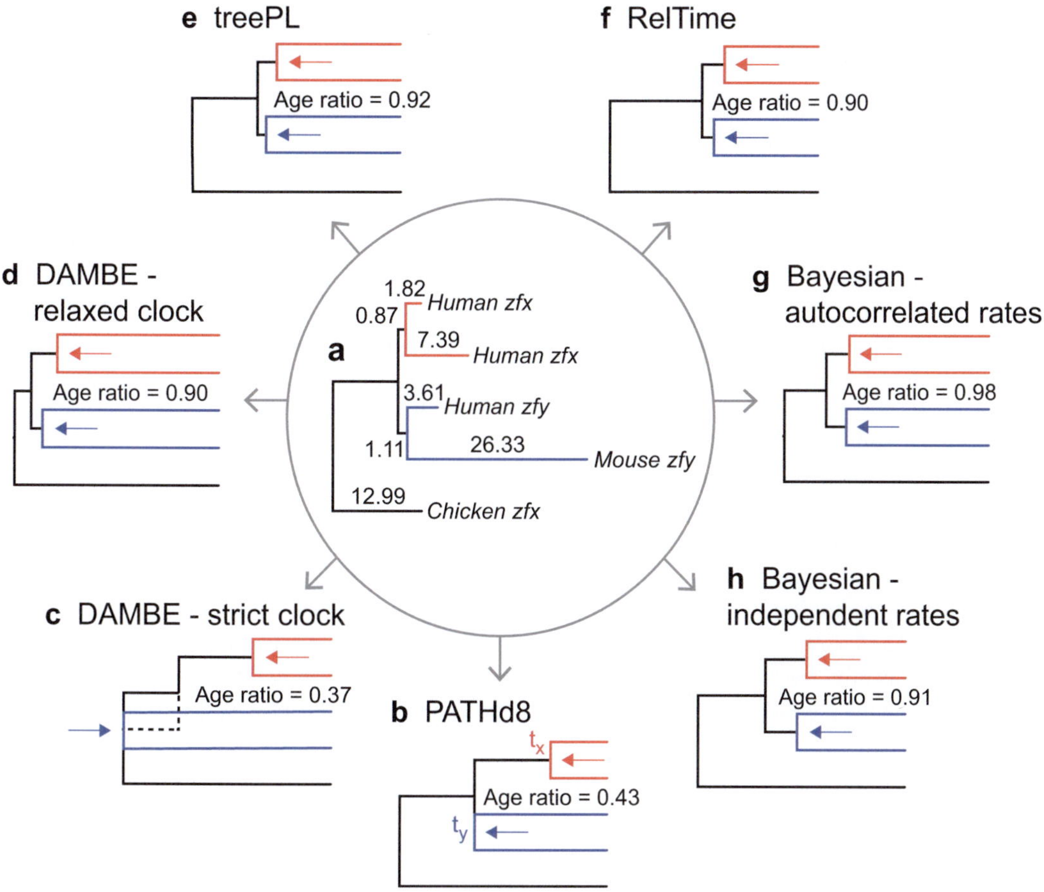

Fig. 12.2 Molecular dating analysis of four DNA sequences. (**a**) An example phylogeny of orthologous DNA sequences of two zinc-finger genes (GenBank accession numbers gi296010876, gi113205066, gi223890138, gi156938288, and gi363728820). The branch lengths are shown in substitutions per 100 base pairs. This is an excellent test case because the expected time for human–mouse species divergence based on gene *zfx* (t_X, red clade) and *zfy* (t_Y, blue clade) should be the same ($t_X/t_Y = 1$), as the gene duplication event occurred prior to the diversification of mammals. Shown are the time-trees produced by (**b**) PATHd8, (**c**) DAMBE with strict clock, (**d**) DAMBE with relaxed clock, (**e**) treePL, (**f**) RelTime, (**g**) MCMCTree (Bayesian) with the autocorrelated branch-rates model, and (**h**) MCMCTree (Bayesian) with the independent branch-rates model. Ratios of node ages for human–mouse divergence based on *zfx* (t_X, red arrow) and *zfy* (t_Y, blue arrow) genes of all resulting time-trees are labelled. One root calibration was used in PATHd8, DAMBE, treePL, and Bayesian analyses. No calibrations were used in the RelTime analysis

Therefore, the use of strict-clock methods is appropriate only if lineages have evolved with a strictly or nearly constant rate. The simplest way to ensure that this condition is valid is to conduct a molecular clock test. An early molecular clock test was proposed by Fitch (1976) for data sets containing two sequences and an outgroup, and was followed by many others (Wu and Li 1985; Muse and Weir 1992; Tajima 1993). For larger data sets, equality of rates on multiple lineages can be evaluated by least squares (Takezaki et al. 1995) and by likelihood-ratio tests (Nei and Kumar 2000). Software packages such as MEGA X (Kumar et al. 2018), LinTre (Takezaki et al. 1995), PAML (Yang 2007), and DAMBE can be used for testing the molecular clock. For example, the difference in log likelihoods with and without assuming the strict clock was

207.33 in PAML for the example data in Fig. 12.2a. The likelihood-ratio test rejects the molecular clock ($P < 10^{-80}$, degrees of freedom = 3) for this data set.

In fact, we expect the hypothesis of the strict molecular clock to be readily rejected for most contemporary data sets, which often consist of many genes and/or genomic segments from many species. Therefore, a practitioner usually needs to use dating methods that do not assume a strict clock. They might choose to apply local clocks that allow different rates in different clades (subtrees) in the phylogeny (Hasegawa et al. 1989; Yoder and Yang 2000). In local-clock methods, a strict clock is assumed to exist within each clade, so one needs to specify clades that show rate homogeneity (clocklike evolution). This is not straightforward to accomplish unless there are clear biological reasons for defining such clades (Sanderson 2002; Ho and Duchêne 2014). Consequently, methods that allow rates to vary throughout the phylogeny are more practical in analyses of real data.

12.2.2 Relaxing the Strict Clock

Relaxed-clock methods allow molecular dating when evolutionary rates vary throughout the tree. We focus on rapid non-Bayesian approaches, as Bayesian approaches have been discussed extensively elsewhere (dos Reis et al. 2016; Nascimento et al. 2017) and in Chaps. 6 and 13. Among the non-Bayesian approaches, penalized likelihood and RelTime are often used. Penalized likelihood estimates divergence times under the statistical criterion that minimizes the squared differences between ancestral and descendent branch rates (Sanderson 1997, 2002). That is, large rate changes are penalized, which is biologically intuitive because an ancestor and its direct descendants are likely to share similar genomic properties, biological attributes, and living environments, and thus will tend to have more similar mutation rates (Gillespie 1984). This property would result in autocorrelation in branch rates (Thorne et al. 1998; Kishino et al. 2001) (Table 12.2).

The penalized-likelihood approach uses a penalty parameter (λ) for penalizing rate changes (Sanderson 2002). A large penalty will favour a strict-clock model, because it will tend to assign very similar rates to ancestor–descendant pairs. Small values of λ will allow rates to vary throughout the tree and will relax the molecular clock. The optimal value of λ depends on the data set being analysed and can be determined by a cross-validation procedure (Sanderson 2002). In this procedure, one terminal branch is removed from the tree at a time, so its immediate ancestral node and other branches are left in place. The rate and node age of the immediate ancestral node is estimated using the remaining branches for a given λ. The optimal value of λ is that which minimizes the difference between the observed substitutions on the ancestral branch and the number of inferred substitutions, which is calculated using the estimated rate and node age. This rate-smoothing approach is effective when applied to the example data in Fig. 12.2a. Penalized likelihood produced an estimate of $t_X/t_Y = 0.92$ (Fig. 12.2e), which is much closer to 1 than that from methods based on a strict clock. The original penalized-likelihood method was implemented in the r8s software (Sanderson 2003) and a faster version is implemented in the treePL software (Smith and O'Meara 2012) and in the R package APE (Paradis 2013). The penalized-likelihood method has also been adopted by Xia and Yang (2011) in their strict-clock method to relax the clock through rate smoothing (Xia 2018a). It produced a time ratio of 0.90 when applied to the example data (Fig. 12.2d).

The RelTime approach is another relaxed-clock method that minimizes differences between the evolutionary rates of ancestral and descendent lineages (Tamura et al. 2012, 2018). An evolutionary lineage consists of a branch and the descendent clade (including all of the taxa and branches). For example, lineage *a* contains three branches in Fig. 12.3, so the length of lineage *a* (L_a) is based on b_1, b_2, and b_5. Tamura et al. (2018) presented a mathematical formulation that produces relative lineage rates purely from the branch lengths in a phylogeny. In their algebraic relative-rate framework, the difference between

Table 12.2 Differences between Bayesian dating methods, penalized likelihood, and RelTime

	Bayesian	Penalized likelihood	RelTime
Framework	Bayesian statistics	Penalized likelihood	Algebra
Rate prior	Independent or autocorrelated rates and probability distributions	Autocorrelated rates and a penalty parameter	Not needed
Tree prior	Birth-death or coalescent process	Not needed	Not needed
Estimate	Node times and branch rates	Node times and branch rates	Node times and lineage rates
Uncertainty	Credibility intervals	Confidence intervals	Confidence intervals
Consider site sampling error	Yes	Yes	Yes
Consider rate variation	Yes	No	Yes
Consider calibrations	Yes; allow the use of boundaries and densities	Yes; allow the use of boundaries	Yes; allow the use of boundaries and densities

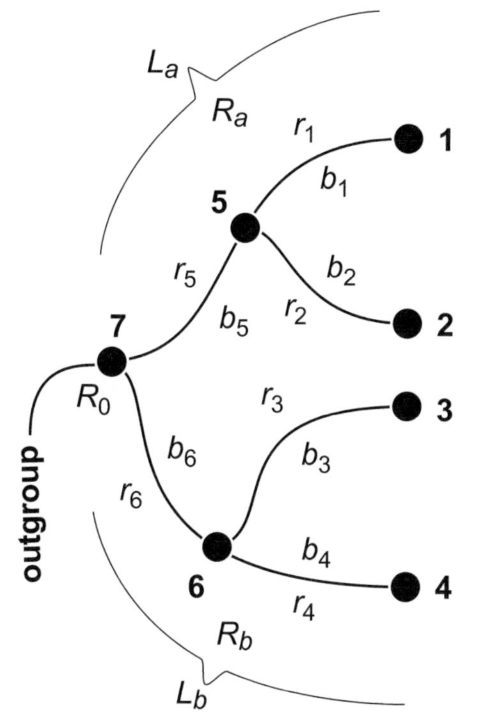

Fig. 12.3 An example phylogeny showing branch lengths (b), branch rates (r), lineage lengths (L), and lineage rates (R). R_a is the rate of the lineage L_a that consists of branches with lengths of b_5, b_1, and b_2, and R_b is the rate of the lineage L_b that consists of branches with lengths of b_6, b_3, and b_4. Lineage rates R_1 to R_4 are the same as branch rates r_1 to r_4, so they are not shown. Relative lineage rates can be computed in MEGA X from branch lengths using Eqs. (6–9, 19–24) for arithmetic means or (28–31, 34–39) for geometric means in Tamura et al. (2018)

rates in ancestral and descendent lineages is minimized and the observed difference in evolutionary rates between sister lineages is accommodated.

The use of lineage rates, rather than the branch rates, is a major difference between RelTime and other relaxed-clock methods (Table 12.2). For example, Bayesian methods use a statistical distribution (e.g., lognormal) as a prior to account for the variation in branch rates across a phylogeny, and the penalized-likelihood method smooths the rate change between ancestral and descendent branch rates using a global penalty function. If we consider node 7 in Fig. 12.3, penalized-likelihood computation will attempt to minimize the difference between branch rates r_5 and r_1 and for other pairs globally. In contrast, RelTime will minimize the difference between lineage rates R_a and R_1 and other pairs individually. Therefore, RelTime does not need to use any penalty functions or distributional priors, which makes it different from penalized-likelihood and Bayesian methods. In the example four-taxon data, RelTime produces a t_X/t_Y ratio of 0.9, which is close to 1.0 (Fig. 12.2f). The RelTime method is available in the MEGA X software. Mello (2018) has provided a detailed protocol for estimating time-trees with RelTime in MEGA X.

Overall, we find that the time ratios produced by non-Bayesian relaxed-clock methods (Fig. 12.2d–f) are similar to the estimate generated by the Bayesian approach when an independent branch-rate (IBR) model was used (0.91, Fig. 12.2h). The use of an autocorrelated branch-rate (ABR) model produced a time ratio of 0.98, an estimate that is very close to 1 (Fig. 12.2g). The ABR model assumes that the branch-specific molecular rates are autocorrelated, such that closely related branches share similar rates and distantly related branches have more different rates (Thorne et al. 1998; Kishino et al. 2001; Ho and Duchêne 2014). The IBR model assumes molecular rates are independent among branches, such that rates on closely related branches do not need to be similar (Drummond et al. 2006; Ho and Duchêne 2014). Results from Bayesian analyses suggest that the ABR model might fit these data better. In fact, Tao et al. (2019) reported the autocorrelation of branch rates to be the dominant pattern in molecular phylogenies for diverse groups of species in an analysis of DNA and amino acid sequences. Therefore, the assumption of autocorrelation is likely to be valid for this example data set.

12.2.3 Performance of Non-Bayesian Relaxed-Clock Methods

Non-Bayesian relaxed-clock methods have been tested extensively in computer simulations with large data sets. Smith and O'Meara (2012) conducted computer simulations under the ABR model on large phylogenies (100–10,000 species) and reported that the penalized-likelihood method can achieve high accuracy in estimating divergence times (see Fig. 1 in Smith and O'Meara 2012). However, they did not test the performance of penalized likelihood using IBR data sets and did not evaluate the accuracy of divergence-time estimates node-by-node; their investigations conducted using the treePL and r8s software were rather limited in scope and depth. In contrast, RelTime has been extensively tested and has a well-justified mathematical foundation (Tamura et al. 2018).

Tamura et al. (2012) conducted extensive simulations under ABR and IBR scenarios on a master time-tree of 446 taxa. In all scenarios, RelTime produced estimates of node ages that were close to the true values (Fig. 12.4a–c; also see Figs. 3 and 5 in Tamura et al. 2012). RelTime estimates were similar to those from the Bayesian method in the IBR case where rate variation was low (Fig. 12.4a). However, the Bayesian method tended to overestimate divergence times (median deviation = 19%) when the rate variation in IBR was larger (Fig. 12.4b). This pattern might relate to the need to specify a single model of branch rates in Bayesian methods. When the specified rate model is not the correct model for the observed rate variation, biased time estimates might be produced. Model averaging can potentially reduce this bias in Bayesian analysis (Li and Drummond 2012). In contrast, RelTime does not need to model branch rates and it performed much better in this case (Fig. 12.4b, median deviation = −5%). RelTime also performed better (median deviation = −2%) than the Bayesian method (median deviation = 14%) for the ABR data sets (Fig. 12.4c). Mello et al. (2021) also reported RelTime to perform as well as Bayesian methods for dating phylogenies that encompass both species and population divergences using simulated data sets.

Apart from the simulation tests, Chernikova et al. (2011) and Gunter et al. (2016) reported that penalized-likelihood methods produced results consistent with those from Bayesian analyses for some data sets. Mello et al. (2017) and Battistuzzi et al. (2018) have also examined many empirical data sets from different groups across the tree of life and found that RelTime produced time estimates that were very similar to those from Bayesian methods, as long as the equivalent calibration boundaries were used. Tao et al. (2020) developed a method for utilizing calibration densities in RelTime and found that RelTime produced not only time estimates but also the surrounding uncertainties that were comparable to those from Bayesian methods in empirical data analyses.

In fact, some studies have found that non-Bayesian methods performed better than Bayesian methods when some priors (e.g.,

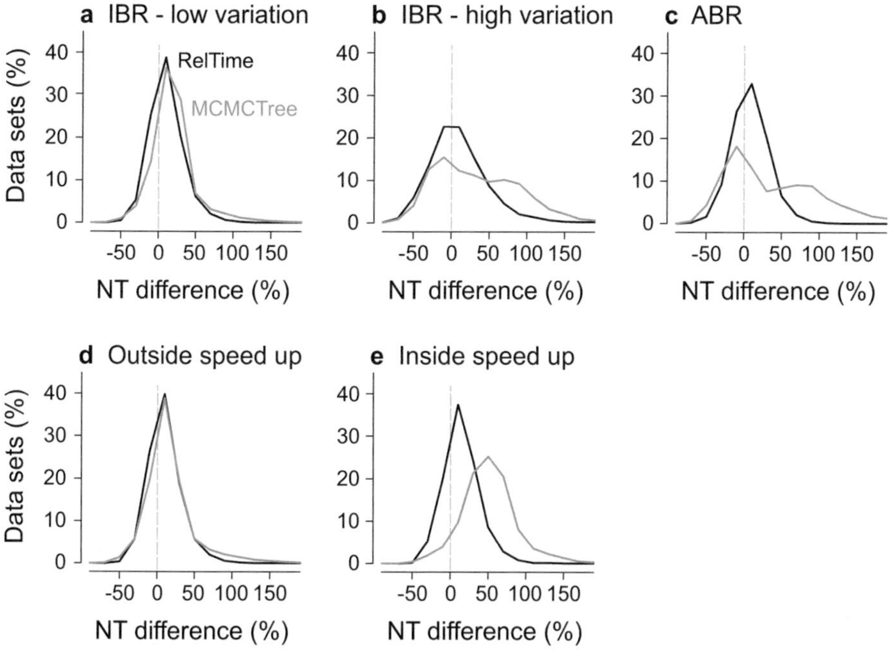

Fig. 12.4 Distributions of the normalized differences between true node times (NT) and estimated times obtained from RelTime and MCMCTree for internal nodes. Comparisons of the performance of RelTime (black curve) and MCMCTree (grey curve) for data sets simulated under (**a**) independent branch-rates (IBR) model with low variation, (**b**) IBR model with high variation, and (**c**) autocorrelated branch-rates (ABR) model. Comparisons of the performance of RelTime (black curve) and MCMCTree (grey curve) for estimating node times (**d**) outside the speed-up clades and (**e**) inside the speed-up clades. Data and results are from Tamura et al. (2012). Dashed grey line indicates the 0% difference in NT

branch-rate model) were incorrectly specified (Tamura et al. 2012, 2018). For example, Tamura et al. (2012) did a simulation test of a clade-specific speed-up, where a random clade of at least 50 taxa was selected to undergo a rate increase while the rest of the branches remained at their original rates simulated under the IBR model. This meant that two different IBR models were applied to the same phylogeny, where one clade had a higher mean rate and the other clade evolved more slowly. The Bayesian method yielded accurate estimates in one clade, but biased estimates in the other clade (Fig. 12.4d–e, also see Fig. 5 in Tamura et al. 2012). This occurred because the single model of branch-rate variation was unable to account for the heterogeneity associated with multiple contrasting clade-specific rate variations. However, RelTime performed well and generated accurate time estimates for both clades (Fig. 12.4d–e), because RelTime does not require the specification of a branch-rate model.

Therefore, the high computational speeds afforded by some of the non-Bayesian dating methods do not come at the expense of accuracy. In fact, whenever possible, it is prudent to analyse data by using methods based on different statistical frameworks to obtain reliable estimates and to assess the potential biases introduced by the assumptions and methods (see Sect. 12.9). However, efficient non-Bayesian methods might be the only feasible option for many users for analysing large data sets containing thousands of genes and species (e.g., Li et al. 2019).

12.2.4 Eliminating Rate Variability Before Molecular Dating

Before proceeding further, let us consider approaches to reduce or eliminate rate variation in data sets containing multiple genes or genomic segments, before applying clock methods. This is important because high rate variation is a key contributor to the uncertainty in time estimates (Zhu et al. 2015; Kumar and Hedges 2016). By reducing the degree of molecular rate variation in a phylogeny, both the accuracy and precision of time estimates might be improved.

We can eliminate (or reduce) evolutionary rate variation by excluding species that have evolved significantly more quickly or slowly than the rest in a sequence alignment, or by excluding genes that fail the molecular clock test. For data sets that contain large numbers of genes and genomic segments, this is a viable option for dating species divergences (Hedges et al. 1996; Smith et al. 2018). In the 1990s, Hedges et al. (1996) and Kumar and Hedges (1998) applied this strategy to date major mammalian and vertebrate divergences, respectively, because relaxed-clock methods were not available at that time. In those early multigene studies, genes and species failing the molecular clock test of Tajima (1993) were removed before divergences were dated using a strict clock. These analyses revealed that major orders of placental mammals and of birds were likely to have originated prior to the K-Pg extinction (Hedges et al. 1996), which challenged the hypothesis of adaptive radiation and founded a very active area of biological research (Kumar and Hedges 1998; Eizirik et al. 2001; dos Reis et al. 2014; Phillips 2015; Prum et al. 2015). Takezaki et al. (1995) presented a statistical approach to detect lineages that evolved at rates that were significantly different from the phylogeny-wide average. Using such gene- and species-elimination approaches, evolutionary timescales were assembled from many large data sets, including those for Hawaiian drosophilids (Russo et al. 1995), diatoms (Kooistra and Medlin 1996), metazoans (Wray et al. 1996), and major eukaryote lineages (Doolittle et al. 1996; Feng et al. 1997).

Smith et al. (2018) proposed a 'gene shopping' approach that extended the original practice of Hedges et al. (1996) to genes that passed the molecular clock test in large phylogenies. Their strategy also requires that the selected genes have a sufficient number of informative sites and that selected gene trees are highly concordant with the species tree. They reported that the application of strict-clock or relaxed-clock methods on the selected clocklike genes improved the precision of time estimates by more than 50%, as the 95% highest posterior density (HPD) intervals became much narrower. The higher precision is achieved by reducing the rate heterogeneity in the phylogeny, which is a key contributor to wide 95% HPD intervals. Higher precision of estimates enables more powerful tests of biological hypotheses and helps to establish evolutionary and ecological patterns more reliably.

Even after 'gene shopping', it is possible that some intrinsic directional rate variation remains in the data set because molecular clock tests are not so powerful when sequences are short or the evolutionary rate is low. This can be remedied by applying a more stringent clock test to exclude genes and species showing even small rate differences (Kumar and Hedges 1998; Hedges and Kumar 2003; Hedges and Shah 2003). We also propose that one should do 'species shopping' to remove species that show evolutionary rates significantly different from others before conducting molecular clock dating, to further reduce rate variation and the uncertainty in time estimates (Takezaki et al. 1995; Hedges et al. 1996). In our view, whenever feasible, a combination of gene shopping and species shopping with relaxed-clock methods is the best strategy when many genes and species are available for estimating divergence times.

12.3 Utility of Relative Divergence Times

All of the non-Bayesian methods can generate relative times directly from a phylogeny in which branch lengths are either provided by the user or inferred from the sequence data using a model of nucleotide or amino acid substitution. The ability to produce relative node ages and rates without using any branch-rate model, speciation model, and even calibration priors can have many benefits (Tamura et al. 2012). First, the relative node ages obtained without any calibrations can be used to identify the calibration constraints or densities that would be expected to have a notable impact on the final time estimation (Marshall 2008). This is because the relative and absolute ('calibrated') node ages should be linearly related when calibration constraints and/or densities do not conflict with the signal from molecular data (Battistuzzi et al. 2015).

Second, the estimates of relative rates can be directly used to identify lineages with significantly lower or higher evolutionary rates, because the standard errors of the relative-rate estimates are available. Those lineages are potentially very interesting because they might indicate the presence of strong selective pressure and other biological factors (Chikina et al. 2016). In addition, the relative rates computed from branch lengths only, without knowing node times, provide insights into evolutionary patterns between the ingroup and outgroup sequences. If the distributions of lineage rates are significantly different, the assumption of the same pattern of rate variation between the ingroup and outgroup taxa might need to be reconsidered.

Third, the relative lineage rates estimated by RelTime can be used for generating new tests of biological hypotheses and for model selection. Tao et al. (2019) used these lineage rates and the machine-learning framework to develop a new statistical test (called CorrTest) that can distinguish between IBR and ABR models, which has been challenging previously (Paradis 2013; Ho et al. 2015a). CorrTest performed better than other methods in detecting the presence of rate autocorrelation in a simulation analysis.

Fourth, the relative divergence times might be useful for detecting clades that have undergone a shift in the rate of diversification, which might indicate the effect of a geological event or the appearance of an ecological niche. Therefore, the knowledge of relative times and rates is useful for discovering exciting biological patterns, developing new methods, and examining the impact of fossil constraints or other prior settings.

12.4 Inferring Absolute Divergence Times

12.4.1 Dating with a Fixed Global Evolutionary Rate

A substantial proportion (12%) of molecular clock studies have been found to use a fixed substitution rate to calibrate the molecular clock (Hipsley and Müller 2014). This is the only choice in cases where no node calibrations are available. An average evolutionary rate from another species group is used to date the divergences in the species group of interest. The estimation of node times is simple in this case: a fixed evolutionary rate is used to convert node heights (in substitutions per site) into divergence times. One just needs to divide all the node heights (in substitutions per site) by the fixed rate of evolution (in substitutions per site per time unit, such as years or million years). Some dating programs (e.g., MEGA X) provide such an option. The use of a fixed rate is only reasonable if there is a good reason to believe that the average evolutionary rates and the biological markers are the same between the species group from which the calibration rate has been derived and the species group to which it is being applied (Wilke et al. 2009). Also, the reliability of the fixed substitution rate depends on the calibrations used in the study from which the rate is obtained (Ho and Phillips 2009).

12.4.2 Dating with a Fixed Node Calibration

A better approach is to derive the clock calibration by using a known divergence time for a node in a phylogeny and then to scale all other node ages in this phylogeny based on this clock calibration. This approach does not require one to assume a molecular clock, because rapid relaxed-clock methods can deal with rate differences among branches and lineages to generate an ultrametric tree. The clock calibration is the relative node height divided by fixed time in the ultrametric tree, and then this calibration sets the scale to convert relative times into absolute times. In MEGA X and other programs (Sanderson 2003; Britton et al. 2007; Smith and O'Meara 2012; Xia 2018a), this can be easily done by assigning a fixed time to a node, which converts all other node heights into times.

For analyses with fixed node calibrations, calibration times can come from biogeography or from ecological/environmental considerations. In fact, a literature survey of molecular dating studies has shown that 15% used times derived from geological events that were associated with geophysical isolation or the appearance of new habitats (Hipsley and Müller 2014). These calibrations can be derived from vicariance, geodispersal, or biological dispersal (Ho et al. 2015b). The geological event is a good source of calibration especially for the species that were directly affected by that event (see Chap. 9). However, it is not appropriate to use those calibrations if the research goal is to test the impact of those geological events (Kodandaramaiah 2011). Similarly, one can use the fossil record to obtain an estimate of a single divergence time in the tree, which is then used to calibrate the clock. Many early studies used a single calibration point because gene-specific alignments generally contained only a few species (e.g., Hedges et al. 1996).

12.4.3 Dating with Multiple Node Calibrations

The most common approach to calibrate a molecular clock is to use many dates derived from the fossil record (Hipsley and Müller 2014). As expected, this practice is particularly common for fossil-rich groups in the tree of life (Ksepka et al. 2015). In fact, studies have been using increasingly large numbers of calibrations, with some contemporary analyses incorporating many tens of calibrations (e.g., Meredith et al. 2011; dos Reis et al. 2015; Barba-Montoya et al. 2018; Morris et al. 2018).

12.4.3.1 Using Multiple Fixed Calibrations or Calibration Constraints

Efficient non-Bayesian relaxed-clock methods allow the use of multiple point calibrations. For example, RelTime uses a linear regression between the relative node heights in the ultrametric tree and all of the user-supplied fixed calibration points. The resulting scaling factor (f) then converts all of the relative times into absolute divergence times. In practice, however, fossil dates do not correspond directly to actual species divergence times, so they are rarely used as fixed calibration points. Instead, the earliest fossil record usually provides a reliable minimum age constraint on a node in the phylogeny (Hedges and Kumar 2004). In some cases, it is possible to place a maximum age constraint, but these are usually difficult to determine (Marshall 2008; Ho and Duchêne 2014; Bromham et al. 2018; Hedges et al. 2018). In practice, despite these difficulties, many researchers prefer to impose both minimum and maximum constraints on multiple nodes in the phylogeny.

RelTime can use all types of constraints in calibrating the molecular clock. It generates a global time factor (f) that produces time estimates that best satisfy the calibration constraints. If there is a range of f values that do not violate the calibration constraints, then the midpoint of that range becomes the estimate of f. When one or more of the absolute times fall outside the calibration constraints, then f is set so that the deviation from the calibration constraints is minimized. After that, times for calibrated nodes are adjusted to ensure that the calibration constraints are fully respected, such that the estimated times for any offending nodes are between the minimum and maximum constraint times specified by the user. This requires altering local evolutionary rates, which prompts re-optimization of all other node times in the tree recursively in the RelTime algorithm (Tamura et al. 2013; Tao et al. 2020).

The penalized-likelihood method adds age constraints in the optimization of the penalty functions of rate smoothing, to ensure that the absolute times are within the calibration constraints imposed by the researcher (see the documentation for the r8s software). PATHd8 also smooths rates to resolve the conflicts between estimated ages and calibrations. However, PATHd8 requires the specification of at least one fixed node age as the anchor calibration, which is used to scale relative dates to absolute dates as the first step. Then, the method smooths rates of sister lineages to fit all calibration constraints (Britton et al. 2007). Also, because PATHd8 is fundamentally a strict-clock method, it has limited power in smoothing the rates compared with the relaxed-clock methods (e.g., penalized likelihood and RelTime). The least-squares-based method (DAMBE) utilizes the calibration bounds during the minimization of the residual sum of squares (RSS) of patristic distances and pairwise distances computed based on the evolutionary rate and predicted divergence times (Xia and Yang 2011). In this case, times used to compute the distance are controlled by the calibration constraints imposed in the RSS minimization. To minimize the RSS, the resulting times will be equal to the maximum or minimum bounds in some cases (Xia and Yang 2011).

12.4.3.2 Using Calibration Constraints with Probability Densities

In addition to minimum and/or maximum constraints, it is becoming commonplace to use probability densities that reflect prior belief about the possible location of the true species divergence time relative to the minimum and/or maximum constraints. Early on, Hedges and Kumar (2004) mentioned several possible distributions (triangular, lognormal, and uniform densities) to model such calibration uncertainty. However, they preferred a uniform distribution for their studies due to a lack of additional information about the true density (Meredith et al. 2011; Morris et al. 2018). With the development of Bayesian methods, it became possible to incorporate any desired probability density in molecular dating (Drummond et al. 2006; Yang and Rannala 2006; Barba-Montoya et al. 2017). Indeed, more recent studies use nonuniform distributions (e.g., Cauchy, lognormal, and exponential distributions) in which a stronger constraint is placed on the minimum time. As expected, the quality of the calibrations and the density assumptions have a major impact on divergence-time estimates in Bayesian analyses, even if a huge amount of molecular data is available (Barba-Montoya et al. 2017; Bromham et al. 2018).

Tao et al. (2020) have developed an approach to incorporate such densities and automatically accommodate the interactions among calibrations in the RelTime method. The new approach resamples calibration constraints from densities many times, to generate a distribution of times for each calibrated node that is analogous to the 'effective prior' in Bayesian approaches, and then derives minimum and maximum bounds (called effective bounds) for use in the RelTime analysis to estimate divergence times and confidence intervals. Confidence intervals produced by this approach overlapped with those reported by the Bayesian analyses and were much narrower than those generated by using the original approach that did not account for interactions among calibrations in RelTime (Tao et al. 2020). The new approach is available in MEGA X for the RelTime method. These effective bounds can also be used in penalized likelihood and other non-Bayesian dating analyses.

12.4.3.3 Using Molecular Dates as Calibrations (Secondary Calibrations)

Many studies use previously published molecular dates to calibrate the clock. These are referred to as secondary calibrations because they are not based on direct fossil or biogeographical data, but rather on inferred molecular dates. A literature survey found that about 15% of studies have used secondary calibrations (Hipsley and Müller 2014). The use of secondary calibrations traces its origins to Kumar and Hedges (1998). They estimated vertebrate divergence times using a secondary mammalian calibration, which was inferred by using the bird-mammal divergence time from the fossil record. This procedure was needed for inferring intra- and interordinal dates using protein sequence alignments that lacked bird sequences. This approach enabled them to increase the number of genes that could be used to infer divergence times. In fact, Hedges and Kumar (2004) suggested that, in some situations, more accurate time estimates might be obtained by using a secondary calibration from a robust source than by using an unreliable primary fossil calibration.

Secondary calibrations continue to be used for groups that have limited fossil records, such as bacteria (Chriki-Adeeb and Chriki 2016) and fungi (Heckman et al. 2001). They have also been used in several recent studies to increase the total number of available calibrations for dating large phylogenies that contain hundreds of species (e.g., dos Reis et al. 2012, 2018). Ultimately, one must use secondary calibrations judiciously, because this practice might produce results significantly different from those produced by using primary calibrations (Graur and Martin 2004; Sauquet et al. 2012; Schenk 2016). However, Hedges and Kumar (2004) found that the inconsistencies in times estimated using the primary and secondary calibrations reported by Graur and Martin (2004) were caused by an incorrect assumption of Gaussian distribution of multigene times and, thus, an incorrect calculation of the time and confidence intervals. The actual distribution should be very skewed because of the small sample size and a large extrapolation. In fact, Morrison (2008) suggested that a lognormal distribution is the most appropriate to model a secondary calibration. The time estimated using the secondary calibration was consistent with the primary time when a skewed distribution was assumed (Hedges and Kumar 2004). Clearly, further research is needed to inform best practices for using secondary calibrations.

12.5 Molecular Dating with Missing Sequence Information

Modern studies often involve large data sets with hundreds of species and genes, due to the growth of public databases and dramatically decreased sequencing costs. However, a disadvantage of building and using such big data sets is that they might contain a large proportion of missing data. For example, the alignment analysed by Barba-Montoya et al. (2018) had 71.4% missing data. Fortunately, both empirical and simulation studies have found that missing data had little impact on divergence-time estimation by both Bayesian and non-Bayesian dating methods, especially when multiple calibrations were used (Douzery et al. 2004; Filipski et al. 2014; Zheng and Wiens 2015). These results indicate that molecular time estimation is robust even when sequences are missing from the majority of genes for most of the species. However, if the data are highly or systematically sparse, resulting in pairs of species with no common genes, then divergence-time estimation can be seriously misled, especially when only a few or no calibrations are used (Filipski et al. 2014; Zheng and Wiens 2015).

Filipski et al. (2014) showed that time estimates for nodes with zero data coverage (i.e., nodes without any common genes for any pair of species in the immediate descendent clades) were unreliable because there were no data to allow the corresponding branch lengths to be estimated. In general, the accuracy of branch-length estimates is low when the overall number of informative characters is small, which would result in poor time estimates (Wiens and Moen 2008; Wiens and Morrill 2011). Limited numbers of informative sites in sequence alignments can reduce the accuracy and precision of time estimates and, thus, lead to spurious changes in diversification rates (Marin and Hedges 2018) and mislead statistical tests of evolutionary rate correlation (Tao et al. 2019). Therefore, it is important to detect nodes with low or zero data coverage before any dating analysis.

One can use MEGA X to visualize data coverage for each node in a phylogeny (Fig. 12.5). The data coverage for each node in the phylogeny is

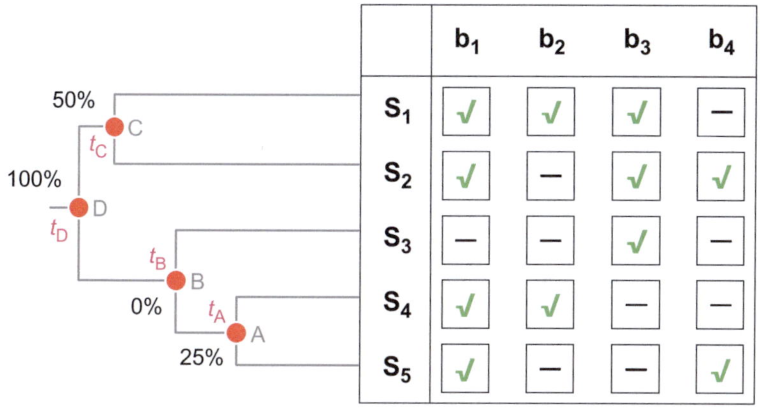

Fig. 12.5 An example of computing node data coverage for a phylogeny containing five species (S) and four nucleotide bases or amino acid residues (b) in the alignment matrix. Node times are given by t_i. Not all bases are available for each species. The available states are designated by check marks and missing ones are indicated by dashes in the matrix. The percentage of data coverage of each internal node is shown

the percentage of positions at which at least one pair of sequences in the descendent clades has a valid nucleotide base or amino acid residue. For example, node A has a data coverage of 25% because only one out of four sites has a valid state between sequences S_4 and S_5. Node B has a data coverage of 0%, because S_3 does not share any positions with a valid state in either S_4 or S_5 (Fig. 12.5). When the data coverage is zero (or low), there is no (or limited) ingroup information to allow the estimation of branch lengths (branch lengths = 0), and RelTime will predict that no time has elapsed on that branch. This results in the age of node B (t_B) becoming the same as the age of node D (t_D) (Filipski et al. 2014). Therefore, dates for nodes with high data coverage are expected to be estimated with higher accuracy.

12.6 Estimation of Confidence Intervals

In Bayesian methods, the credibility intervals or HPD intervals of node ages can be derived from the posterior distributions of times. Although the Bayesian credibility intervals and HPD intervals are not the same as the traditional analytical confidence intervals used in frequentist statistics (Jaynes and Kempthorne 1976), many researchers interpret them in a similar way. However, for non-Bayesian methods, the calculation of confidence intervals is complex. This is because it is difficult to generate analytical equations to account for the variance in node times introduced by the stochastic error in branch-length estimation, the rate heterogeneity among branches, and the uncertainty of calibrations. Therefore, many non-Bayesian methods (e.g., penalized likelihood) compute confidence intervals for divergence times by using the bootstrap approach, in which only sites or genes of molecular sequences are resampled. This leads to overly narrow confidence intervals because the site-bootstrapping approach only captures errors associated with the estimation of branch lengths in the tree. It cannot account for the variance introduced by evolutionary rate differences among lineages, which can have a big impact on the precision of time estimation (Kumar and Hedges 2016) (Table 12.2).

Tamura et al. (2013) suggested a method to generate confidence intervals encompassing the error due to branch-length estimation and rate variation for the RelTime method. Tao et al. (2020) improved this method and presented the analytical equations to compute confidence intervals for RelTime reliably, which is available in MEGA X. Simulation analyses showed that RelTime performed better than Bayesian methods and produced confidence intervals with high probabilities of containing the true values ($\geq$94%) for both small and large data sets when a minimum number of calibrations was used.

The uncertainty in calibrations is also an important source of estimation error in the inference of divergence times. Therefore, reliable and well-constrained calibrations can be very effective in reducing the widths of confidence intervals. Bayesian methods use different probability densities to accommodate the uncertainty in calibrations and to account automatically for the interaction among calibrations. Tao et al. (2020) have developed a new method for use in the RelTime framework to derive calibration boundaries from probability densities that account for their interactions (mentioned above). The resulting confidence intervals are comparable to the HPD intervals generated from Bayesian methods in empirical analyses (Tao et al. 2020). This method, with modifications, can also be used for other non-Bayesian methods (e.g., penalized likelihood).

12.7 Dating with Non-contemporaneous Molecular Data

In some studies, molecular sequences are obtained from biological samples that have been acquired at different times. This is common in the analysis of DNA and protein sequences from fast-evolving pathogens and those generated from ancient samples (Rambaut 2000; Stadler and Yang 2013; Biek et al. 2015). This makes the tips of the evolutionary tree asynchronous. Several rapid dating methods have been developed for this type of sequence data (see also Chap. 10). As with the evolution of methods for dating analyses of contemporaneous data, the first approaches to be developed were based on a strict clock. In the single-rate dated tips (SRDT) method, the slope of a linear regression between the root-to-tip distances (or pairwise distances from the outgroup sequence) and the sampling dates is used to determine the global rate and the dates for the internal nodes (Li et al. 1988; Bollyky and Holmes 1999; Rambaut 2000). SRDT is a very fast method and has been implemented in the TipDate software (Rambaut 2000). Some UPGMA-like methods, such as serial-sampled UPGMA (Drummond and Rodrigo 2000) and TREBLE (Yang et al. 2007), were also developed under the strict-clock model. The least-squares method of Xia and Yang (2011), implemented in the DAMBE software, can also be modified to analyse non-contemporaneous data to minimize the residual sum of squares under a global clock (Xia 2018b).

Non-Bayesian methods that relax the assumption of rate constancy have also been developed, and they do not require the specification of many priors as in Bayesian approaches (To et al. 2016; Miura et al. 2020). Maximum-likelihood methods have been developed to estimate substitution rates and node dates under local and discrete clocks (Physher; Fourment and Holmes 2014) and under a relaxed clock (TreeTime; Sagulenko et al. 2018). TreeTime uses a normal prior to control the rate variation to be more autocorrelated-like or independent-like. The penalized-likelihood method implemented in r8s can also be used for dating non-contemporaneous data (Sanderson 2003). To et al. (2016) developed a least-squares dating (LSD) method that assumes the noise in molecular rates to be normal-like to account for independent rate variation across branches. Volz and Frost (2017) combined the maximum-likelihood and least-squares criteria to develop treedater. Miura et al. (2020) developed a method based on the RelTime approach, called RelTime with Dated Tips (RTDT), and the method is available in MEGA X.

Many of these non-Bayesian methods have been evaluated using data sets simulated under IBR models. They perform as well as Bayesian methods in estimating substitution rates and the root age (Fourment and Holmes 2014; To et al. 2016; Volz and Frost 2017; Sagulenko et al. 2018). Miura et al. (2020) conducted a benchmark study to assess the performance of various Bayesian and non-Bayesian methods in estimating divergence times for a large collection of simulated data sets, which were simulated under ABR and IBR models, using different tree shapes, and with strong and weak temporal signals. For data sets with moderate or strong temporal signals, RTDT performed better than other non-Bayesian methods because it produced good node-by-node time estimates and reliable confidence intervals that often contained the true values. Other non-Bayesian methods (e.g., LSD and TreeTime) performed well for IBR data sets, but not for ABR data sets. When there was a weak temporal signal in the data, Bayesian methods provided better estimates than non-Bayesian methods, as long as the correct rate model was specified. Tong et al. (2018) also suggested that non-Bayesian methods produced reliable rate estimates when the evolutionary rate was high, but that Bayesian methods generated slightly better estimates when there was a low evolutionary rate and weak temporal signal.

Non-Bayesian methods also allow the data to have missing sampling dates or to have uncertainties in the sampling dates (Volz and Frost 2017; Sagulenko et al. 2018; Miura et al. 2020). All of these non-Bayesian methods are orders of magnitude faster than Bayesian methods (Volz and Frost 2017; Miura et al. 2020), so they provide the feasibility of dating phylogenies with thousands of tips and sampling dates, which are expected to become increasingly common in molecular epidemiology. Miura et al. (2020) provided brief guidelines for users to select the most appropriate method for tip-dating analysis, based on the characteristics of the data set being analysed.

12.8 Phylogenetic Uncertainty

In the above, we focused on the application of non-Bayesian methods for estimating divergence times and confidence intervals for a given topology, because molecular dating is frequently done after inferring a phylogenetic tree. Ideally, one would obtain a reliable tree topology using maximum likelihood and other methods, and then estimate divergence times and their uncertainties based on this fixed topology. If the inferred tree is inaccurate, divergence times estimated for many of the nodes will be meaningless, because they would not correspond to actual evolutionary divergence events. The placement of calibrations can also become complicated when the phylogenetic tree is not well established. The presence of uncertainty in the tree topology is expected to inflate the uncertainty of divergence-time estimates (Ho 2009).

In some situations, however, one might fix the nodes of interest and allow the rest of the phylogeny to be inferred from the data. In this case, it is possible to apply a chosen non-Bayesian method to each alternative topology and report the mean time estimate, the standard deviation, and a summary confidence interval around the meantime of the node of interest across all of the candidate topologies. For example, it is of great interest to date the origin of a set of pathogenic strains in tip-dating analyses. The accuracy of time estimates for this node has been tested in simulation analyses by using phylogenies inferred from the sequence alignment, rather than fixing the topology (To et al. 2016; Volz and Frost 2017; Sagulenko et al. 2018; Miura et al. 2020). The results of these analyses have been very encouraging, with RTDT and other non-Bayesian methods producing reliable estimates for this node. Similar procedures can be applied to dating species and divergences between duplicated genes by using relaxed non-Bayesian methods.

12.9 Concluding Remarks

We anticipate that RelTime, penalized likelihood, and other non-Bayesian methods will become more widely used for a number of reasons. First, the computational speed and reliable inferences offered by these non-Bayesian methods allow one to use larger data sets for dating the tree of life or for testing biological hypotheses. Because Bayesian methods often demand large amounts of computational time and memory, many researchers adopt a divide-and-conquer approach by running the Bayesian methods on small partitions and gluing the results together (Misof et al. 2014; Oliveros et al. 2019). Alternatively, researchers might filter the genes until the size of the remaining data set is feasible for Bayesian inference (Hughes et al. 2018). This situation is going to become more acute, because progress in sequencing technology has been a boon for molecular systematics and biodiversity research, leading to a two-dimensional expansion of data sets (sites and species) available for dating studies (e.g., Zeng et al. 2014; Testo and Sundue 2016; Zheng and Wiens 2016; Barba-Montoya et al. 2018; Hughes et al. 2018). For this reason, faster Bayesian implementations are also being developed (Åkerborg et al. 2008; Lartillot et al. 2013).

Second, the use of efficient and reliable methods will enhance scientific rigour by allowing an assessment of the robustness of estimates to the assumptions made in dating analyses. Such analyses might involve studying the effects of using different combinations of genes, species, calibrations, and priors. Owing to computational time requirements, such explorations can be difficult for large data sets. Rapid non-Bayesian methods provide researchers with a toolkit to test the sensitivity of molecular time estimates and to improve downstream investigations of the biological process.

Third, the computational time requirements imposed by Bayesian methods make it challenging to examine the accuracy and precision of their estimates for large data sets, whereas rapid non-Bayesian methods have been tested on data sets with hundreds to thousands of species (Smith and O'Meara 2012; Tamura et al. 2012, 2018). A high computational burden also discourages independent evaluation of Bayesian date estimates by others interested in reproducing the results. Many practitioners are frustrated by the fact that independent attempts to simply reproduce the results of Bayesian dating can take weeks to months, and can only be pursued by research groups with access to extensive computing resources. This delays, and even impedes, scientific discourse and progress. The presence of reliable, efficient non-Bayesian methods is very useful and makes molecular dating accessible to all, including those without ready access to high-performance computing infrastructure.

Admittedly, Bayesian methods are useful when one wishes to incorporate some other information into divergence-time inference (e.g., biogeographic data) or to get a joint inference of some other phylogenetic features (e.g., population dynamics parameters). However, whether the inclusion of additional information or the joint inference will improve the accuracy of divergence-time estimation requires more extensive study, because appropriate settings for priors are usually unknown.

In fact, we suggest that users apply both Bayesian and non-Bayesian methods to obtain estimates of divergence times and their confidence intervals for molecular data sets, where possible. This would allow us to detect potential biases introduced by the assumptions and methods. Nevertheless, it is important to note that concordance between time estimates from Bayesian and non-Bayesian approaches should not be taken to suggest that the estimated times are correct. This is because the estimation of absolute divergence times highly depends on the calibration constraints used, and all methods will be negatively affected if the calibration constraints or densities used are incorrect (Battistuzzi et al. 2015; Hedges et al. 2018). For example, the use of an exponential density indicates a very high probability that the node age is close to the minimum constraint (Hedges and Kumar 2004; Ho and Duchêne 2014). Without proper justification and prior independent data, the choice of calibration density is largely subjective (Heath 2012; Bromham et al. 2018), which can adversely affect molecular date estimates. Different density distributions, even with the same minimum and maximum bounds, can produce different posterior time estimates in Bayesian methods (dos Reis et al. 2015; Barba-Montoya et al. 2017; Warnock et al. 2017; Morris et al. 2018). In addition, there are concerns about the imposition of maximum constraints on node times, because the fossil record only provides reliable minimum constraints (Battistuzzi et al. 2015; Bromham et al. 2018; Hedges et al. 2018). Therefore, one needs to examine the reliability of calibrations before conducting dating analyses (Andújar et al. 2014; Battistuzzi et al. 2015; Hedges et al. 2018).

In general, we see no reason for avoiding non-Bayesian methods for constructing time-trees, given that they are computationally efficient and produce estimates of divergence times and their surrounding uncertainties that are scientifically rigorous and reproducible. In particular, efficient non-Bayesian methods might be the only feasible option for many users for analysing large data sets containing thousands of genes and species.

References

Åkerborg Ö, Sennblad B, Lagergren J (2008) Birth-death prior on phylogeny and speed dating. BMC Evol Biol 8:77

Andújar C, Soria-Carrasco V, Serrano J, Gómez-Zurita J (2014) Congruence test of molecular clock calibration hypotheses based on Bayes factor comparisons. Methods Ecol Evol 5:226–242

Barba-Montoya J, dos Reis M, Yang Z (2017) Comparison of different strategies for using fossil calibrations to generate the time prior in Bayesian molecular clock dating. Mol Phylogenet Evol 114:386–400

Barba-Montoya J, dos Reis M, Schneider H, Donoghue PCJ, Yang Z (2018) Constraining uncertainty in the timescale of angiosperm evolution and the veracity of a Cretaceous Terrestrial Revolution. New Phytol 218:819–834

Battistuzzi FU, Billing-Ross P, Murillo O, Filipski A, Kumar S (2015) A protocol for diagnosing the effect of calibration priors on posterior time estimates: A case study for the Cambrian explosion of animal phyla. Mol Biol Evol 32:1907–1912

Battistuzzi FU, Tao Q, Jones L, Tamura K, Kumar S (2018) RelTime relaxes the strict molecular clock throughout the phylogeny. Genome Biol Evol 10:1631–1636

Bhatnagar N, Bogdanov A, Mossel E (2011) The computational complexity of estimating MCMC convergence time. In: Goldberg LA, Jansen K, Ravi R, Rolim JDP (eds) Approximation, randomization, and combinatorial optimization. Algorithms and techniques. Springer, Heidelberg, pp 424–435

Biek R, Pybus OG, Lloyd-Smith JO, Didelot X (2015) Measurably evolving pathogens in the genomic era. Trends Ecol Evol 30:306–313

Bollyky PL, Holmes EC (1999) Reconstructing the complex evolutionary history of hepatitis B virus. J Mol Evol 49:130–141

Britton T, Oxelman B, Vinnersten A, Bremer K (2002) Phylogenetic dating with confidence intervals using mean path lengths. Mol Phylogenet Evol 24:58–65

Britton T, Anderson CL, Jacquet D, Lundqvist S, Bremer K (2007) Estimating divergence times in large phylogenetic trees. Syst Biol 56:741–752

Bromham L, Duchêne S, Hua X, Ritchie AM, Duchêne DA, Ho SYW (2018) Bayesian molecular dating: opening up the black box. Biol Rev 93:1165–1191

Chernikova D, Motamedi S, Csürös M, Koonin EV, Rogozin IB (2011) A late origin of the extant eukaryotic diversity: divergence time estimates using rare genomic changes. Biol Direct 6:26

Chikina M, Robinson JD, Clark NL (2016) Hundreds of genes experienced convergent shifts in selective pressure in marine mammals. Mol Biol Evol 33:2182–2192

Chriki-Adeeb R, Chriki A (2016) Estimating divergence times and substitution rates in Rhizobia. Evol Bioinform 12:87–97

Crosby RW, Williams TL (2017) Fast algorithms for computing phylogenetic divergence time. BMC Bioinform 18:514

Doolittle RF, Feng DF, Tsang S, Cho G, Little E (1996) Determining divergence times of the major kingdoms of living organisms with a protein clock. Science 271:470–477

dos Reis M, Inoue J, Hasegawa M, Asher RJ, Donoghue PC, Yang Z (2012) Phylogenomic datasets provide both precision and accuracy in estimating the timescale of placental mammal phylogeny. Proc R Soc B 279:3491–3500

dos Reis M, Donoghue PC, Yang Z (2014) Neither phylogenomic nor palaeontological data support a Palaeogene origin of placental mammals. Biol Lett 10:20131003

dos Reis M, Thawornwattana Y, Angelis K, Telford MJ, Donoghue PC, Yang Z (2015) Uncertainty in the timing of origin of animals and the limits of precision in molecular timescales. Curr Biol 25:1–12

dos Reis M, Donoghue PC, Yang Z (2016) Bayesian molecular clock dating of species divergences in the genomics era. Nat Rev Genet 17:71–80

dos Reis M, Gunnell GF, Barba-Montoya J, Wilkins A, Yang Z, Yoder AD (2018) Using phylogenomic data to explore the effects of relaxed clocks and calibration strategies on divergence time estimation: primates as a test case. Syst Biol 67:594–615

Douzery EJP, Snell EA, Bapteste E, Delsuc F, Philippe H (2004) The timing of eukaryotic evolution: does a relaxed molecular clock reconcile proteins and fossils? Proc Natl Acad Sci USA 101:15386–15391

Drummond A, Rodrigo AG (2000) Reconstructing genealogies of serial samples under the assumption of a molecular clock using serial-sample UPGMA. Mol Biol Evol 17:1807–1815

Drummond AJ, Ho SYW, Phillips MJ, Rambaut A (2006) Relaxed phylogenetics and dating with confidence. PLOS Biol 4:e88

Eizirik E, Murphy W, O'Brien S (2001) Molecular dating and biogeography of the early placental mammal radiation. J Hered 92:212–219

Faria NR, Rambaut A, Suchard MA, Baele G, Bedford T, Ward MJ, Tatem AJ, Sousa JD, Arinaminpathy N, Pépin J, Posada D, Peeters M, Pybus OG, Lemey P (2014) The early spread and epidemic ignition of HIV-1 in human populations. Science 346:56–61

Feng DF, Cho G, Doolittle RF (1997) Determining divergence times with a protein clock: update and reevaluation. Proc Natl Acad Sci USA 94:13028–13033

Filipski A, Murillo O, Freydenzon A, Tamura K, Kumar S (2014) Prospects for building large timetrees using molecular data with incomplete gene coverage among species. Mol Biol Evol 31:2542–2550

Fitch WM (1976) Molecular evolutionary clocks. In: Ayala FJ (ed) Molecular evolution. Sinauer, Sunderland, MA, pp 160–178

Fourment M, Holmes EC (2014) Novel non-parametric models to estimate evolutionary rates and divergence times from heterochronous sequence data. BMC Evol Biol 14:163

Gillespie JH (1984) The molecular clock may be an episodic clock. Proc Natl Acad Sci USA 81:8009–8013

Graur D, Martin W (2004) Reading the entrails of chickens: molecular timescales of evolution and the illusion of precision. Trends Genet 20:80–86

Gunter NL, Weir TA, Slipinksi A, Bocak L, Cameron SL (2016) If dung beetles (Scarabaeidae: Scarabaeinae) arose in association with dinosaurs, did they also suffer a mass co-extinction at the K-Pg boundary? PLOS ONE 11:e0153570

Hasegawa M, Kishino H, Yano T (1989) Estimation of branching dates among primates by molecular clocks of nuclear DNA which slowed down in Hominoidea. J Hum Evol 18:461–476

Heath TA (2012) A hierarchical Bayesian model for calibrating estimates of species divergence times. Syst Biol 61:793–809

Heckman DS, Geiser DM, Eidell BR, Stauffer RL, Kardos NL, Hedges SB (2001) Molecular evidence for the early colonization of land by fungi and plants. Science 293:1129–1133

Hedges SB, Kumar S (2003) Genomic clocks and evolutionary timescales. Trends Genet 19:200–206

Hedges SB, Kumar S (2004) Precision of molecular time estimates. Trends Genet 20:242–247

Hedges SB, Kumar S (2009) The timetree of life. Oxford University Press, Oxford, UK

Hedges SB, Shah P (2003) Comparison of mode estimation methods and application in molecular clock analysis. BMC Bioinform 4:31

Hedges SB, Parker PH, Sibley CG, Kumar S (1996) Continental breakup and the ordinal diversification of birds and mammals. Nature 381:226–229

Hedges SB, Marin J, Suleski M, Paymer M, Kumar S (2015) Tree of life reveals clock-like speciation and diversification. Mol Biol Evol 32:835–845

Hedges SB, Tao Q, Walker M, Kumar S (2018) Accurate timetrees require accurate calibrations. Proc Natl Acad Sci USA 115:E9510–E9511

Hipsley CA, Müller J (2014) Beyond fossil calibrations: realities of molecular clock practices in evolutionary biology. Front Genet 5:138

Ho SYW (2009) An examination of phylogenetic models of substitution rate variation among lineages. Biol Lett 5:421–424

Ho SYW (2014) The changing face of the molecular evolutionary clock. Trends Ecol Evol 29:496–503

Ho SYW, Duchêne S (2014) Molecular-clock methods for estimating evolutionary rates and timescales. Mol Ecol 23:5947–5965

Ho SYW, Phillips MJ (2009) Accounting for calibration uncertainty in phylogenetic estimation of evolutionary divergence times. Syst Biol 58:367–380

Ho SYW, Duchêne S, Duchêne D (2015a) Simulating and detecting autocorrelation of molecular evolutionary rates among lineages. Mol Ecol Resour 15:688–696

Ho SYW, Tong KJ, Foster CS, Ritchie AM, Lo N, Crisp MD (2015b) Biogeographic calibrations for the molecular clock. Biol Lett 11:20150194

Huerta-Cepas J, Gabaldón T (2011) Assigning duplication events to relative temporal scales in genome-wide studies. Bioinformatics 27:38–45

Hughes LC, Ortí G, Huang Y, Sun Y, Baldwin CC, Thompson AW, Arcila D, Betancur-R R, Li C, Becker L, Bellora N, Zhao X, Li X, Wang M, Fang C, Xie B, Zhou Z, Huang H, Chen S, Venkatesh B, Shi Q (2018) Comprehensive phylogeny of ray-finned fishes (Actinopterygii) based on transcriptomic and genomic data. Proc Natl Acad Sci USA 115:6249–6254

Jaynes ET, Kempthorne O (1976) Confidence intervals vs Bayesian intervals. In: Harper WL, Hooker CA (eds) Foundations of probability theory, statistical inference, and statistical theories of science. Springer, Dordrecht, pp 175–257

Jiao Y, Wickett NJ, Ayyampalayam S, Chanderbali AS, Landherr L, Ralph PE, Tomsho LP, Hu Y, Liang H, Soltis PS, Soltis DE, Clifton SW, Schlarbaum SE, Schuster SC, Ma H, Leebens-Mack J, dePamphilis CW (2011) Ancestral polyploidy in seed plants and angiosperms. Nature 473:97–100

Kishino H, Thorne JL, Bruno WJ (2001) Performance of a divergence time estimation method under a probabilistic model of rate evolution. Mol Biol Evol 18:352–361

Kodandaramaiah U (2011) Tectonic calibrations in molecular dating. Curr Zool 57:116–124

Kooistra WH, Medlin LK (1996) Evolution of the diatoms (Bacillariophyta): IV. A reconstruction of their age from small subunit rRNA coding regions and the fossil record. Mol Phylogenet Evol 6:391–407

Ksepka DT, Parham JF, Allman JF, Benton MJ, Carrano MT, Cranston KA, Donoghue PC, Head JJ, Hermsen EJ, Irmis RB, Joyce WG, Kohli M, Lamm KD, Leehr D, Patané JL, Polly D, Phillips MJ, Smith NA,

Smith ND, Van Tuinen M, Ware JL, Warnock RCM (2015) The Fossil Calibration Database, a new resource for divergence dating. Syst Biol 64:853–859

Kumar S (2005) Molecular clocks: four decades of evolution. Nat Rev Genet 6:654–662

Kumar S, Hedges SB (1998) A molecular timescale for vertebrate evolution. Nature 392:917–920

Kumar S, Hedges SB (2016) Advances in time estimation methods for molecular data. Mol Biol Evol 33:863–869

Kumar S, Stecher G, Li M, Knyaz C, Tamura K (2018) MEGA X: molecular evolutionary genetics analysis across computing platforms. Mol Biol Evol 35:1547–1549

Langley CH, Fitch WM (1974) An examination of the constancy of the rate of molecular evolution. J Mol Evol 3:161–177

Lartillot N, Rodrigue N, Stubbs D, Richer J (2013) PhyloBayes MPI: phylogenetic reconstruction with infinite mixtures of profiles in a parallel environment. Syst Biol 62:611–615

Li WLS, Drummond AJ (2012) Model averaging and Bayes factor calculation of relaxed molecular clocks in Bayesian phylogenetics. Mol Biol Evol 29:751–761

Li WH, Tanimura M, Sharp PM (1988) Rates and dates of divergence between AIDS virus nucleotide sequences. Mol Biol Evol 5:313–330

Li H-T, Yi T-S, Gao L-M, Ma P-F, Zhang T, Yang J-B, Gitzendanner MA, Fritsch PW, Cai J, Luo Y, Wang H, van der Bank M, Zhang S-D, Wang Q-F, Wang J, Zhang Z-R, Fu C-N, Yang J, Hollingsworth PMN, Chase MW, Soltis DE, Soltis PS, Li D-Z (2019) Origin of angiosperms and the puzzle of the Jurassic gap. Nat Plants 5:461–470

Louca S, Shih PM, Pennell MW, Fischer WW, Parfrey LW, Doebeli M (2018) Bacterial diversification through geological time. Nat Ecol Evol 2:1458–1467

Lu L-M, Mao L-F, Yang T, Ye J-F, Liu B, Li H-L, Sun M, Miller JT, Mathews S, Hu H-H, Niu Y-T, Peng D-X, Chen Y-H, Smith SA, Chen M, Xiang K-L, Le C-T, Dang V-C, Soltis PS, Soltis DE, Li J-H, Chen Z-D (2018) Evolutionary history of the angiosperm flora of China. Nature 554:234–238

Marin J, Hedges SB (2018) Undersampling genomes has biased time and rate estimates throughout the tree of life. Mol Biol Evol 35:2077–2084

Marin J, Battistuzzi FU, Brown AC, Hedges SB (2017) The timetree of prokaryotes: new insights into their evolution and speciation. Mol Biol Evol 34:437–446

Marshall CR (2008) A simple method for bracketing absolute divergence times on molecular phylogenies using multiple fossil calibration points. Am Nat 171:726–742

Mello B (2018) Estimating timetrees with MEGA and the TimeTree resource. Mol Biol Evol 35:2334–2342

Mello B, Tao Q, Tamura K, Kumar S (2017) Fast and accurate estimates of divergence times from big data. Mol Biol Evol 34:45–50

Mello B, Tao Q, Kumar S (2021) Molecular dating for phylogenies containing a mix of populations and species by using Bayesian and RelTime approaches. Mol Ecol Resour (in press)

Meredith RW, Janečka JE, Gatesy J, Ryder OA, Fisher CA, Teeling EC, Goodbla A, Eizirik E, Simão TL, Stadler T, Rabosky DL, Honeycutt RL, Flynn JJ, Ingram CM, Steiner C, Williams TL, Robinson TJ, Burk-Herrick A, Westerman M, Ayoub NA, Springer MS, Murphy WJ (2011) Impacts of the Cretaceous Terrestrial Revolution and KPg extinction on mammal diversification. Science 334:521–524

Metsky HC, Matranga CB, Wohl S, Schaffner SF, Freije CA, Winnicki SM, West K, Qu J, Baniecki ML, Gladden-Young A, Lin AE, Tomkins-Tinch CH, Ye SH, Park DJ, Luo CY, Barnes KG, Shah RR, Chak B, Barbosa-Lima G, Delatorre E, Vieira YR, Paul LM, Tan AL, Barcellona CM, Porcelli MC, Vasquez C, Cannons AC, Cone MR, Hogan KN, Kopp EW, Anzinger JJ, Garcia KF, Parham LA, Ramírez RMG, Montoya MCM, Rojas DP, Brown CM, Hennigan S, Sabina B, Scotland S, Gangavarapu K, Grubaugh ND, Oliveira G, Robles-Sikisaka R, Rambaut A, Gehrke L, Smole S, Halloran ME, Villar L, Mattar S, Lorenzana I, Cerbino-Neto J, Valim C, Degrave W, Bozza PT, Gnirke A, Andersen KG, Isern S, Michael SF, Bozza FA, Souza TML, Bosch I, Yozwiak NL, MacInnis BL, Sabeti PC (2017) Zika virus evolution and spread in the Americas. Nature 546:411–415

Misof B, Liu S, Meusemann K, Peters RS, Donath A, Mayer C, Frandsen PB, Ware J, Flouri T, Beutel RG, Niehuis O, Petersen M, Izquierdo-Carrasco F, Wappler T, Rust J, Aberer AJ, Aspöck U, Aspöck H, Bartel D, Blanke A, Berger S, Böhm A, Buckley TR, Calcott B, Chen J, Friedrich F, Fukui M, Fujita M, Greve C, Grobe P, Gu S, Huang Y, Jermiin LS, Kawahara AY, Krogmann L, Kubiak M, Lanfear R, Letsch H, Li Y, Li Z, Li J, Lu H, Machida R, Mashimo Y, Kapli P, McKenna DD, Meng G, Nakagaki Y, Navarrete-Heredia JL, Ott M, Ou Y, Pass G, Podsiadlowski L, Pohl H, von Reumont BM, Schütte K, Sekiya K, Shimizu S, Slipinski A, Stamatakis A, Song W, Su X, Szucsich NU, Tan M, Tan X, Tang M, Tang J, Timelthaler G, Tomizuka S, Trautwein M, Tong X, Uchifune T, Walzl MG, Wiegmann BM, Wilbrandt J, Wipfler B, Wong TK, Wu Q, Wu G, Xie Y, Yang S, Yang Q, Yeates DK, Yoshizawa K, Zhang Q, Zhang R, Zhang W, Zhang Y, Zhao J, Zhou C, Zhou L, Ziesmann T, Zou S, Li Y, Xu X, Zhang Y, Yang H, Wang J, Wang J, Kjer KM, Zhou X (2014) Phylogenomics resolves the timing and pattern of insect evolution. Science 346:763–767

Miura S, Tamura K, Tao Q, Huuki LA, Pond SLK, Priest J, Deng J, Kumar S (2020) A new method for inferring timetrees from temporally sampled molecular sequences. PLOS Comput Biol 16:e1007046

Morris JL, Puttick MN, Clark JW, Edwards D, Kenrick P, Pressel S, Wellman CH, Yang Z, Schneider H, Donoghue PC (2018) The timescale of early land

plant evolution. Proc Natl Acad Sci USA 115:E2274–E2283
Morrison DA (2008) How to summarize estimates of ancestral divergence times. Evol Bioinform 4:75–95
Muse SV, Weir BS (1992) Testing for equality of evolutionary rates. Genetics 132:269–276
Nascimento FF, dos Reis M, Yang Z (2017) A biologist's guide to Bayesian phylogenetic analysis. Nat Ecol Evol 1:1446–1454
Nei M, Kumar S (2000) Molecular evolution and phylogenetics. Oxford University Press, Oxford, UK
Oliveros CH, Field DJ, Ksepka DT, Barker FK, Aleixo A, Andersen MJ, Alström P, Benz BW, Braun EL, Braun MJ, Bravo GA, Brumfield RT, Chesser RT, Claramunt S, Cracraft J, Cuervo AM, Derryberry EP, Glenn TC, Harvey MG, Hosner PA, Joseph L, Kimball RT, Mack AL, Miskelly CM, Peterson AT, Robbins MB, Sheldon FH, Silveira LF, Smith BT, White ND, Moyle RG, Faircloth BC (2019) Earth history and the passerine superradiation. Proc Natl Acad Sci USA 116:7916–7925
Paradis E (2013) Molecular dating of phylogenies by likelihood methods: a comparison of models and a new information criterion. Mol Phylogenet Evol 67:436–444
Phillips MJ (2015) Geomolecular dating and the origin of placental mammals. Syst Biol 65:546–557
Prum RO, Berv JS, Dornburg A, Field DJ, Townsend JP, Lemmon EM, Lemmon AR (2015) A comprehensive phylogeny of birds (Aves) using targeted next-generation DNA sequencing. Nature 526:569–578
Rambaut A (2000) Estimating the rate of molecular evolution: incorporating non-contemporaneous sequences into maximum likelihood phylogenies. Bioinformatics 16:395–399
Russo C, Takezaki N, Nei M (1995) Molecular phylogeny and divergence times of drosophilid species. Mol Biol Evol 12:391–404
Sagulenko P, Puller V, Neher RA (2018) TreeTime: maximum-likelihood phylodynamic analysis. Virus Evol 4:vex042
Sanderson MJ (1997) A nonparametric approach to estimating divergence times in the absence of rate constancy. Mol Biol Evol 14:1218–1231
Sanderson MJ (2002) Estimating absolute rates of molecular evolution and divergence times: a penalized likelihood approach. Mol Biol Evol 19:101–109
Sanderson MJ (2003) r8s: inferring absolute rates of molecular evolution and divergence times in the absence of a molecular clock. Bioinformatics 19:301–302
Sarich VM, Wilson AC (1967) Immunological time scale for hominid evolution. Science 158:1200–1203
Sarich VM, Wilson AC (1973) Generation time and genomic evolution in primates. Science 179:1144–1147
Sauquet H, Ho SYW, Gandolfo MA, Jordan GJ, Wilf P, Cantrill DJ, Bayly MJ, Bromham L, Brown GK, Carpenter RJ, Lee DM, Murphy DJ, Sniderman JM, Udovicic F (2012) Testing the impact of calibration on molecular divergence times using a fossil-rich group: the case of *Nothofagus* (Fagales). Syst Biol 61:289–313
Schenk JJ (2016) Consequences of secondary calibrations on divergence time estimates. PLOS ONE 11: e0148228
Smith SA, O'Meara BC (2012) treePL: divergence time estimation using penalized likelihood for large phylogenies. Bioinformatics 28:2689–2690
Smith SA, Brown JW, Walker JF (2018) So many genes, so little time: a practical approach to divergence-time estimation in the genomic era. PLOS ONE 13: e0197433
Stadler T, Yang Z (2013) Dating phylogenies with sequentially sampled tips. Syst Biol 62:674–688
Tajima F (1993) Simple methods for testing the molecular evolutionary clock hypothesis. Genetics 135:599–607
Takezaki N, Rzhetsky A, Nei M (1995) Phylogenetic test of the molecular clock and linearized trees. Mol Biol Evol 12:823–833
Tamura K, Battistuzzi FU, Billing-Ross P, Murillo O, Filipski A, Kumar S (2012) Estimating divergence times in large molecular phylogenies. Proc Natl Acad Sci USA 109:19333–19338
Tamura K, Stecher G, Peterson D, Filipski A, Kumar S (2013) MEGA6: molecular evolutionary genetics analysis version 6.0. Mol Biol Evol 30:2725–2729
Tamura K, Tao Q, Kumar S (2018) Theoretical foundation of the RelTime method for estimating divergence times from variable evolutionary rates. Mol Biol Evol 35:1170–1182
Tao Q, Tamura K, Battistuzzi FU, Kumar S (2019) A machine learning method for detecting autocorrelation of evolutionary rates in large phylogenies. Mol Biol Evol 36:811–824
Tao Q, Tamura K, Mello B, Kumar S (2020) Reliable confidence intervals for RelTime estimates of evolutionary divergence times. Mol Biol Evol 37:280–290
Testo W, Sundue M (2016) A 4000-species dataset provides new insight into the evolution of ferns. Mol Phylogenet Evol 105:200–211
Thorne JL, Kishino H, Painter IS (1998) Estimating the rate of evolution of the rate of molecular evolution. Mol Biol Evol 15:1647–1657
To T-H, Jung M, Lycett S, Gascuel O (2016) Fast dating using least-squares criteria and algorithms. Syst Biol 65:82–97
Tong KJ, Duchêne DA, Duchêne S, Geoghegan JL, Ho SYW (2018) A comparison of methods for estimating substitution rates from ancient DNA sequence data. BMC Evol Biol 18:70
Volz E, Frost S (2017) Scalable relaxed clock phylogenetic dating. Virus Evol 3:vex025
Warnock RCM, Yang Z, Donoghue PCJ (2017) Testing the molecular clock using mechanistic models of fossil preservation and molecular evolution. Proc R Soc B 284:20170227
Wiens JJ, Moen DS (2008) Missing data and the accuracy of Bayesian phylogenetics. J Syst Evol 46:307–314

Wiens JJ, Morrill MC (2011) Missing data in phylogenetic analysis: reconciling results from simulations and empirical data. Syst Biol 60:719–731

Wilke T, Schultheiß R, Albrecht C (2009) As time goes by: a simple fool's guide to molecular clock approaches in invertebrates. Am Malacol Bull 27:25–45

Worobey M, Han G-Z, Rambaut A (2014) A synchronized global sweep of the internal genes of modern avian influenza virus. Nature 508:254–257

Wray GA, Levinton JS, Shapiro LH (1996) Molecular evidence for deep Precambrian divergences among metazoan phyla. Science 274:568–573

Wu C-I, Li W-H (1985) Evidence for higher rates of nucleotide substitution in rodents than in man. Proc Natl Acad Sci USA 82:1741–1745

Xia X (2018a) DAMBE7: New and improved tools for data analysis in molecular biology and evolution. Mol Biol Evol 35:1550–1552

Xia X (2018b) Bioinformatics and the cell: modern computational approaches in genomics, proteomics and transcriptomics. Springer International, New York

Xia X, Yang Q (2011) A distance-based least-square method for dating speciation events. Mol Phylogenet Evol 59:342–353

Yang Z (2007) PAML 4: phylogenetic analysis by maximum likelihood. Mol Biol Evol 24:1586–1591

Yang Z, Rannala B (2006) Bayesian estimation of species divergence times under a molecular clock using multiple fossil calibrations with soft bounds. Mol Biol Evol 23:212–226

Yang Z, O'Brien JD, Zheng X, Zhu H-Q, She Z-S (2007) Tree and rate estimation by local evaluation of heterochronous nucleotide data. Bioinformatics 23:169–176

Yoder AD, Yang Z (2000) Estimation of primate speciation dates using local molecular clocks. Mol Biol Evol 17:1081–1090

Yu Y, Xiang Q, Manos PS, Soltis DE, Soltis PS, Song B-H, Cheng S, Liu X, Wong G (2017) Whole-genome duplication and molecular evolution in *Cornus* L. (Cornaceae) – Insights from transcriptome sequences. PLOS ONE 12:e0171361

Zeng L, Zhang Q, Sun R, Kong H, Zhang N, Ma H (2014) Resolution of deep angiosperm phylogeny using conserved nuclear genes and estimates of early divergence times. Nat Commun 5:4956

Zheng Y, Wiens JJ (2015) Do missing data influence the accuracy of divergence-time estimation with BEAST? Mol Phylogenet Evol 85:41–49

Zheng Y, Wiens JJ (2016) Combining phylogenomic and supermatrix approaches, and a time-calibrated phylogeny for squamate reptiles (lizards and snakes) based on 52 genes and 4162 species. Mol Phylogenet Evol 94:537–547

Zhu T, dos Reis M, Yang Z (2015) Characterization of the uncertainty of divergence time estimation under relaxed molecular clock models using multiple loci. Syst Biol 64:267–280

Zuckerkandl E, Pauling L (1962) Molecular disease, evolution, and genic heterogeneity. In: Kasha M, Pullman B (eds) Horizons in biochemistry. Academic, New York, pp 189–225

Bayesian Phylogenomic Dating

13

Sandra Álvarez-Carretero and Mario dos Reis

Abstract

The development of divergence-time estimation methods has been an active area of research since the early 1960s, when the molecular clock was first postulated by Zuckerkandl and Pauling. Thanks to technological and computational improvements, more powerful and cutting-edge techniques and algorithms have been developed to better understand species evolution at the molecular level. These have led to improved methods for molecular clock dating of speciation events. During the past two decades, the approaches for DNA sequencing have substantially advanced and their costs have decreased, thus enabling large-scale genome-sequencing projects that aim to sequence all species in the tree of life. Being able to access thousands of complete genomes, however, has brought new biological and computational challenges to phylogenomic analyses. We might have more data, but also new questions to answer. Inferring reliable phylogenies and accurately dating them is now the main goal of phylogenomic analyses. Although new computational tools that implement more complex evolutionary models have been developed, there remain challenges in dealing with issues such as polytomies, incomplete lineage sorting, and the uncertainty in the fossil record. This chapter aims to guide the reader through the steps of Bayesian phylogenomic dating analyses, from data collection and processing up to the inference of the species tree and subsequent clock dating analysis. We pay close attention to the Bayesian paradigm in molecular clock dating, focusing on the effects that the prior and the likelihood can have on the estimated divergence times when using phylogenomic data. We describe strategies to speed up computation when using large genomic data sets, such as the approximate-likelihood method, which produces speed-ups of up to 1000× in time-tree inference. We also discuss strategies to improve the efficiency of Markov chain Monte Carlo sampling.

Keywords

Bayesian inference · Molecular clock · Divergence times · Phylogeny · Approximate likelihood · Model selection

13.1 Introduction

In the early 1960s, Zuckerkandl and Pauling's realization that globin chains evolve at an approximately constant rate led them to postulate the

S. Álvarez-Carretero (✉) · M. dos Reis
School of Biological and Chemical Sciences, Queen Mary University of London, London, UK
e-mail: s.alvarez-carretero@ucl.ac.uk; sandra.ac93@gmail.com

S. Y. W. Ho (ed.), *The Molecular Evolutionary Clock*, https://doi.org/10.1007/978-3-030-60181-2_13

existence of a molecular clock (Zuckerkandl and Pauling 1962). This rate constancy means that molecular phylogenies can be calibrated to geological time by using information from the fossil record, thus allowing the inference of species divergence times. It is now acknowledged that molecular rate constancy only applies to closely related species (e.g., Goodman et al. 1971; Ohta and Kimura 1971; Jukes and Holmquist 1972; Langley and Fitch 1974; Fitch and Langley 1976; Li et al. 1987), leading to the development of 'relaxed-clock' methods that allow estimation of divergence times in phylogenies with more distantly related species (Takezaki et al. 1995; Sanderson 1997; Rambaut and Bromham 1998; Thorne et al. 1998; Huelsenbeck et al. 2000; Kishino et al. 2001; Drummond et al. 2006; Yang and Rannala 2006). Continual advances in technology and computation, as well as the development of new and more efficient methods and techniques (particularly the use of the Bayesian method), have changed the way molecular and palaeontological data are used for divergence-time estimation (dos Reis et al. 2016).

The availability of molecular data for different species has increased exponentially during the six decades since the proposal of the molecular clock hypothesis. We have moved from sequencing a unique DNA fragment in one run with the Sanger method (Sanger et al. 1977) to high-throughput approaches that allow the sequencing of millions of DNA bases in parallel (Zhang et al. 2011; Slatko et al. 2018). Thus, over 50,000 prokaryotic and eukaryotic genomes are, as of 2020, readily available for analysis in GenBank (Fig. 13.1). The phylogenetic analysis of complete genomes, phylogenomics, is a breakthrough that is expected to bring deeper insights into the evolutionary history of species (Eisen and Fraser 2003; Delsuc et al. 2005).

Nevertheless, just as the number of complete genomes has been continually increasing, the challenges encountered when working with them have also increased. In the beginnings of phylogenomics, the aim was not only to develop computational tools that could better differentiate between orthologues and paralogues to resolve the incongruence among phylogenetic trees (Gee 2003; Rokas et al. 2003; Delsuc et al. 2005). More importantly, phylogenomics was expected to be the panacea to resolve the polytomies in the tree of life (Delsuc et al. 2005; Rokas and Carroll 2006). These expectations do not seem to have been fulfilled yet, and thus the challenges described above are still unresolved in current phylogenomic analyses, demanding even more efficient and effective algorithms.

In the first phylogenomic analyses, reconstructing and resolving incongruences in inferred phylogenies were the first issues to be

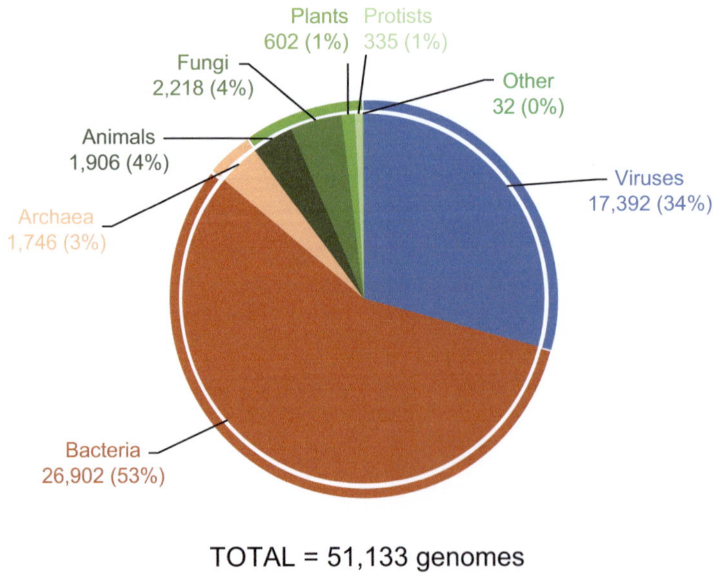

Fig. 13.1 Number of genome sequences available in GenBank as of April 2020. The outer circles correspond to Bacteria (dark orange), Archaea (light orange), Eukarya (green), and viruses (dark blue)

encountered when developing computational methods. Different statistical approaches were proposed for this purpose: distance methods such as minimum evolution using least-squares optimality (e.g., Rzhetsky and Nei 1992; Bryant and Waddell 1998) and neighbour-joining (Saitou and Nei 1987), maximum parsimony (Camin and Sokal 1965; Fitch 1970), maximum likelihood (Felsenstein 1981, 1988), and the Bayesian approach (Rannala and Yang 1996; Mau et al. 1999; Larget and Simon 1999). In addition, it was observed that phylogenies inferred for different genes did not always have the same topology, leading to the realization that different genes can have different evolutionary histories (e.g., Camin and Sokal 1965; Takahata 1989; Huelsenbeck et al. 2001; Thorne and Kishino 2002; dos Reis et al. 2016; Zhang et al. 2016). This observation encouraged the development of software that could incorporate the inference of coalescence events using phylogenomic data (e.g., Rannala and Yang 2003; Liu and Pearl 2007; Heled and Drummond 2010; Yang and Rannala 2010; Bryant et al. 2012; Mirarab et al. 2014; Bouckaert et al. 2019).

As phylogenies for different taxa started to be resolved, a parallel methodological and computational challenge arose: dating the inferred phylogeny with phylogenomic data. The methodology used to estimate divergence times is based on the combination of molecular data (genetic and/or genome sequences) and the fossil record (continuous and/or discrete morphological data, and the ages of fossils), with the latter used to calibrate the inferred phylogeny with geological times (Benton and Donoghue 2007; Chap. 8). Unfortunately, the fossil record is not complete, adding a level of uncertainty to the analysis that needs to be taken into account (e.g., dos Reis et al. 2015).

The Bayesian paradigm is an interesting approach to tackle this problem: it can integrate different sources of information and account for uncertainty on model parameters through the use of probability distributions (see Chap. 6). Despite Bayes's theorem dating back to the 1700s (Bayes 1763), it was not until 1998 that it was introduced by Thorne et al. (1998) for divergence-time estimation. Since then, Bayesian clock dating methods have been extended to integrate different kinds of data (molecular and morphological) and to estimate model parameters such as evolutionary rates or times of speciation (e.g., Kishino et al. 2001; Thorne and Kishino 2002; Rannala and Yang 2003; Drummond et al. 2006; Yang and Rannala 2006; dos Reis et al. 2012, 2016, 2018; Ronquist et al. 2012a; Heath et al. 2014; Zhu et al. 2015; Davín et al. 2018; Bouckaert et al. 2019).

Nevertheless, there are three main drawbacks associated with Bayesian clock dating. First, establishing prior probability distributions for model parameters is somewhat subjective, in particular when modelling fossil uncertainties (e.g., see Tavaré et al. 2002; Drummond et al. 2006; Yang and Rannala 2006; Benton et al. 2015). Second, the phylogenetic likelihood calculation is computationally expensive, and thus alternative approaches to approximate the likelihood have been explored (e.g., Thorne et al. 1998; Beaumont et al. 2002; Guindon 2010; dos Reis and Yang 2011; dos Reis et al. 2012). Third, the uncertainty in posterior estimates can become small but will never disappear in clock dating of extant species, because divergence times and evolutionary rates are unidentifiable in the likelihood function (Rannala and Yang 2007; dos Reis and Yang 2013; Zhu et al. 2015).

Faster dating methods using penalized likelihood have been developed (Sanderson 2002; Yang and Yoder 2003; Yang 2004; Smith and O'Meara 2012), increasing computational efficiency compared with Bayesian methods (see Chap. 12). These fast methods, however, do not account for fossil or branch-length uncertainty (Thorne and Kishino 2005). Despite these challenges, the Bayesian approach has become the preferred method for phylogenomic dating (e.g., Clarke et al. 2011; dos Reis et al. 2012, 2018; Springer et al. 2012; Jarvis et al. 2014; Misof et al. 2014; Zheng and Wiens 2016; Barba-Montoya et al. 2018).

This chapter aims to guide the reader through the different steps of Bayesian phylogenomic dating analyses, from data collection and processing up to the inference of the species tree and subsequent clock dating analysis. We pay close attention to the Bayesian paradigm in molecular clock dating, focusing on the effects that the prior and the likelihood can have on the estimated divergence times when using phylogenomic data. We discuss strategies to speed up computation when using large genomic data sets, such as the approximate-likelihood method, which produces speed-ups of up to 1000× in time-tree inference, as well as strategies to improve the efficiency of Markov chain Monte Carlo (MCMC) sampling. General reviews of Bayesian phylogenetic inference are available elsewhere (Holder and Lewis 2003; Yang 2014; Nascimento et al. 2017).

13.2 Bayesian Phylogenomic Dating

In this section, we will go through the different steps involved in a Bayesian phylogenomic dating analysis. We start with data preparation, then explain how to set up the prior distributions for rates and times, set up the likelihood model, and summarize the posterior distribution.

13.2.1 Preparing the Data

Before we can perform Bayesian inference of divergence times using phylogenomic data, there are several steps that need to be carried out: (1) data collection, (2) sequence alignment, (3) alignment partitioning, (4) species-tree inference, and (5) model selection. The general approach is illustrated in Fig. 13.2.

The first step, data collection, can start with the user (1) collecting biological samples from the species of interest, (2) sequencing and bioinformatically processing the raw sequence data from the samples, or (3) directly downloading orthologous genes for the species of interest from a database such as GenBank (Benson et al. 2018) or Ensembl (Zerbino et al. 2018). Starting at one point or another will depend on the design of the research project. Here, we give a summary of each of these steps.

13.2.1.1 Obtaining and Generating the Molecular Data

Collecting samples suitable for DNA sequencing from the species of interest (e.g., blood or tissue) is the first step in many studies. For instance, studies on unsequenced non-model organisms might start with gathering samples for DNA extractions. Once the samples are collected, they need to be processed using different molecular techniques and sent for sequencing. At the end of this stage, sequencing reads are obtained, the quality of which will have a direct impact on the subsequent analyses. For example, if the samples have been contaminated or there have been problems during sequencing, the quality of the raw sequence reads will be poor and will affect the downstream in silico analyses (e.g., Kircher and Kelso 2010; Strong et al. 2014; Kebschull and Zador 2015; Ballenghien et al. 2017). Even though there might be systematic and random errors that cannot be avoided nor controlled for in this stage, these can be minimized if the protocols for each technique carried out during the experimental work are cautiously followed (e.g., Cheung et al. 2011; Benjamini and Speed 2012; Wong et al. 2012; Ross et al. 2013). This will help to increase the quality of raw sequence data.

13.2.1.2 Processing the Molecular Sequence Data

The raw sequence data must be processed through a pipeline of bioinformatics software to filter, assemble, and annotate the data. Quality-control checks should be done before each step (Guo et al. 2014). During the filtering step, the raw sequence data are trimmed by removing the ends of the reads. This is done to remove the adapter sequences used during sequencing. In addition, the user can specify parameters such as the Phred score, the minimum read length, or the minimum base quality to further restrict the trimming process (see some trimmer software examples in Table 13.1). After trimming, read quality needs to be checked for any errors before the genome assembly is generated (Laehnemann

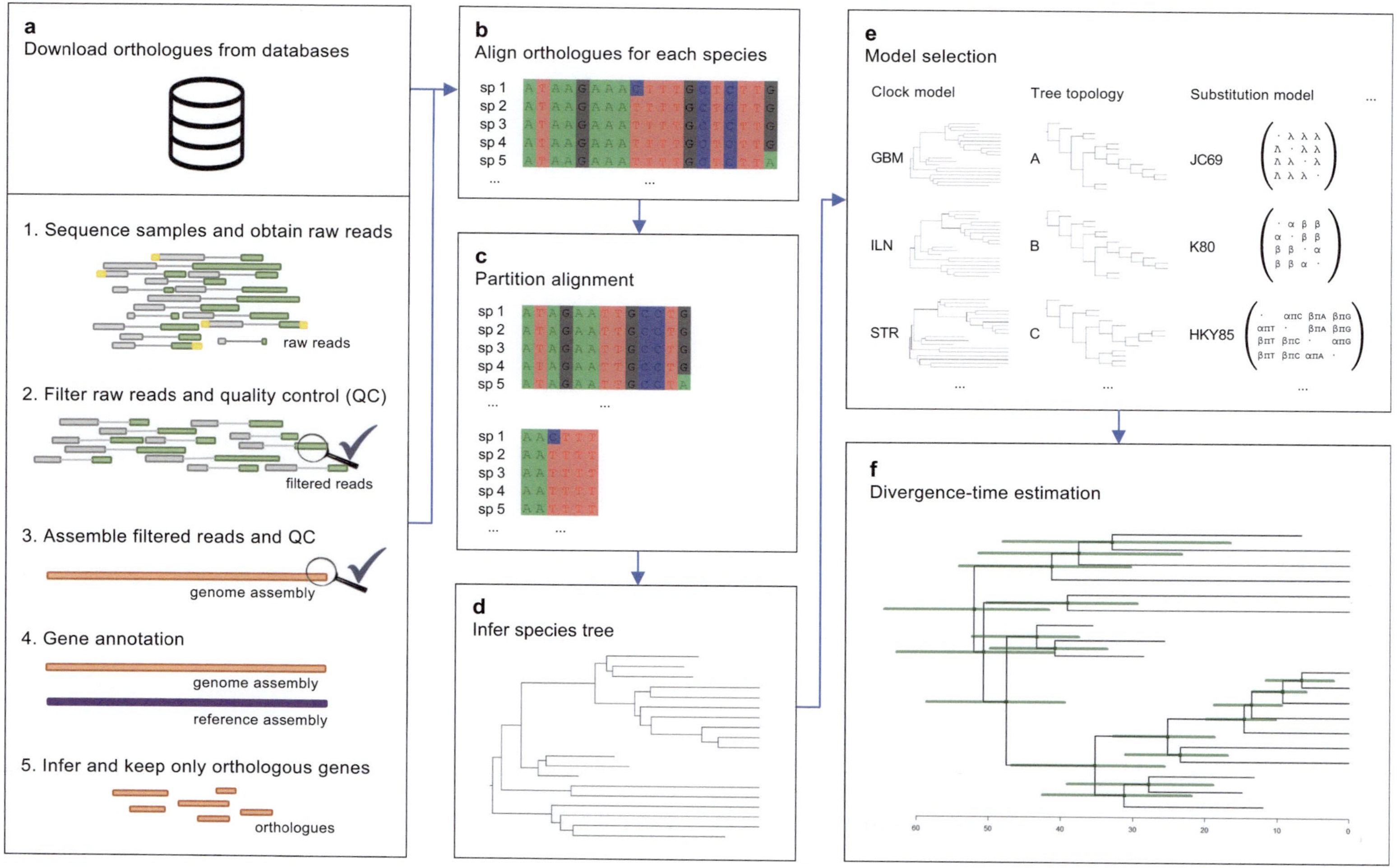

Fig. 13.2 Steps from data collection to Bayesian phylogenomic dating. (**a**) Data collection. We start with raw sequence data for the species of interest, or with already filtered orthologous genes downloaded from specific databases. Raw data will need to be filtered and assembled. The quality of the raw sequencing reads after each step is measured using quality-control software before genes in the genome assembly can be annotated. The final annotated assembly is screened for orthologues. (**b**) Sequence

Fig. 13.2 (continued) alignment. Software is used to align the orthologues collected for the species of interest. (**c**) Alignment partitioning. The alignment is partitioned according to a specific scheme, such as concatenation of codon positions of protein-coding genes and/or grouping genes with similar evolutionary rates or nucleotide compositions. (**d**) Species-tree inference. The partitioned alignment is used to infer the species tree using an approach such as maximum likelihood or Bayesian inference. (**e**) Model selection. The evolutionary model for the data needs to be chosen. There are different models that can be tested, such as different clock models, tree topologies, and substitution models. Methods such as Bayes factors can be used to select the best model for the data. (**f**) Divergence-time estimation. The last step uses the partitioned phylogenomic alignment, the inferred species tree, and the selected model to infer the species divergence times with the preferred Bayesian MCMC dating software

Table 13.1 Selection of software used to filter, assemble, and annotate raw sequence data

Software	Data	Reads	Adapter removal	Quality filtering trimming (QFT)	Read-error correction	Assembly	Annotation	Parallelization	Citations
Trimming tools									
Cutadapt v2.1	454/Roche, ABI SOLiD (*colorspace*), Illumina	SE, PE	Y	Y	N	N	N	Multi-core support	Martin (2011)
Trimmomatic v0.39	Illumina	SE, PE	Y	Y	N	N	N	Multithread support	Bolger et al. (2014)
PrinSeq • lite v0.20.4 • web v0.20.1	Designed for 454/Roche, but works with other data	SE, PE	Y	Y	N	N	N	No	Schmieder and Edwards (2011)
Assembly tools									
ABySS v2.1.5	Not specified	SE, PE, MP	N	Y	N[a]	Y	N	MPI and multithread support	Simpson et al. (2009), Jackman et al. (2017)
Canu v1.8	PacBio RS II or Oxford Nanopore MinION	LRSM	N	Y[b]	Y	Y	N	Grid environment support	Koren et al. (2017)
MaSuRCA v3.3.1	Illumina, PacBio/MinION. Other data need to be converted to Celera Assembler compatible frg files	SE, PE, MP	Y	Y[c]	Y	Y	N	Grid environment support	Zimin et al. (2013)
Ray Meta v2.3.1	Single genomes, metagenomes, transcriptomes[d]	SE, PE	N	Y	N	Y	Y[e]	MPI support	Boisvert et al. (2012)
SGA v0.10.15	Illumina	SE, PE	N	Y[b]	Y	Y	N	No	Simpson and Durbin (2012)
SOAPdenovo2 v2.04-r241	Illumina GA short reads	SE, PE, MP	N	Y[b]	Y	Y	N	Multithread support	Luo et al. (2012)
SPAdes & MetaSPAdes v3.13	Designed for Illumina and IonTorrent, but works with hybrid assemblies using PacBio, Oxford Nanopore, and Sanger reads	SE, PE, MP	N	Y[b]	Y[f]	Y	N	Multithread support	Bankevich et al. (2012)
Trinity v2.8.4	Transcripts from Illumina RNA-Seq data	SE, PE, MP	Y[g]	Y[h]	Y	Y	N	Multithread support[i]	Grabherr et al. (2011)

(continued)

Table 13.1 (continued)

Software	Data	Reads	Adapter removal	Quality filtering trimming (QFT)	Read-error correction	Assembly	Annotation	Parallelization	Citations
Velvet v1.2.10	Not specified	SE, PE, MP	N	Y[b]	Y	Y	N	Multithread support	Zerbino and Birney (2008)
Annotation tools									
Prokka v1.13	Bacterial, archaeal, and viral assemblies	–	N	N	N	N	Y	Multithread support	Seemann (2014)
MAKER v2.31.10	Any assembly	–	N	N	N	N	Y	MPI support	Cantarel et al. (2008)

SE single-end reads, *PE* paired-end reads, *MP* mate-pairs reads, *LRSM* long-read single-molecule sequences, *QFT* quality filtering trimming
[a]Does not do error correction, but uses linked-read libraries to correct assembly errors
[b]After read-error correction
[c]Before and after read-error correction
[d]Ray can work with transcriptomes, but it has not been extensively tested
[e]There is an optional taxonomic profiling with coloured de Bruijn graphs
[f]There is an optional correction for mismatches and short indels
[g]Yes, if QFT enabled with Trimmomatic
[h]Optional pre-QFT with Trimmomatic; additional QFT after read-error correction
[i]Requires a script with the list of commands to execute

et al. 2016). Quality-control software such as FastQC (Andrews 2010) can be used for this purpose.

Once reads have been processed, the assembly step takes place. This involves aligning and stacking the short sequence reads to construct large stretches of sequenced genome known as contigs (e.g., Nagarajan and Pop 2013; Ekblom and Wolf 2014). There are several assembler tools available, each using its own specific algorithms (see Table 13.1). The output of genome assemblies can depend on the assembler used, whether a reference genome was available, and the type of species for which sequence reads are being assembled. Therefore, it is important to decide which assembly is to be kept (e.g., Magoc et al. 2013). The quality of a genome assembly can be measured with QUAST (Gurevich et al. 2013), a program that not only measures assembly quality but also compares multiple assemblies generated with different approaches. Measures such as the cumulative length or the number of misassemblies are included in the QUAST report.

The next step, genome annotation, is a critical component of the bioinformatics pipeline (e.g., Miller et al. 2010; Schatz et al. 2010; Ekblom and Wolf 2014). If genes are not properly annotated, they might be assigned to the wrong orthologue or paralogue groups, thus compromising the subsequent analyses (Phillippy et al. 2008; Florea et al. 2011). There are several tools that use a reference genome to annotate genome assemblies (see Table 13.1) as well as databases that already contain annotated genomes, such as Ensembl or GenBank. Even though standard practices have been established to annotate the genomes (Madupu et al. 2010; Klimke et al. 2011), automatic annotation is not always accurate. Thus, manual curation of the annotated genomes may be required (Yandell and Ence 2012).

13.2.1.3 Inferring Orthologues

Most molecular dating software cannot deal with the computational costs associated with using whole-genome alignments. Thus, it is common practice to select sets of orthologous genes from genome assemblies to be used in the dating analysis. Orthologues are genes related via speciation, while paralogues have evolved via duplication events (Koonin 2005). In phylogenomic dating analyses, the interest lies in estimating species divergence times in the phylogeny of interest, and thus orthologues must be used (Siu-Ting et al. 2019).

Different methodologies have been developed for orthology inference and several orthology databases have been created (see Table 13.2). Benchmarking has been proposed to assess the performance of orthology-inference methods and to compare databases of orthologues (see Hulsen et al. 2006; Chen et al. 2007; Altenhoff and Dessimoz 2009; Boeckmann et al. 2011; Altenhoff et al. 2016).

13.2.1.4 Generating the Phylogenomic Alignment

Once the user has a set of orthologous genes for each species of interest, the sequences must then be aligned (the second step in Fig. 13.2). A range of bioinformatics tools can be used to obtain a multiple sequence alignment (see Table 13.3). Biological processes such as insertions, deletions, or nucleotide substitutions can be considered when aligning the sequences (e.g., Vingron and von Haeseler 1997; Chowdhury and Garai 2017). They are modelled and integrated in a scoring function to evaluate the quality of an alignment (e.g., likelihood function, hidden Markov models, mismatch, gap-opening penalty, and gap-extension penalty). If the wrong scoring function is used or if it is not correctly optimized, however, the resulting alignment can contain errors, thus affecting the subsequent analyses (for an example of the impact of alignment errors on testing for positive selection, see Fletcher and Yang 2010).

Aligners based on progressive alignment algorithms tend to be very popular because of their speed. Unfortunately, the guide tree used might influence the generated alignment and hence the inferred phylogeny (Lake 1991; Thorne and Kishino 1992; Redelings and Suchard 2005). In contrast, dynamic programming can resolve this issue because it can sum over all possible pairwise alignments, hence accounting for all

Table 13.2 Selection of software and databases used for orthology inference

Software/ database	Brief description	Citations
OMA Orthology Database in 2018	Database and method used to infer orthologues in complete genomes. The stand-alone OMA pipeline can be used on custom genomic/transcriptomic data, which can be combined with precomputed data exported from parts of the OMA database.	Roth et al. (2008), Dalquen et al. (2013), Altenhoff et al. (2018)
OrthoFinder v2.2.7	Comprehensive platform for comparative genomics. Not only finds orthogroups and orthologues, but also infers rooted gene trees and gene-duplication events. The latter can also be mapped onto an inferred rooted species tree.	Emms and Kelly (2015)
Eggnog v4.5.1	Database with orthologous groups predicted with a hierarchical functional annotation EggNOG pipeline.	Huerta-Cepas et al. (2016)
HieranoiDB Hieranoid2	Interactive database with hierarchical groups of orthologues from InParanoid aggregated by Hieranoid2.	Kaduk et al. (2017), Kaduk and Sonnhammer (2017)
PANTHER v14.1	Classification system of proteins (and their genes) generated after combining bioinformatics algorithms and manual curation. Part of the Gene Ontology Phylogenetic Annotation Project.	Mi et al. (2019)
MetaPhOrs	Database with phylogeny-based orthology and paralogy predictions computed after relying on publicly available homology-prediction servers. Each prediction comes with a quality assessment using a consistency score and an evidence level.	Pryszcz et al. (2011)
InParanoid8	Uses pairwise similarity scores between complete proteomes to infer orthology groups. Proteomes with sequences similar to those already in the group are included, i.e., in-paralogues. The relatedness between in-paralogue members and the seed orthologue is provided with a confidence value.	Sonnhammer and Östlund (2015)
Ensembl	Interactive platform used for comparative genomics analyses. Genes and genomes in this database have undergone manual curation, and are tagged as orthologous based on identity thresholds, which depend on the most recent common ancestor of the species pair.	Vilella et al. (2009)
OrthoDB v9.1 OrthoDB pipeline v2.4.4	Comprehensive hierarchical catalogue of orthologues. The pipeline software is the OrthoDB stand-alone version that can be used to find and cluster in-paralogues from the Fasta files provided by the user.	Zdobnov et al. (2017)

Table 13.3 Selection of software used to generate multiple sequence alignments for phylogenomic data sets

Software	Brief description	Citations
Clustal Ω	In default mode, uses the input sequence to generate the guide tree with mBED (Blackshields et al. 2010). Uses a hidden Markov model (Söding 2005) to generate the alignment.	Larkin et al. (2007), Sievers et al. (2011)
Muscle	Uses a guide tree, which subsequently is refined.	Edgar (2004)
MAFFT	Same approach as in MUSCLE.	Katoh and Standley (2013)
ProbCons	Same approach as in MUSCLE but computes the expected accuracy and the probabilistic consistency transformation for the refinement steps to improve the alignment.	Do et al. (2005)
PRANK	Uses a guide tree, but it is not subsequently refined. Uses the tree to score each alignment column.	Löytynoja and Goldman (2005, 2008), Löytynoja (2014)
FSA	Does not use a guide tree. Uses an alignment accuracy metric that penalizes the wrong alignments. Can estimate the branch lengths, indel rates, and substitution rates.	Bradley et al. (2009)

gene evolutionary histories in the genome sequence. Nevertheless, it becomes computationally intractable for phylogenomic data in terms of time and memory requirements (Redelings and Suchard 2009), in which case using approximate methods becomes an appealing solution. For instance, methods based on MCMC and hidden Markov models have been implemented in the software BAli-Phy (Suchard and Redelings 2006) to jointly estimate the sequence alignment and the tree, with promising results. Computational limitations in this software, however, restrict the number of taxa that can be included in the alignment and the length of the molecular sequences.

Once the phylogenomic sequence alignment is obtained, it might need to be partitioned (the third step in Fig. 13.2). There are several partitioning approaches that can be used, and the best partitioning scheme for a specific alignment can be selected using bioinformatics tools such as PartitionFinder (Lanfear et al. 2012, 2017) or ClockstaR (Duchêne and Ho 2014). Despite extensive benchmarking analyses (e.g., Brown and Lemmon 2007; Duchêne et al. 2011; Duchêne and Ho 2014; Foster and Ho 2017; Angelis et al. 2018), no standard practice has yet been established for deciding on the best partitioning scheme. Nevertheless, the choice of partitioning scheme can affect the accuracy of phylogenomic analysis, so the user should cautiously evaluate different partitioning schemes prior to Bayesian inference.

13.2.1.5 Inferring the Species Tree

The last step before estimating the species divergence times consists of inferring the species tree using the phylogenomic data (the fourth step in Fig. 13.2). As pointed out in Sect. 13.1, however, studies have shown that different genes can lead to trees with different topologies for the same taxa. These observations suggest that each gene could have its own evolutionary history. As a result, researchers have developed software based on the multispecies coalescent to better estimate species trees that accommodate gene-tree versus species-tree discrepancies, such as BPP (Yang and Yoder 2003; Yang and Rannala 2010), *BEAST (Heled and Drummond 2010; Bouckaert et al. 2019), SNAPP (Bryant et al. 2012), BEST (Liu and Pearl 2007), and ASTRAL (Mirarab et al. 2014).

Some of the issues that might cause conflict among gene trees include hidden paralogy, hybridization, recombination, horizontal gene transfer, incomplete lineage sorting, and substitution saturation (e.g., Galtier and Daubin 2008; Smith et al. 2015). In addition, some of the taxa included in a phylogeny might not have a complete genome available. This means that some of the genes included in the phylogenomic alignment might not be present in all of the taxa, which can result in an alignment with large amounts of missing data. Apart from that, not all of the genes evolve at the same evolutionary rate on each lineage (i.e., rate heterogeneity or heterotachy). Missing data and heterotachy can lead to long-branch attraction, an artefact that involves the clustering of long branches even though the species are not actually closely related (Felsenstein 1978; Huelsenbeck 1998). If this is left uncorrected, using the wrong tree topology will negatively affect the inference of the branch lengths and the model parameters, such as evolutionary rates and divergence times. Sampling taxa to include slow-evolving species (Aguinaldo et al. 1997) and to root deep-level trees (Brinkmann and Philippe 1999; Brinkmann et al. 2005) might help to reduce long-branch attraction.

13.2.2 Bayesian Divergence-Time Inference

Once a suitable species tree and sequence alignment are available, we can infer the species divergence times (Table 13.4). In theory, it is possible to infer the tree topology and the divergence times simultaneously (e.g., Drummond et al. 2006). For very large phylogenomic data sets, however, this is impractical because the computation is too expensive. Therefore, it is customary to fix the topology in the analysis, or at least to provide constraints on the monophyly of clades

Table 13.4 Selection of software for Bayesian inference of species divergence times. Based on Table 1 in dos Reis et al. (2016)

Software	Brief description	Citations
BEAST	Comprehensive suite of models. Particularly aimed at analyses of infectious diseases using molecular, phenotypic, and epidemiological data.	Suchard et al. (2018)
BEAST 2	Comprehensive suite of models. Especially strong for the analysis of serially sampled sequence data. Includes models of morphological traits.	Bouckaert et al. (2019)
MCMCtree	Comprehensive suite of models of rate variation. Fast approximate-likelihood method that allows the estimation of time-trees using genome alignments. Brownian motion model to analyse quantitative morphological data.	Yang (2007)
MrBayes	Large suite of models for morphological and molecular evolutionary analysis. Comprehensive suite of models of rate variation.	Ronquist et al. (2012b)
MultiDivtime	First Bayesian clock-dating program. Introduced the geometric Brownian model (autocorrelated-rates model) and the approximate-likelihood approach.	Thorne et al. (1998), Thorne and Kishino (2002)
RevBayes	Interactive suite of models. Uses probabilistic graphical models and its own language Rev. Requires the user to fully specify the model for the analysis. Models to analyse quantitative data implemented but not exhaustively tested.	Höhna et al. (2016)
PhyloBayes	Broad suite of models. Uses data augmentation to speed up likelihood calculation.	Lartillot et al. (2009)

(Drummond et al. 2006; Yang and Rannala 2006).

The posterior distribution of times (**t**) and rates (**r**) given the phylogenomic alignment (D) is

$$f(\mathbf{t},\mathbf{r}|D) = \frac{1}{z} f(\mathbf{t}) f(\mathbf{r}|\mathbf{t}) f(D|\mathbf{t},\mathbf{r}) \qquad (13.1)$$

where $f(\mathbf{t})$ is the prior on times, $f(\mathbf{r}|\ \mathbf{t})$ is the prior on the molecular rates, and $f(D|\ \mathbf{t}, \mathbf{r})$ is the likelihood of the phylogenomic alignment given **t** and **r**. The constant z is set so that the posterior distribution integrates to 1 and so it is a proper probability distribution. In Bayesian clock dating, z usually cannot be calculated analytically and thus MCMC sampling is needed (e.g., Thorne et al. 1998; Yang and Rannala 2006).

A review of the Bayesian theory of clock dating is given in Chap. 6 and in previous publications (Chaps. 6 and 7 in Yang 2014; Heath and Moore 2014). Here, we focus on how calculation of the prior and the likelihood affect the phylogenomic analysis. We discuss the strategies that can be followed to speed up the computation of the posterior and thus improve the computational efficiency of MCMC sampling on large phylogenomic alignments.

13.2.2.1 Approximating the Likelihood

Approximating the likelihood is perhaps the most important strategy to speed up clock dating in phylogenomic alignments. Calculation of the likelihood in a phylogeny is proportional to the number of distinct configurations of characters in an alignment column (site patterns). For phylogenomic alignments with millions of sites, likelihood computation can be very expensive, and thus a typical phylogenomic MCMC might take several months to complete. Thorne et al. (1998) suggested using the Taylor expansion of the log likelihood as an approximation to speed up the computation. This approximation allows Bayesian estimation of divergence times on large phylogenomic alignments that would otherwise be intractable.

Let $\mathbf{b} = \{b_i = t_i r_i\}$ be the vector of branch lengths (in substitutions per site) in the tree, where t_i is the time duration of the ith branch, and r_i is the molecular rate on the branch. Let $\ell(\mathbf{b}) = \log f(D|\ \mathbf{t}, \mathbf{r})$ be the log likelihood written as a function of the branch lengths. The Taylor expansion of $\ell(\mathbf{b})$ around the maximum-likelihood estimates of the branch lengths, $\widehat{\mathbf{b}}$, is

$$\ell(\mathbf{b}) \approx \ell\left(\widehat{\mathbf{b}}\right) + \mathbf{g}^T \Delta\mathbf{b} + \frac{1}{2}\Delta\mathbf{b}^T \mathbf{H} \Delta\mathbf{b} \qquad (13.2)$$

where $\Delta\mathbf{b} = \mathbf{b} - \widehat{\mathbf{b}}$, and $\mathbf{g} = \{g_i\}$ and $\mathbf{H} = \{H_{ij}\}$ are the gradient (the vector of first derivatives) and the Hessian (the matrix of second derivatives) of the likelihood function evaluated at the maximum-likelihood estimates. Thus, to use the approximation, a two-step procedure is followed. First, $\widehat{\mathbf{b}}$, $\mathbf{g}$, and $\mathbf{H}$ are estimated by maximum likelihood for each partition in the phylogenomic alignment. Note that the substitution model is chosen at this step. Then, once $\widehat{\mathbf{b}}$, $\mathbf{g}$, and $\mathbf{H}$ are obtained, they are used to approximate the likelihood in Eq. (13.2) during MCMC sampling.

We note the following about the likelihood approximation:

1. When the maximum-likelihood estimates of the branch lengths are inside the parameter space (i.e., when they are not 0 or ∞), the gradient is zero and the second term in Eq. (13.2) is null. In this case, the approximated likelihood is simply $L = \exp(\ell) \approx L\left(\widehat{\mathbf{b}}\right) \times \exp\left(\frac{1}{2}\Delta\mathbf{b}^T \mathbf{H} \Delta\mathbf{b}\right)$, which is proportional to the multivariate normal density with mean $\widehat{\mathbf{b}}$ and covariance matrix $-\mathbf{H}^{-1}$. When some branch lengths are zero, the second term dominates the approximation for those branches. When some branch lengths are infinite, the approximation does not work well. We suggest that taxa and/or partitions with infinite branches be removed from the analysis.
2. The approximation can be improved by using transformations on the branch lengths, $\mathbf{u} = h(\mathbf{b})$ (dos Reis and Yang 2011). The approximate log likelihood on the transformed parameter space is

$$\ell(\mathbf{u}) \approx \ell(\widehat{\mathbf{u}}) + \Delta\mathbf{u}^T \mathbf{g}_u + \frac{1}{2}\Delta\mathbf{u}^T \mathbf{H}_u \Delta\mathbf{u},$$

where $\mathbf{g}_u$ and $\mathbf{H}_u$ are the gradient and Hessian of the transformed likelihood. The square-root and arcsine-based transforms appear to provide very good approximations (Fig. 13.3).
3. The approximation improves as the number of sites in the alignment is increased. Analyses of empirical data have shown that as few as 10,000 sites are enough to provide an excellent approximation under a relaxed molecular clock on a mitochondrial mammal phylogeny (Fig. 13.4). Nevertheless, the approximation does not work well under the strict clock (Fig. 13.4). This is because, when the clock is violated (as usually happens in real data), proposed branch lengths during MCMC sampling will tend to be very far away from the maximum-likelihood estimates, i.e., in the left and right tails of the likelihood curve, where the approximation is poor (Fig. 13.3). Therefore, we do not recommend using approximate likelihood calculation under the strict clock unless it is used in the analysis of closely related organisms, when the clock is not violated.
4. Because we must estimate $\widehat{\mathbf{b}}$, $\mathbf{g}$, and $\mathbf{H}$ on a fixed tree before we carry out MCMC sampling of the posterior, it is not possible to use the approximate method to coestimate the tree topology and divergence times. Nonetheless, if there are uncertain nodes in the tree, it is possible instead to run a separate analysis on each topology. In this case, one set of $\widehat{\mathbf{b}}$, $\mathbf{g}$, and $\mathbf{H}$ must be estimated for each tree topology separately.

Bayesian inference with approximate likelihood calculation is orders of magnitude quicker than with exact calculation (Battistuzzi et al. 2011; Tamura et al. 2012; Mello et al. 2017). For instance, Battistuzzi et al. (2011) found speed-ups of up to 1000× when comparing the approximation with the exact method, without loss of accuracy. The approximate method is implemented in the programs MCMCtree (dos Reis and Yang 2011) and MultiDivTime (Thorne et al. 1998). The implementation in MCMCtree

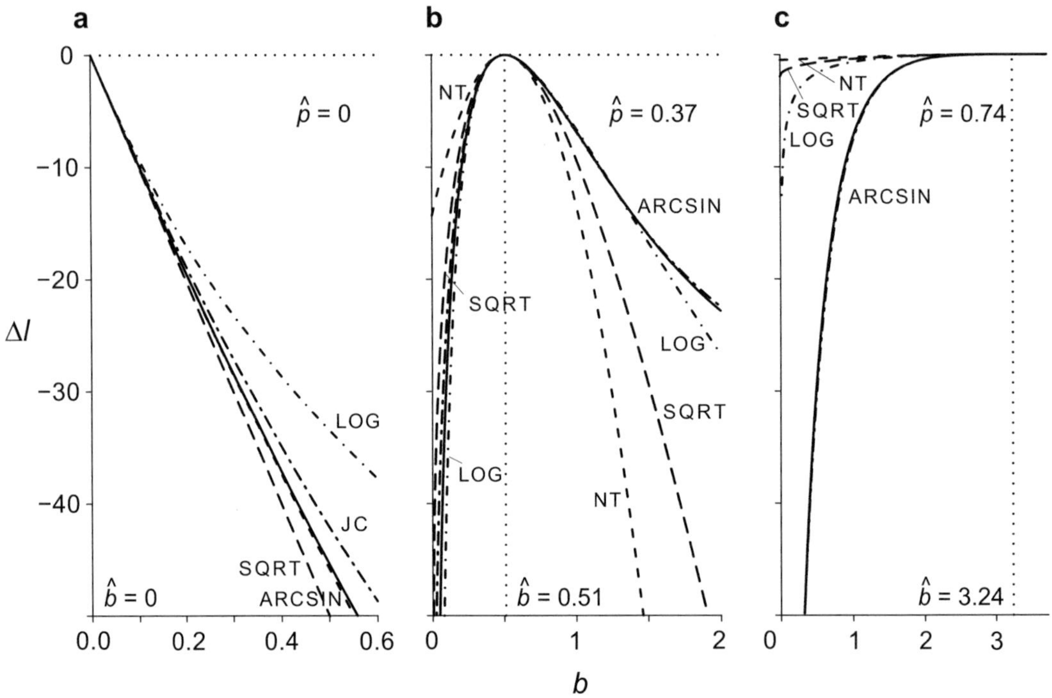

Fig. 13.3 The approximate-likelihood method for molecular clock dating. Log-likelihood curves are calculated exactly (solid line) and approximately (dashed lines) as a function of the molecular distance (the branch length, b) between two sequences of length $n = 100$. The JC69 substitution model is used. (**a**) Two identical sequences, $x = 0$ nucleotide differences, in which case the maximum-likelihood estimate of the branch length is $\hat{b} = 0$. (**b**) Two sequences with $x = 37$ nucleotide differences, in which case $0 < b < \infty$. (**c**) Two sequences with $x = 74$ nucleotide differences, in which case the alignment is almost saturated. The approximate likelihood is calculated using four transforms: NT, SQRT, LOG, and ARCSIN. The ARCSIN transform appears to provide the best approximation overall. Redrawn from Fig. 1 in dos Reis and Yang (2011)

has successfully been used to estimate divergence times on phylogenomic alignments of various data sets, including mammals (Meredith et al. 2011; dos Reis et al. 2012), birds (Jarvis et al. 2014), metazoans (dos Reis et al. 2015), and plants (Barba-Montoya et al. 2018; Morris et al. 2018); and to study the origin of life on Earth (Betts et al. 2018). A tutorial for approximate likelihood calculation with MCMCtree is given by dos Reis and Yang (2019).

13.2.2.2 The Rate Prior

The realization of the violation of the molecular clock motivated the search for methods that could accommodate rate heterogeneity across lineages (Welch and Bromham 2005), resulting in the development of relaxed-clock models (Thorne et al. 1998; Drummond et al. 2006). These approaches allow the evolutionary rate to vary across branches of the phylogeny, making it possible to determine which lineages evolve quickly or slowly. The adequacy of clock models has been assessed in several studies, in which different probability distributions are used to model rate evolution among lineages (e.g., Heath et al. 2012; Ho 2014; Zhu et al. 2015; dos Reis et al. 2016, 2018). The log-normal independent-rates model (Drummond et al. 2006; Rannala and Yang 2007) and the log-normal autocorrelated-rates model (Thorne et al. 1998; Rannala and Yang 2007) are the most commonly used. Using different rate models on the same data set can lead to different posterior estimates of node times (e.g., Aris-Brosou and Yang 2002;

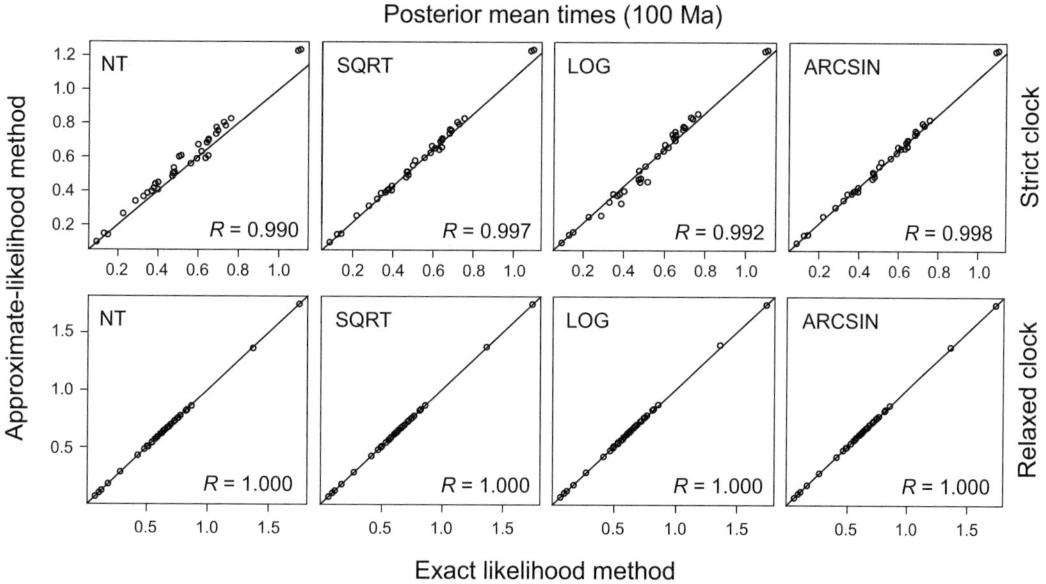

Fig. 13.4 Estimated posterior mean times under the strict clock and the autocorrelated-rates models for a mammal data set using approximate and exact likelihood calculations. The approximate likelihood is calculated using four transforms: NT, SQRT, LOG, and ARCSIN. The approximation does not work well under the strict clock when the clock is seriously violated. Redrawn from Fig. 3 in dos Reis and Yang (2011)

Beaulieu et al. 2015; dos Reis et al. 2015, 2018; Barba-Montoya et al. 2018). Therefore, selection of the rate model is a very important step before proceeding with divergence-time estimation.

Besides selecting the best rate model (which accounts for rate variation among lineages), it might also be important to select an appropriate model of rate heterogeneity among sites in genes (Gillespie 1991; Welch and Bromham 2005). For instance, rate heterogeneity among sites or alignment partitions can be accommodated by discretizing the data set into different categories. Current practice is to use a probability distribution such as the gamma (Yang 1994) or the Dirichlet (dos Reis et al. 2014) so that sites or loci assigned to the same category share the same evolutionary rate. Usually, a gamma distribution divides sites within a locus into several categories according to different evolutionary rates under a nucleotide substitution model. A similar approach can be used to model the prior on the mean rate among loci, first using a gamma distribution and then the Dirichlet distribution to partition rates among loci (dos Reis et al. 2014). Natural selection does not have the same impact on different codon positions in protein-coding genes, with first and second codon positions usually evolving much more slowly and having a smaller transition/transversion bias than third codon positions (Bofkin and Goldman 2007). Such different evolutionary patterns can be easily accommodated by separating the first and second codon positions from the third codon positions into two different data partitions.

Deciding on the best partitioning scheme is not easy for the user, but there is software that can automatically perform this task, such as ClockstaR (Duchêne and Ho 2014) and PartitionFinder (Lanfear et al. 2012, 2017). The algorithms of these two programs differ in their approach to assigning data subsets to partitions: ClockstaR uses a tree-distance metric to generate clusters using the partitioning-along-medoids method (Kaufman and Rousseeuw 1990), while PartitionFinder relies on a hierarchical clustering approach based on information-theoretic measures (Bayesian information criterion, Akaike information criterion, or Akaike information criterion with small-sample correction). Alternatively, the use of principal-components analysis to find better strategies to partition the data has also been explored (dos Reis et al. 2012). Several benchmarking analyses have been carried out with both simulated and real data sets with the aim of defining the best partitioning scheme to accommodate rate heterogeneity (Brown and Lemmon 2007; Duchêne et al. 2011; Duchêne and Ho 2014; Foster and Ho 2017; Angelis et al. 2018). Simulation results show no consensus on a best partitioning scheme, hence the importance of carefully choosing a suitable partitioning scheme before proceeding with Bayesian inference.

When the molecular clock is violated and the posterior times and rates are inferred, the concatenation approach tends to result in less precise estimates and wider credibility intervals (Angelis et al. 2018). Increasing the data size in the same partitioning block (concatenating sites) does not result in more informative data nor in more certain posterior estimates (dos Reis and Yang 2013; dos Reis et al. 2014; Zhu et al. 2015). Note that rate heterogeneity along branches is confounded with the divergence times in the molecular branch lengths. Thus, information on patterns of rate heterogeneity among loci is lost if the different loci are concatenated. Nevertheless, if the alignment is correctly partitioned, the process of rate evolution can then be thought of as being replicated in each partition. Consequently, when the molecular clock is relaxed, partitioning the molecular data can add information that might lead to posterior estimates with smaller variances. Figure 13.5 shows the resulting posterior divergence times estimated with different partitioning schemes and the two relaxed-clock models implemented in MCMCtree (i.e., the independent-rates and the autocorrelated-rates models) for a plant data set (Angelis et al. 2018). These results are just one example of the dramatic effect that changing the partitioning scheme can have on the estimated posterior times.

Dating software such as MCMCtree (Yang 2007), BEAST (Suchard et al. 2018), BEAST 2 (Bouckaert et al. 2019), MrBayes (Ronquist et al. 2012b), and RevBayes (Höhna et al. 2016) have implemented different relaxed-clock models, the use of partitioned data sets, and discrete probability distributions to model the rate prior. For programs that do not currently implement approximate likelihood, phylogenomic data can be divided into smaller subsets of genes and/or species for the dating analysis (e.g., Misof et al. 2014; Upham et al. 2019). The process can be repeated for further subsamples of genes and/or species and the results collated to provide consolidated estimates. This procedure, however, does not make efficient use of all of the data and does not benefit from asymptotic reduction of estimated variances as when using the complete data sets.

13.2.2.3 The Time Prior and Fossil Calibrations

The posterior divergence times estimated in a dating analysis will not be calibrated to geological time (i.e., absolute ages) unless fossil or geological information is incorporated. Common practice is to use probability distributions to model node ages based on fossil evidence (see Chap. 8). These distributions can account for fossil uncertainty, such as uncertainty related to the age of the fossil or its assignment to a particular lineage. In many cases, calibration distributions are set to comply with minimum and maximum node ages as informed by the fossil evidence (Benton and Donoghue 2007). Deciding which probability distribution and/or which maximum and minimum bounds to use, however, involves some subjectivity. This means that different studies analysing the same data set might use different calibration distributions, thus resulting in different posterior time estimates. There is ongoing debate on how calibrations should be constructed (e.g., Tavaré et al. 2002; Yang and Rannala 2006; Benton and Donoghue 2007; Heled and Drummond 2012; Heath et al. 2012; Parham et al. 2012; Nowak et al. 2013).

In addition to these difficulties, some of the models used by different dating software might not be implemented in the same way (e.g., rate models), which makes it difficult to compare the results obtained by different tools. Therefore, it is very important for the user to first run their preferred dating software without data to verify the

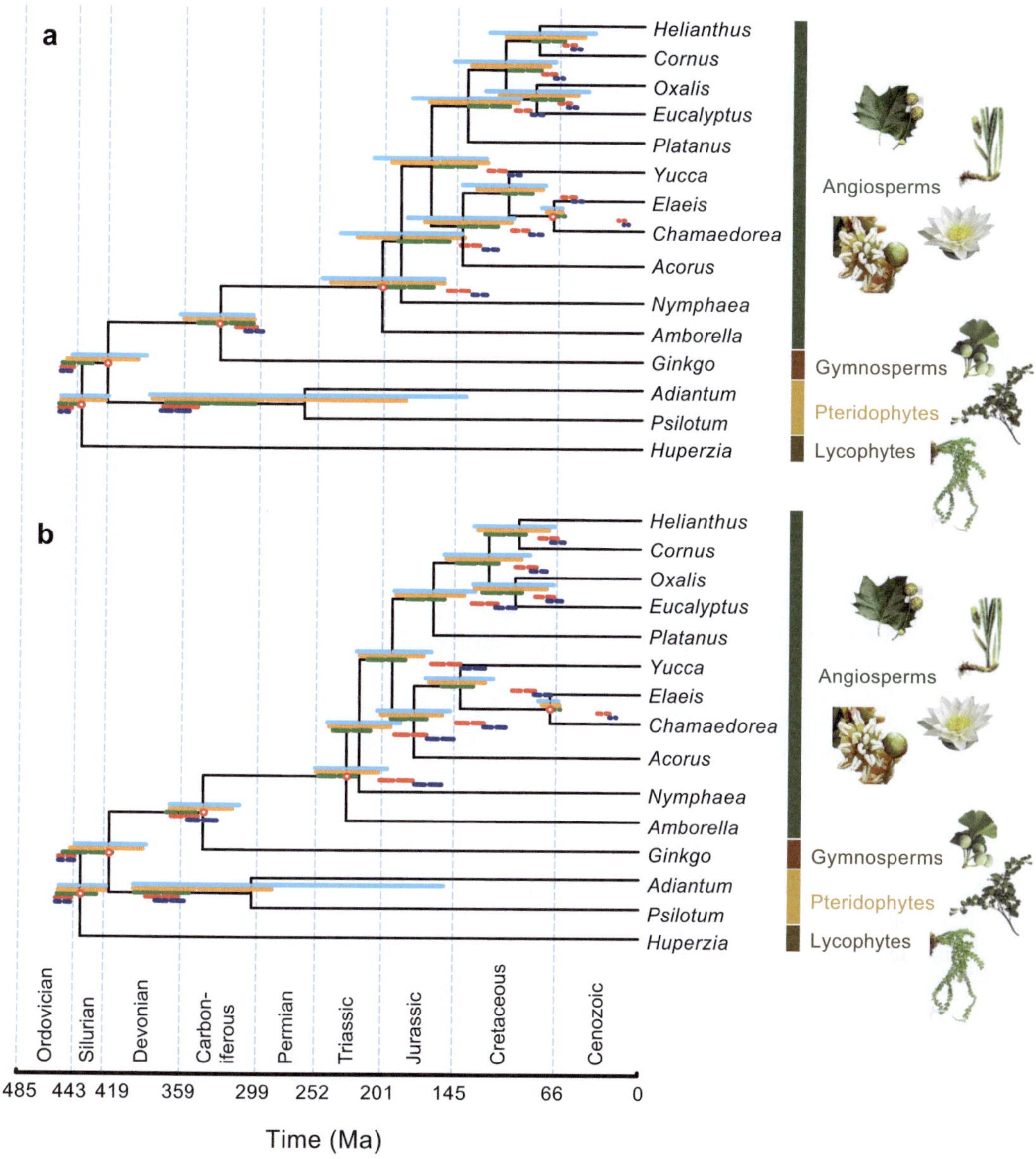

Fig. 13.5 Effect of partitioning schemes on divergence-time estimates in a plant phylogeny under the (**a**) independent-rates model and (**b**) autocorrelated-rates model. Coloured bars show the estimates from five different partitioning schemes. From top to bottom, the schemes applied are: (1) concatenation and ClockstaR, (2) codon position, (3) PartitionFinder, (4) gene, and (5) gene codon positions. The choice of partitioning scheme has a substantial effect on the divergence-time estimates. Modified from Fig. 4 in Angelis et al. (2018)

rate and time prior distributions applied in the dating analysis. For instance, there are some dating programs that do not require the user to input a prior for the root age. This might be troublesome if users decide to accept the default parameters, which is quite often the case among users with less technical background. Using the default time prior for the root age, which might not be an appropriate prior, might have a big impact on the posterior time estimates.

Although using time constraints on small data sets to estimate divergence times is straightforward and computationally tractable, it becomes very inefficient with phylogenomic data sets. This issue was tackled by dos Reis et al. (2012, 2018) through their sequential Bayesian procedure. In this approach, the user first estimates the posterior times of a small sample from the phylogenomic data set (i.e., few species and large alignment) and then uses the posterior times from this analysis to build the priors in a subsequent analysis (i.e., with many species but with a short alignment). If a phylogenomic data set, D, can be split into two independent (and thus non-overlapping) subsets, $D = (D_1, D_2)$, then the posterior distribution of θ (e.g., divergence times, molecular rates, and model parameters) can be written as

$$\begin{aligned} f(\theta|D) &\propto f(\theta)f(D_1|\theta)f(D_2|\theta) \\ &\propto f(\theta|D_1)f(D_2|\theta) \end{aligned} \tag{13.3}$$

In other words, the posterior when using the first data partition, $f(\theta| D_1) \propto f(\theta)f(D_1| \theta)$, can be used as the prior for the subsequent analysis of the second subset. This strategy of partitioning data into independent subsets is well known in Bayesian data analysis (Gelman et al. 2013). We note that this approach is different from the use of secondary calibrations (see Graur and Martin 2004): the sequential Bayesian analysis is a method that uses two non-overlapping and independent partitions, D_1 and D_2 (dos Reis et al. 2018). When using MCMCtree, this methodology reduces the computational burden in phylogenomic dating analyses because it avoids the computational cost of sampling rates for lineages with missing data in a partition. In practice, we would first run a dating analysis of the first partition and collect an MCMC sample of times and rates. Then, a statistical distribution (such as the skew-t, Wilkinson et al. 2011) can be used to approximate the posterior density for each node age. The fitted distribution is then used to construct the calibrations for the corresponding nodes when analysing the second subset. We note that, when using this method with fitted skew-t distributions, the posterior correlation among node ages is ignored, in which case this approach gives an approximation to the true posterior.

Even though large amounts of data can be used for phylogenomic dating analyses, there is uncertainty regarding the molecular branch lengths: the times and rates are confounded and cannot be estimated separately. Consequently, as the amount of molecular data increases and tends to infinity, the uncertainty in posterior time estimates does not converge to zero (Yang and Rannala 2006; Rannala and Yang 2007). Instead, the uncertainty converges to a limiting distribution that depends on the fossil uncertainties. This is known as the infinite-sites theory (Yang and Rannala 2006; Rannala and Yang 2007). An example is given in Fig. 13.6, where the 95% credibility interval (CI) widths for each estimated node age, i.e., a measure of the uncertainty in the estimates, are plotted against different data sizes. The uncertainty in the estimates does not converge to zero despite the data size increasing substantially. Therefore, there is a limit to the amount of data that is informative in phylogenomic analysis. How informative a molecular alignment is on the divergence times can be assessed by using the so-called infinite-sites plot, in which the posterior CI widths are plotted against the posterior mean node times (Fig. 13.7). If the data points fall on a straight line, the limit of uncertainty has been reached; including further data in the analysis will not improve the time estimates.

13.2.2.4 Running the MCMC and Summarizing the Posterior

Once the data are prepared and the appropriate model and prior distributions have been chosen, Bayesian inference of evolutionary rate and divergence times can proceed. To obtain our posterior estimates, we need to run the MCMC so that we can obtain a sample from the posterior distribution. This simulation technique is necessary because the normalizing constant of the Bayes equation (z in Eq. 13.1) cannot be obtained analytically.

When analysing phylogenomic data, the MCMC chain might take too long to run before convergence is reached. Therefore, poor choices of proposal mechanisms and/or step lengths can lead to poor mixing and problems with

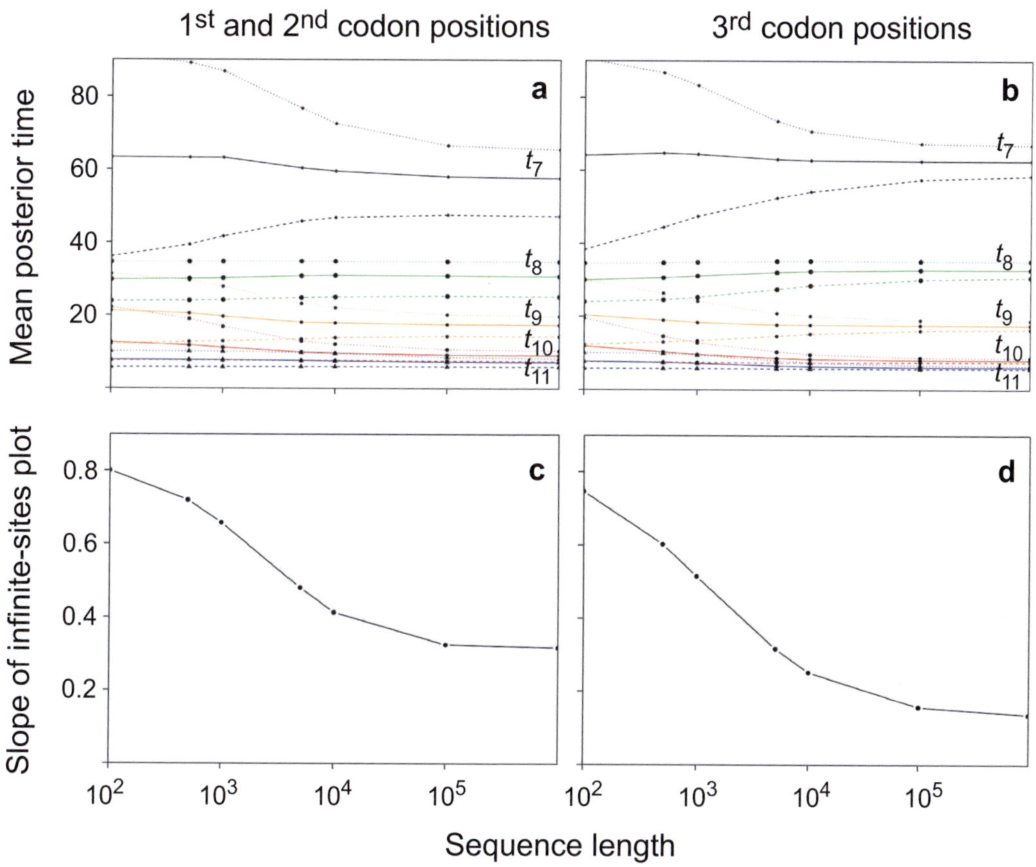

Fig. 13.6 Effect of sequence length on uncertainty in estimated divergence times. Top panels (**a**) and (**b**): the 95% credibility interval (CI, dotted line) is plotted along the mean posterior time (solid line) for the five divergence times in a five-species primate phylogeny for (**a**) 1st and 2nd and (**b**) 3rd codon positions. Note that, for a limiting sequence length, the uncertainty (measured as the CI width) in the time estimates cannot be further reduced. Bottom panels (**c**) and (**d**): the slope of the infinite-sites plot (see Fig. 13.7) indicates how much uncertainty remains in a dating analysis. For instance, in (**c**) and for 100 nucleotides, the slope of the infinite-sites plot is 0.8, meaning that 0.8 million years of uncertainty are added to the posterior CI width for every 1 million years of divergence. For large data, the slope does not go to zero but converges to a limit. Redrawn from Fig. 9 in dos Reis and Yang (2013)

convergence (Yang and Rodríguez 2013). The user can diagnose these issues by plotting the sampled parameters of interest against the MCMC sampling iteration, known as an MCMC trace plot. Software such as Tracer (Rambaut et al. 2018) can be used, although R (R Core Team 2019) can also be used for this purpose. Figure 13.8 shows an example of a healthy (or efficient) MCMC chain in comparison with a chain that is inefficient and has not converged.

If convergence or mixing issues are observed in the MCMC traces, there are a number of strategies that can be used to improve the efficiency of the MCMC. For instance, we can increase the number of MCMC iterations if the chain has not achieved convergence and needs to run longer. Alternatively, we might increase the sampling frequency but keep the same number of samples to be collected during the MCMC (Nascimento et al. 2017). This has the effect of

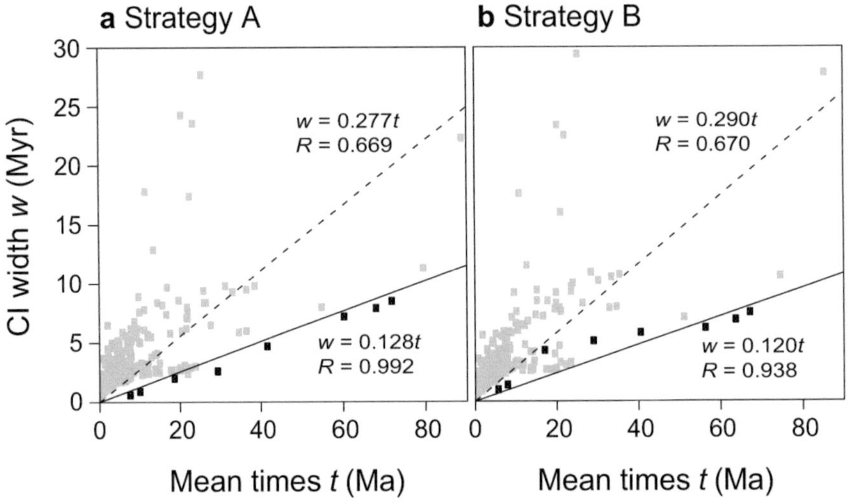

Fig. 13.7 The infinite-sites plot of credibility interval widths (w) against posterior mean times (t). The data are a primate phylogeny of 372 species, with eight of these species having complete genomes in the analysis. Two fossil calibration strategies were used: (**a**) Strategy A and (**b**) Strategy B (for details, see dos Reis et al. 2018). An autocorrelated-rates model was used in both cases. For the genome-scale data (black), the dots fall close to a straight line, meaning that the uncertainty in divergence times for these nodes cannot be further reduced by simply adding more genomic data. For the remaining species (grey), which have relatively short alignments, the dots are far from a straight line, indicating that including more molecular data for these species in the analysis might improve the divergence-time estimates. Redrawn from Fig. 6 in dos Reis et al. (2018)

lengthening the MCMC chain without producing excessively large files. Another option is for the user to run several chains using overdispersed starting points, and later combine the samples to generate a unified posterior summary. As a general rule of thumb, the user should aim to collect an effective sample size between 1000 and 10,000; although effective sample sizes as low as 200 are commonly found in phylogenetic analyses. For more details on MCMC efficiency and effective sample sizes, see Nascimento et al. (2017).

13.3 Further Considerations

The initial idea of phylogenomic data as the ultimate solution to the issues faced in phylogenetic analyses, such as polytomies or gene-tree conflict, was overestimated. Not only have these problems persisted, but new and more complex computational challenges have arisen with the increased availability of whole-genome data. There is a constant search for new computational methods to improve the modelling of evolutionary processes. For instance, some software packages have implemented the multispecies coalescent model, which integrates incomplete lineage sorting and can potentially allow hybridization. Unfortunately, other evolutionary processes such as recombination are more complex to model and difficult to integrate into the multispecies coalescent model.

In addition to computational limitations, it is worth paying attention to the selection of the best model for the data. The user has the responsibility of understanding their data, and thus should be familiar with the steps involved in a complete Bayesian clock dating analysis starting from the way the data are preprocessed and including the Bayesian inference. These steps include deciding which fossils are best suited to calibrate the phylogeny, deciding on the most appropriate partitioning scheme, and choosing appropriate parameters for the Bayesian model (e.g., molecular clock, prior on the root age, prior on the evolutionary rate, nucleotide substitution model, fossilized birth–death process, and the tree

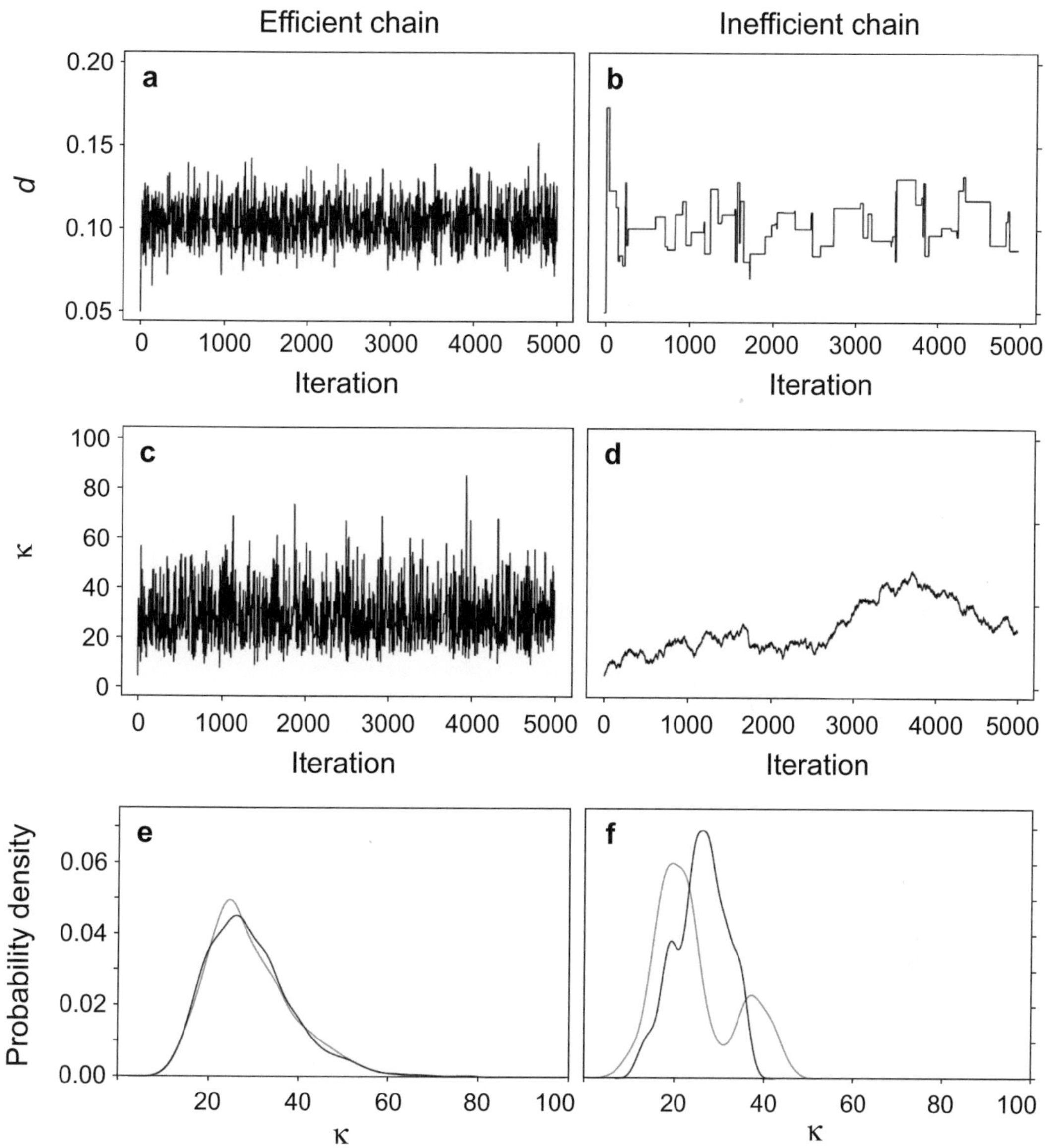

Fig. 13.8 Trace plots and histograms used to diagnose the convergence and mixing efficiency of Markov chain Monte Carlo (MCMC) sampling. The trace plots for efficient MCMC chains can be distinguished by their caterpillar shape (**a**, **c**), while the inefficient chains (**b**, **d**) show mixing problems because they either become stuck at the same values or traverse the space slowly. The histograms of an efficient chain (**e**) clearly show a distribution with an identifiable mean, while those of the inefficient chain (**f**) do not. Redrawn from Fig. 2 in Nascimento et al. (2017)

topology). These ultimately depend on the data set being analysed. Unfortunately, users with a lack of statistical or computational background often select default software settings as an easy-to-follow protocol. This is poor practice and should not be followed. We would also like to stress again how important it is for the user to run their preferred MCMC software without data to verify the prior that will be used in the Bayesian dating analysis. In fact, many common MCMC clock dating problems, such as misspecified fossil calibrations, can be easily diagnosed when analysing the prior.

While the ages of fossils are carefully used to select the most accurate probability distributions to calibrate the phylogeny, fossils are also informative about another important feature: their morphology. These characters can also be used in dating analyses because it is possible to infer morphological distances between extinct and extant species in a phylogeny (see Chap. 7). Discrete morphological characters have been widely used in models that can treat fossils as dated tips, combined with molecular data to infer species divergence times (Nylander et al. 2004; Pyron 2011; Ronquist et al. 2012b). The most commonly used model of discrete character evolution is the Mk model (Lewis 2001). Nevertheless, there are some disadvantages: the Mk model assumes that rates of change are equal among character states and, even though this assumption can be relaxed (e.g., symmetrical and all-rates-different models; Paradis et al. 2004), character correlation is not accounted for (Felsenstein 2005) and a correction for acquisition bias is still needed (Lewis 2001; Leaché et al. 2015).

Continuous morphological characters can help to resolve the issues with discrete characters, and thus are a promising data source for Bayesian dating analyses. Recent studies have shown how morphometric data can be used as quantitative characters under the Brownian diffusion model (Felsenstein 1973) to estimate phylogenies (Parins-Fukuchi 2018a, b) and infer the evolutionary rate and species divergence times (Álvarez-Carretero et al. 2019). The latter approach has been implemented in the dating software MCMCtree to account for character correlation and variation in character measurements within a population. Using these new dating approaches, with phylogenomic data sets combined with thousands of aligned morphometric landmarks, offers very interesting prospects and may lead to a new area of research: morphogenomic dating.

In summary, improvement in phylogenomic dating seems to be tied to parallel progress in computational and technological equipment. As soon as more efficient algorithms and more realistic models are included in Bayesian dating analyses, the estimates of species divergence times should improve.

References

Aguinaldo AMA, Turbeville JM, Linford LS, Rivera MC, Garey JR, Raff RA, Lake JA (1997) Evidence for a clade of nematodes, arthropods and other moulting animals. Nature 387:489–493

Altenhoff AM, Dessimoz C (2009) Phylogenetic and functional assessment of orthologs inference projects and methods. PLOS Comput Biol 5:e1000262

Altenhoff AM, Boeckmann B, Capella-Gutierrez S, Dalquen DA, DeLuca T, Forslund K, Huerta-Cepas J, Linard B, Pereira C, Pryszcz LP, Schrieber F, da Silva AS, Szklarczyk D, Train C-M, Bork P, Lecompte O, von Mering C, Xenarios I, Sjölander K, Jensen LJ, Martin MJ, Muffato M, Quest for Orthologs Consortium, Gabaldón T, Lewis SE, Thomas PD, Sonnhammer E, Dessimoz C (2016) Standardized benchmarking in the quest for orthologs. Nat Methods 13:425–430

Altenhoff AM, Glover NM, Train C-M, Kaleb K, Vesztrocy AW, Dylus D, de Farias TM, Zile K, Stevenson C, Long J, Redestig H, Gonnet GH, Dessimoz C (2018) The OMA orthology database in 2018: retrieving evolutionary relationships among all domains of life through richer web and programmatic interfaces. Nucleic Acids Res 46:D477–D485

Álvarez-Carretero S, Goswami A, Yang Z, dos Reis M (2019) Bayesian estimation of species divergence times using correlated quantitative characters. Syst Biol 68:967–986

Andrews S (2010) FastQC: a quality control tool for high throughput sequence data. http://www.bioinformatics.babraham.ac.uk/projects/fastqc

Angelis K, Álvarez-Carretero S, dos Reis M, Yang Z (2018) An evaluation of different partitioning strategies for Bayesian estimation of species divergence times. Syst Biol 67:61–77

Aris-Brosou S, Yang Z (2002) Effects of models of rate evolution on estimation of divergence dates with special reference to the metazoan 18S ribosomal RNA phylogeny. Syst Biol 51:703–714

Ballenghien M, Faivre N, Galtier N (2017) Patterns of cross-contamination in a multispecies population genomic project: detection, quantification, impact, and solutions. BMC Biol 15:25

Bankevich A, Nurk S, Antipov D, Gurevich AA, Dvorkin M, Kulikov AS, Lesin VM, Nikolenko SI, Pham S, Prjibelski AD, Pyshkin AV, Sirotkin AV, Vyahhi N, Tesler G, Alekseyev MA, Pevzner PA (2012) SPAdes: a new genome assembly algorithm and its applications to single-cell sequencing. J Comput Biol 19:455–477

Barba-Montoya J, dos Reis M, Schneider H, Donoghue PCJ, Yang Z (2018) Constraining uncertainty in the timescale of angiosperm evolution and the veracity of a Cretaceous Terrestrial Revolution. New Phytol 218:819–834

Battistuzzi FU, Billing-Ross P, Paliwal A, Kumar S (2011) Fast and slow implementations of relaxed-clock methods show similar patterns of accuracy in estimating divergence times. Mol Biol Evol 28:2439–2442

Bayes T (1763) An essay towards solving a problem in the doctrine of chances. By the late Rev. Mr. Bayes, F. R. S. communicated by Mr. Price, in a letter to John Canton, A. M. F. R. S. Philos Trans R Soc 53:370–418

Beaulieu JM, O'Meara BC, Crane P, Donoghue MJ (2015) Heterogeneous rates of molecular evolution and diversification could explain the Triassic Age estimate for angiosperms. Syst Biol 64:869–878

Beaumont MA, Zhang W, Balding DJ (2002) Approximate Bayesian computation in population genetics. Genetics 162:2025–2035

Benjamini Y, Speed TP (2012) Summarizing and correcting the GC content bias in high-throughput sequencing. Nucleic Acids Res 40:e72

Benson DA, Cavanaugh M, Clark K, Karsch-Mizrachi I, Ostell J, Pruitt KD, Sayers EW (2018) GenBank. Nucleic Acids Res 46:D41–D47

Benton MJ, Donoghue PCJ (2007) Paleontological evidence to date the Tree of Life. Mol Biol Evol 24:26–53

Benton MJ, Donoghue PCJ, Asher RJ, Friedman M, Near TJ, Vinther J (2015) Constraints on the timescale of animal evolutionary history. Palaeontol Electron 18:1.1FC

Betts HC, Puttick MN, Clark JW, Williams TA, Donoghue PCJ, Pisani D (2018) Integrated genomic and fossil evidence illuminates life's early evolution and eukaryote origin. Nat Ecol Evol 2:1556–1562

Blackshields G, Sievers F, Shi W, Wim A, Higgins DG (2010) Sequence embedding for fast construction of guide trees for multiple sequence alignment. Algorithms Mol Biol 5:21

Boeckmann B, Robinson-Rechavi M, Xenarios I, Dessimoz C (2011) Conceptual framework and pilot study to benchmark phylogenomic databases based on reference gene trees. Br Bioinform 12:423–435

Bofkin L, Goldman N (2007) Variation in evolutionary processes at different codon positions. Mol Biol Evol 24:513–521

Boisvert S, Raymond F, Godzaridis É, Laviolette F, Corbeil J (2012) Ray Meta: scalable de novo metagenome assembly and profiling. Genome Biol 13:R122

Bolger AM, Lohse M, Usadel B (2014) Trimmomatic: a flexible trimmer for Illumina sequence data. Bioinformatics 30:2114–2120

Bouckaert R, Vaughan TG, Barido-Sottani J, Duchêne S, Fourment M, Gavryushkina A, Heled J, Jones G, Kühnert D, De Maio N, Matschiner M, Mendes FK, Müller NF, Ogilvie HA, du Plessis L, Popinga A, Rambaut A, Rasmussen D, Siveroni I, Suchard MA, Wu C-H, Xie D, Zhang C, Stadler T, Drummond AJ (2019) BEAST 2.5: an advanced software platform for Bayesian evolutionary analysis. PLOS Comput Biol 15:e1006650

Bradley RK, Roberts A, Smoot M, Juvekar S, Do J, Dewey C, Holmes I, Pachter L (2009) Fast statistical alignment. PLOS Comput Biol 5:e1000392

Brinkmann H, Philippe H (1999) Archaea sister group of Bacteria? Indications from tree reconstruction artifacts in ancient phylogenies. Mol Biol Evol 16:817–825

Brinkmann H, van der Giezen M, Zhou Y, de Raucourt GP, Philippe H (2005) An empirical assessment of long-branch attraction artefacts in deep eukaryotic phylogenomics. Syst Biol 54:743–757

Brown JM, Lemmon AR (2007) The importance of data partitioning and the utility of Bayes factors in Bayesian phylogenetics. Syst Biol 56:643–655

Bryant D, Waddell P (1998) Rapid evaluation of least-squares and minimum-evolution criteria on phylogenetic trees. Mol Biol Evol 15:1346–1359

Bryant D, Bouckaert R, Felsenstein J, Rosenberg NA, RoyChoudhury A (2012) Inferring species trees directly from biallelic genetic markers: bypassing gene trees in a full coalescent analysis. Mol Biol Evol 29:1917–1932

Camin JH, Sokal RR (1965) A method for deducing branching sequences in phylogeny. Evolution 19:311–326

Cantarel BL, Korf I, Robb SMC, Parra G, Ross E, Moore B, Holt C, Sánchez Alvarado A, Yandell M (2008) MAKER: an easy-to-use annotation pipeline designed for emerging model organism genomes. Genome Res 18:188–196

Chen F, Mackey AJ, Vermunt JK, Roos DS (2007) Assessing performance of orthology detection strategies applied to eukaryotic genomes. PLOS ONE 2:e383

Cheung M-S, Down TA, Latorre I, Ahringer J (2011) Systematic bias in high-throughput sequencing data and its correction by BEADS. Nucleic Acids Res 39: e103

Chowdhury B, Garai G (2017) A review on multiple sequence alignment from the perspective of genetic algorithm. Genomics 109:419–431

Clarke JT, Warnock RCM, Donoghue PCJ (2011) Establishing a time-scale for plant evolution. New Phytol 192:266–301

Dalquen DA, Altenhoff AM, Gonnet GH, Dessimoz C (2013) The impact of gene duplication, insertion, deletion, lateral gene transfer and sequencing error on orthology inference: a simulation study. PLOS ONE 8:e56925

Davín AA, Tannier E, Williams TA, Boussau B, Daubin V, Szöllősi GJ (2018) Gene transfers can date the Tree of Life. Nat Ecol Evol 2:904–909

Delsuc F, Brinkmann H, Philippe H (2005) Phylogenomics and the reconstruction of the Tree of Life. Nat Rev Genet 6:361–375

Do CB, Mahabhashyam MSP, Brudno M, Batzoglou S (2005) ProbCons: probabilistic consistency-based multiple sequence alignment. Genome Res 15:330–340 dos Reis M, Yang Z (2011) Approximate likelihood calculation on a phylogeny for Bayesian estimation of divergence times. Mol Biol Evol 28:2161–2172

dos Reis M, Yang Z (2013) The unbearable uncertainty of Bayesian divergence time estimation. J Syst Evol 51:30–43

dos Reis M, Yang Z (2019) Bayesian molecular clock dating using genome-scale datasets. In: Anisimova M (ed) Evolutionary genomics. Springer, New York, pp 309–330

dos Reis M, Inoue J, Hasegawa M, Asher RJ, Donoghue PCJ, Yang Z (2012) Phylogenomic datasets provide both precision and accuracy in estimating the timescale of placental mammal phylogeny. Proc R Soc B 279:3491–3500

dos Reis M, Zhu T, Yang Z (2014) The impact of the rate prior on Bayesian estimation of divergence times with multiple loci. Syst Biol 63:555–565

dos Reis M, Thawornwattana Y, Angelis K, Telford MJ, Donoghue PCJ, Yang Z (2015) Uncertainty in the timing of origin of animals and the limits of precision in molecular timescales. Curr Biol 25:2939–2950

dos Reis M, Donoghue PCJ, Yang Z (2016) Bayesian molecular clock dating of species divergences in the genomics era. Nat Rev Genet 17:71–80

dos Reis M, Gunnell GF, Barba-Montoya J, Wilkins A, Yang Z, Yoder AD (2018) Using phylogenomic data to explore the effects of relaxed clocks and calibration strategies on divergence time estimation: primates as a test case. Syst Biol 67:594–615

Drummond AJ, Ho SYW, Phillips MJ, Rambaut A (2006) Relaxed phylogenetics and dating with confidence. PLOS Biol 4:e88

Duchêne S, Ho SYW (2014) Using multiple relaxed-clock models to estimate evolutionary timescales from DNA sequence data. Mol Phylogenet Evol 77:65–70

Duchêne S, Archer FI, Vilstrup J, Caballero S, Morin PA (2011) Mitogenome phylogenetics: the impact of using single regions and partitioning schemes on topology, substitution rate and divergence time estimation. PLOS ONE 6:e27138

Edgar RC (2004) MUSCLE: multiple sequence alignment with high accuracy and high throughput. Nucleic Acids Res 32:1792–1797

Eisen JA, Fraser CM (2003) Phylogenomics: intersection of evolution and genomics. Science 300:1706–1707

Ekblom R, Wolf JBW (2014) A field guide to whole-genome sequencing, assembly and annotation. Evol Appl 7:1026–1042

Emms DM, Kelly S (2015) OrthoFinder: solving fundamental biases in whole genome comparisons dramatically improves orthogroup inference accuracy. Genome Biol 16:157

Felsenstein J (1973) Maximum-likelihood estimation of evolutionary trees from continuous characters. Am J Hum Genet 25:471–492

Felsenstein J (1978) Cases in which parsimony or compatibility methods will be positively misleading. Syst Zool 27:401–410

Felsenstein J (1981) Evolutionary trees from DNA sequences: a maximum likelihood approach. J Mol Evol 17:368–376

Felsenstein J (1988) Phylogenies from molecular sequences: inference and reliability. Annu Rev Genet 22:521–565

Felsenstein J (2005) Using the quantitative genetic threshold model for inferences between and within species. Philos Trans R Soc B 360:1427–1434

Fitch WM (1970) Distinguishing homologous from analogous proteins. Syst Biol 19:99–113

Fitch WM, Langley CH (1976) Evolutionary rates in proteins: neutral mutations and the molecular clock. In: Goodman M, Tashian RE, Tashian JH (eds) Molecular anthropology. Springer, Boston, MA, pp 197–219

Fletcher W, Yang Z (2010) The effect of insertions, deletions, and alignment errors on the branch-site test of positive selection. Mol Biol Evol 27:2257–2267

Florea L, Souvorov A, Kalbfleisch TS, Salzberg S (2011) Genome assembly has a major impact on gene content: a comparison of annotation in two *Bos taurus* assemblies. PLOS ONE 6:e21400

Foster CSP, Ho SYW (2017) Strategies for partitioning clock models in phylogenomic dating: application to the angiosperm evolutionary timescale. Genome Biol Evol 9:2752–2763

Galtier N, Daubin V (2008) Dealing with incongruence in phylogenomic analyses. Philos Trans R Soc B 363:4023–4029

Gee H (2003) Ending incongruence. Nature 425:782

Gelman A, Carlin JB, Stern HS, Dunson DB, Vehtari A, Rubin DB (2013) Bayesian data analysis, 3rd edn. Chapman and Hall/CRC, Boca Raton, FL

Gillespie JH (1991) The causes of molecular evolution. Oxford University Press, Oxford, UK

Goodman M, Barnabas J, Matsuda G, Moore GW (1971) Molecular evolution in the descent of man. Nature 233:604–613

Grabherr MG, Haas BJ, Yassour M, Levin JZ, Thompson DA, Amit I, Adiconis X, Fan L, Raychowdhury R, Zeng Q, Chen Z, Mauceli E, Hacohen N, Gnirke A, Rhind N, di Palma F, Birren BW, Nusbaum C, Lindblad-Toh K, Friedman N, Regev A (2011) Full-length transcriptome assembly from RNA-Seq data without a reference genome. Nat Biotechnol 29:644–652

Graur D, Martin W (2004) Reading the entrails of chickens: molecular timescales of evolution and the illusion of precision. Trends Genet 20:80–86

Guindon S (2010) Bayesian estimation of divergence times from large sequence alignments. Mol Biol Evol 27:1768–1781

Guo Y, Ye F, Sheng Q, Clark T, Samuels DC (2014) Three-stage quality control strategies for DNA re-sequencing data. Br Bioinform 15:879–889

Gurevich A, Saveliev V, Vyahhi N, Tesler G (2013) QUAST: quality assessment tool for genome assemblies. Bioinformatics 29:1072–1075

Heath TA, Moore BR (2014) Bayesian inference of species divergence times. In: Chen M-H, Kuo L, Lewis

PO (eds) Bayesian phylogenetics: methods, algorithms, and applications. Chapman and Hall/CRC, Boca Raton, FL, pp 277–318

Heath TA, Holder MT, Huelsenbeck JP (2012) A Dirichlet process prior for estimating lineage-specific substitution rates. Mol Biol Evol 29:939–955

Heath TA, Huelsenbeck JP, Stadler T (2014) The fossilized birth-death process for coherent calibration of divergence-time estimates. Proc Natl Acad Sci USA 111:E2957–E2966

Heled J, Drummond AJ (2010) Bayesian inference of species trees from multilocus data. Mol Biol Evol 27:570–580

Heled J, Drummond AJ (2012) Calibrated tree priors for relaxed phylogenetics and divergence time estimation. Syst Biol 61:138–149

Ho SYW (2014) The changing face of the molecular evolutionary clock. Trends Ecol Evol 29:496–503

Höhna S, Landis MJ, Heath TA, Boussau B, Lartillot N, Moore BR, Huelsenbeck JP, Ronquist F (2016) RevBayes: Bayesian phylogenetic inference using graphical models and an interactive model-specification language. Syst Biol 65:726–736

Holder M, Lewis PO (2003) Phylogeny estimation: traditional and Bayesian approaches. Nat Rev Genet 4:275–284

Huelsenbeck JP (1998) Systematic bias in phylogenetic analysis: is the Strepsiptera problem solved? Syst Biol 47:519–537

Huelsenbeck JP, Larget B, Swofford D (2000) A compound Poisson process for relaxing the molecular clock. Genetics 154:1879–1892

Huelsenbeck JP, Ronquist F, Nielsen R, Bollback JP (2001) Bayesian inference of phylogeny and its impact on evolutionary biology. Science 294:2310–2314

Huerta-Cepas J, Szklarczyk D, Forslund K, Cook H, Heller D, Walter MC, Rattei T, Mende DR, Sunagawa S, Kuhn M, Jensen LJ, von Mering C, Bork P (2016) eggNOG 4.5: a hierarchical orthology framework with improved functional annotations for eukaryotic, prokaryotic and viral sequences. Nucleic Acids Res 44:D286–D293

Hulsen T, Huynen MA, de Vlieg J, Groenen PMA (2006) Benchmarking ortholog identification methods using functional genomics data. Genome Biol 7:R31

Jackman SD, Vandervalk BP, Mohamadi H, Chu J, Yeo S, Hammond SA, Jahesh G, Khan H, Coombe L, Warren RL, Birol I (2017) ABySS 2.0: resource-efficient assembly of large genomes using a Bloom filter. Genome Res 27:768–777

Jarvis ED, Mirarab S, Aberer AJ, Li B, Houde P, Li C, Ho SYW, Faircloth BC, Nabholz B, Howard JT, Suh A, Weber CC, da Fonseca RR, Li J, Zhang F, Li H, Zhou L, Narula N, Liu L, Ganapathy G, Boussau B, Bayzid MS, Zavidovych V, Subramanian S, Gabaldon T, Capella-Gutierrez S, Huerta-Cepas J, Rekepalli B, Munch K, Schierup M, Lindow B, Warren WC, Ray D, Green RE, Bruford MW, Zhan X, Dixon A, Li S, Li N, Huang Y, Derryberry EP, Bertelsen MF, Sheldon FH, Brumfield RT, Mello CV, Lovell PV, Wirthlin M, Schneider MPC, Prosdocimi F, Samaniego JA, Velazquez AMV, Alfaro-Nunez A, Campos PF, Petersen B, Sicheritz-Ponten T, Pas A, Bailey T, Scofield P, Bunce M, Lambert DM, Zhou Q, Perelman P, Driskell AC, Shapiro B, Xiong Z, Zeng Y, Liu S, Li Z, Liu B, Wu K, Xiao J, Yinqi X, Zheng Q, Zhang Y, Yang H, Wang J, Smeds L, Rheindt FE, Braun M, Fjeldsa J, Orlando L, Barker FK, Jonsson KA, Johnson W, Koepfli K-P, O'Brien S, Haussler D, Ryder OA, Rahbek C, Willerslev E, Graves GR, Glenn TC, McCormack J, Burt D, Ellegren H, Alstrom P, Edwards SV, Stamatakis A, Mindell DP, Cracraft J, Braun EL, Warnow T, Jun W, Gilbert MTP, Zhang G (2014) Whole-genome analyses resolve early branches in the Tree of Life of modern birds. Science 346:1320–1331

Jukes TH, Holmquist R (1972) Evolutionary clock: nonconstancy of rate in different species. Science 177:530–532

Kaduk M, Sonnhammer E (2017) Improved orthology inference with Hieranoid 2. Bioinformatics 33:1154–1159

Kaduk M, Riegler C, Lemp O, Sonnhammer ELL (2017) HieranoiDB: a database of orthologs inferred by Hieranoid. Nucleic Acids Res 45:D687–D690

Katoh K, Standley DM (2013) MAFFT multiple sequence alignment software version 7: improvements in performance and usability. Mol Biol Evol 30:772–780

Kaufman L, Rousseeuw PJ (1990) Finding groups in data: an introduction to cluster analysis. Wiley, Hoboken, NJ

Kebschull JM, Zador AM (2015) Sources of PCR-induced distortions in high-throughput sequencing data sets. Nucleic Acids Res 43:e143

Kircher M, Kelso J (2010) High-throughput DNA sequencing – concepts and limitations. BioEssays 32:524–536

Kishino H, Thorne JL, Bruno WJ (2001) Performance of a divergence time estimation method under a probabilistic model of rate evolution. Mol Biol Evol 18:352–361

Klimke W, O'Donovan C, White O, Brister JR, Clark K, Fedorov B, Mizrachi I, Pruitt KD, Tatusova T (2011) Solving the problem: genome annotation standards before the data deluge. Stand Genomic Sci 5:168–193

Koonin EV (2005) Orthologs, paralogs, and evolutionary genomics. Annu Rev Genet 39:309–338

Koren S, Walenz BP, Berlin K, Miller JR, Bergman NH, Phillippy AM (2017) Canu: scalable and accurate long-read assembly via adaptive *k*-mer weighting and repeat separation. Genome Res 27:722–736

Laehnemann D, Borkhardt A, McHardy AC (2016) Denoising DNA deep sequencing data – high-throughput sequencing errors and their correction. Br Bioinform 17:154–179

Lake JA (1991) The order of sequence alignment can bias the selection of tree topology. Mol Biol Evol 8:378–385

Lanfear R, Calcott B, Ho SYW, Guindon S (2012) PartitionFinder: combined selection of partitioning schemes and substitution models for phylogenetic analyses. Mol Biol Evol 29:1695–1701

Lanfear R, Frandsen PB, Wright AM, Senfeld T, Calcott B (2017) PartitionFinder 2: new methods for selecting partitioned models of evolution for molecular and morphological phylogenetic analyses. Mol Biol Evol 34:772–773

Langley CH, Fitch WM (1974) An examination of the constancy of the rate of molecular evolution. J Mol Evol 3:161–177

Larget B, Simon DL (1999) Markov chain Monte Carlo algorithms for the Bayesian analysis of phylogenetic trees. Mol Biol Evol 16:750–759

Larkin MA, Blackshields G, Brown NP, Chenna R, McGettigan PA, McWilliam H, Valentin F, Wallace IM, Wilm A, Lopez R, Thompson JD, Gibson TJ, Higgins DG (2007) Clustal W and Clustal X version 2.0. Bioinformatics 23:2947–2948

Lartillot N, Lepage T, Blanquart S (2009) PhyloBayes 3: a Bayesian software package for phylogenetic reconstruction and molecular dating. Bioinformatics 25:2286–2288

Leaché AD, Banbury BL, Felsenstein J, Nieto-Montes de Oca A, Stamatakis A (2015) Short tree, long tree, right tree, wrong tree: new acquisition bias corrections for inferring SNP phylogenies. Syst Biol 64:1032–1047

Lewis PO (2001) A likelihood approach to estimating phylogeny from discrete morphological character data. Syst Biol 50:913–925

Li W-H, Tanimura M, Sharp PM (1987) An evaluation of the molecular clock hypothesis using mammalian DNA sequences. J Mol Evol 25:330–342

Liu L, Pearl DK (2007) Species trees from gene trees: reconstructing Bayesian posterior distributions of a species phylogeny using estimated gene tree distributions. Syst Biol 56:504–514

Löytynoja A (2014) Phylogeny-aware alignment with PRANK. Methods Mol Biol 1079:155–170

Löytynoja A, Goldman N (2005) An algorithm for progressive multiple alignment of sequences with insertions. Proc Natl Acad Sci USA 102:10557–10562

Löytynoja A, Goldman N (2008) Phylogeny-aware gap placement prevents errors in sequence alignment and evolutionary analysis. Science 320:1632–1635

Luo R, Liu B, Xie Y, Li Z, Huang W, Yuan J, He G, Chen Y, Pan Q, Liu Y, Tang J, Wu G, Zhang H, Shi Y, Liu Y, Yu C, Wang B, Lu Y, Han C, Cheung DW, Yiu SM, Peng S, Zhu X, Liu G, Liao X, Li Y, Yang H, Wang J, Lam TW, Wang J (2012) SOAPdenovo2: an empirically improved memory-efficient short-read de novo assembler. Gigascience 1:18

Madupu R, Brinkac LM, Harrow J, Wilming LG, Böhme U, Lamesch P, Hannick LI (2010) Meeting report: a workshop on best practices in genome annotation. Database 2010:baq001

Magoc T, Pabinger S, Canzar S, Liu X, Su Q, Puiu D, Tallon LJ, Salzberg SL (2013) GAGE-B: an evaluation of genome assemblers for bacterial organisms. Bioinformatics 29:1718–1725

Martin M (2011) Cutadapt removes adapter sequences from high-throughput sequencing reads. EMBnet J 17:10–12

Mau B, Newton MA, Larget B (1999) Bayesian phylogenetic inference via Markov chain Monte Carlo methods. Biometrics 55:1–12

Mello B, Tao Q, Tamura K, Kumar S (2017) Fast and accurate estimates of divergence times from big data. Mol Biol Evol 34:45–50

Meredith RW, Janečka JE, Gatesy J, Ryder OA, Fisher CA, Teeling EC, Goodbla A, Eizirik E, Simão TL, Stadler T, Rabosky DL, Honeycutt RL, Flynn JJ, Ingram CM, Steiner C, Williams TL, Robinson TJ, Burk-Herrick A, Westerman M, Ayoub NA, Springer MS, Murphy WJ (2011) Impacts of the Cretaceous Terrestrial Revolution and KPg extinction on mammal diversification. Science 334:521–524

Mi H, Muruganujan A, Ebert D, Huang X, Thomas PD (2019) PANTHER version 14: more genomes, a new PANTHER GO-slim and improvements in enrichment analysis tools. Nucleic Acids Res 47:D419–D426

Miller JR, Koren S, Sutton G (2010) Assembly algorithms for next-generation sequencing data. Genomics 95:315–327

Mirarab S, Reaz R, Bayzid MS, Zimmermann T, Swenson MS, Warnow T (2014) ASTRAL: genome-scale coalescent-based species tree estimation. Bioinformatics 30:i541–i548

Misof B, Liu S, Meusemann K, Peters RS, Donath A, Mayer C, Frandsen PB, Ware J, Flouri T, Beutel RG, Niehuis O, Petersen M, Izquierdo-Carrasco F, Wappler T, Rust J, Aberer AJ, Aspöck U, Aspöck H, Bartel D, Blanke A, Berger S, Böhm A, Buckley TR, Calcott B, Chen J, Friedrich F, Fukui M, Fujita M, Greve C, Grobe P, Gu S, Huang Y, Jermiin LS, Kawahara AY, Krogmann L, Kubiak M, Lanfear R, Letsch H, Li Y, Li Z, Li J, Lu H, Machida R, Mashimo Y, Kapli P, McKenna DD, Meng G, Nakagaki Y, Navarrete-Heredia JL, Ott M, Ou Y, Pass G, Podsiadlowski L, Pohl H, von Reumont BM, Schütte K, Sekiya K, Shimizu S, Slipinski A, Stamatakis A, Song W, Su X, Szucsich NU, Tan M, Tan X, Tang M, Tang J, Timelthaler G, Tomizuka S, Trautwein M, Tong X, Uchifune T, Walzl MG, Wiegmann BM, Wilbrandt J, Wipfler B, Wong TK, Wu Q, Wu G, Xie Y, Yang S, Yang Q, Yeates DK, Yoshizawa K, Zhang Q, Zhang R, Zhang W, Zhang Y, Zhao J, Zhou C, Zhou L, Ziesmann T, Zou S, Li Y, Xu X, Zhang Y, Yang H, Wang J, Wang J, Kjer KM, Zhou X (2014) Phylogenomics resolves the timing and pattern of insect evolution. Science 346:763–767

Morris JL, Puttick MN, Clark JW, Edwards D, Kenrick P, Pressel S, Wellman CH, Yang Z, Schneider H, Donoghue PCJ (2018) The timescale of early land plant evolution. Proc Natl Acad Sci USA 115:E2274–E2283

Nagarajan N, Pop M (2013) Sequence assembly demystified. Nat Rev Genet 14:157–167

Nascimento FF, dos Reis M, Yang Z (2017) A biologist's guide to Bayesian phylogenetic analysis. Nat Ecol Evol 1:1446–1454

Nowak MD, Smith AB, Simpson C, Zwickl DJ (2013) A simple method for estimating informative node age priors for the fossil calibration of molecular divergence time analyses. PLOS ONE 8:e66245

Nylander JAA, Ronquist F, Huelsenbeck JP, Nieves-Aldrey J (2004) Bayesian phylogenetic analysis of combined data. Syst Biol 53:47–67

Ohta T, Kimura M (1971) On the constancy of the evolutionary rate of cistrons. J Mol Evol 1:18–25

Paradis E, Claude J, Strimmer K (2004) APE: analyses of phylogenetics and evolution in R language. Bioinformatics 20:289–290

Parham JF, Donoghue PCJ, Bell CJ, Calway TD, Head JJ, Holroyd PA, Inoue JG, Irmis RB, Joyce WG, Ksepka DT, Patané JSL, Smith ND, Tarver JE, van Tuinen M, Yang Z, Angielczyk KD, Greenwood JM, Hipsley CA, Jacobs L, Makovicky PJ, Müller J, Smith KT, Theodor JM, Warnock RCM, Benton MJ (2012) Best practices for justifying fossil calibrations. Syst Biol 61:346–359

Parins-Fukuchi C (2018a) Bayesian placement of fossils on phylogenies using quantitative morphometric data. Evolution 72:1801–1814

Parins-Fukuchi C (2018b) Use of continuous traits can improve morphological phylogenetics. Syst Biol 67:328–339

Phillippy AM, Schatz MC, Pop M (2008) Genome assembly forensics: finding the elusive mis-assembly. Genome Biol 9:R55

Pryszcz LP, Huerta-Cepas J, Gabaldón T (2011) MetaPhOrs: orthology and paralogy predictions from multiple phylogenetic evidence using a consistency-based confidence score. Nucleic Acids Res 39:e32

Pyron RA (2011) Divergence time estimation using fossils as terminal taxa and the origins of Lissamphibia. Syst Biol 60:466–481

R Core Team (2019) R: a language and environment for statistical computing. R Foundation for Statistical Computing, Vienna

Rambaut A, Bromham L (1998) Estimating divergence dates from molecular sequences. Mol Biol Evol 15:442–448

Rambaut A, Drummond AJ, Xie D, Baele G, Suchard MA (2018) Posterior summarization in Bayesian phylogenetics using Tracer 1.7. Syst Biol 67:901–904

Rannala B, Yang Z (1996) Probability distribution of molecular evolutionary trees: a new method of phylogenetic inference. J Mol Evol 43:304–311

Rannala B, Yang Z (2003) Bayes estimation of species divergence times and ancestral population sizes using DNA sequences from multiple loci. Genetics 164:1645–1656

Rannala B, Yang Z (2007) Inferring speciation times under an episodic molecular clock. Syst Biol 56:453–466

Redelings BD, Suchard MA (2005) Joint Bayesian estimation of alignment and phylogeny. Syst Biol 54:401–418

Redelings BD, Suchard MA (2009) Robust inferences from ambiguous alignments. In: Rosenberg MS (ed) Sequence alignment: methods, models, concepts, and strategies. University of California Press, Berkeley, CA

Rokas A, Carroll SB (2006) Bushes in the tree of life. PLOS Biol 4:e352

Rokas A, Williams BL, King N, Carroll SB (2003) Genome-scale approaches to resolving incongruence in molecular phylogenies. Nature 425:798–804

Ronquist F, Klopfstein S, Vilhelmsen L, Schulmeister S, Murray DL, Rasnitsyn AP (2012a) A total-evidence approach to dating with fossils, applied to the early radiation of the Hymenoptera. Syst Biol 61:973–999

Ronquist F, Teslenko M, van der Mark P, Ayres DL, Darling A, Höhna S, Larget B, Liu L, Suchard MA, Huelsenbeck JP (2012b) MrBayes 3.2: efficient Bayesian phylogenetic inference and model choice across a large model space. Syst Biol 61:539–542

Ross MG, Russ C, Costello M, Hollinger A, Lennon NJ, Hegarty R, Nusbaum C, Jaffe DB (2013) Characterizing and measuring bias in sequence data. Genome Biol 14:R51

Roth AC, Gonnet GH, Dessimoz C (2008) Algorithm of OMA for large-scale orthology inference. BMC Bioinformatics 9:518

Rzhetsky A, Nei M (1992) Statistical properties of the ordinary least-squares, generalized least-squares, and minimum-evolution methods of phylogenetic inference. J Mol Evol 35:367–375

Saitou N, Nei M (1987) The neighbor-joining method: a new method for reconstructing phylogenetic trees. Mol Biol Evol 4:406–425

Sanderson MJ (1997) A nonparametric approach to estimating divergence times in the absence of rate constancy. Mol Biol Evol 14:1218–1231

Sanderson MJ (2002) Estimating absolute rates of molecular evolution and divergence times: a penalized likelihood approach. Mol Biol Evol 19:101–109

Sanger F, Nicklen S, Coulson AR (1977) DNA sequencing with chain-terminating inhibitors. Proc Natl Acad Sci USA 74:5463–5467

Schatz MC, Delcher AL, Salzberg SL (2010) Assembly of large genomes using second-generation sequencing. Genome Res 20:1165–1173

Schmieder R, Edwards R (2011) Quality control and preprocessing of metagenomic datasets. Bioinformatics 27:863–864

Seemann T (2014) Prokka: rapid prokaryotic genome annotation. Bioinformatics 30:2068–2069

Sievers F, Wilm A, Dineen D, Gibson TJ, Karplus K, Li W, Lopez R, McWilliam H, Remmert M, Söding J, Thompson JD, Higgins DG (2011) Fast, scalable generation of high-quality protein multiple sequence alignments using Clustal Omega. Mol Syst Biol 7:539

Simpson JT, Durbin R (2012) Efficient de novo assembly of large genomes using compressed data structures. Genome Res 22:549–556

Simpson JT, Wong K, Jackman SD, Schein JE, Jones SJM, Birol I (2009) ABySS: a parallel assembler for short read sequence data. Genome Res 19:1117–1123

Siu-Ting K, Torres-Sánchez M, San Mauro D, Wilcockson D, Wilkinson M, Pisani D, O'Connell MJ, Creevey CJ (2019) Inadvertent paralog inclusion drives artifactual topologies and timetree estimates in phylogenomics. Mol Biol Evol 36:1344–1356

Slatko BE, Gardner AF, Ausubel FM (2018) Overview of next-generation sequencing technologies. Curr Protoc Mol Biol 122:e59

Smith SA, O'Meara BC (2012) treePL: divergence time estimation using penalized likelihood for large phylogenies. Bioinformatics 28:2689–2690

Smith SA, Moore MJ, Brown JW, Yang Y (2015) Analysis of phylogenomic datasets reveals conflict, concordance, and gene duplications with examples from animals and plants. BMC Evol Biol 15:150

Söding J (2005) Protein homology detection by HMM-HMM comparison. Bioinformatics 21:951–960

Sonnhammer ELL, Östlund G (2015) InParanoid 8: orthology analysis between 273 proteomes, mostly eukaryotic. Nucleic Acids Res 43:D234–D239

Springer MS, Meredith RW, Gatesy J, Emerling CA, Park J, Rabosky DL, Stadler T, Steiner C, Ryder OA, Janečka JE, Fisher CA, Murphy WJ (2012) Macroevolutionary dynamics and historical biogeography of primate diversification inferred from a species supermatrix. PLOS ONE 7:e49521

Strong MJ, Xu G, Morici L, Bon-Durant SS, Baddoo M, Lin Z, Fewell C, Taylor CM, Flemington EK (2014) Microbial contamination in next generation sequencing: implications for sequence-based analysis of clinical samples. PLOS Pathog 10:e1004437

Suchard MA, Redelings BD (2006) BAli-Phy: simultaneous Bayesian inference of alignment and phylogeny. Bioinformatics 22:2047–2048

Suchard MA, Lemey P, Baele G, Ayres DL, Drummond AJ, Rambaut A (2018) Bayesian phylogenetic and phylodynamic data integration using BEAST 1.10. Virus Evol 4:vey016

Takahata N (1989) Gene genealogy in three related populations: consistency probability between gene and population trees. Genetics 122:957–966

Takezaki N, Rzhetsky A, Nei M (1995) Phylogenetic test of the molecular clock and linearized trees. Mol Biol Evol 12:823–833

Tamura K, Battistuzzi FU, Billing-Ross P, Murillo O, Filipski A, Kumar S (2012) Estimating divergence times in large molecular phylogenies. Proc Natl Acad Sci USA 109:19333–19338

Tavaré S, Marshall CR, Will O, Soligo C, Martin RD (2002) Using the fossil record to estimate the age of the last common ancestor of extant primates. Nature 416:726–729

Thorne JL, Kishino H (1992) Freeing phylogenies from artifacts of alignment. Mol Biol Evol 9:1148–1162

Thorne JL, Kishino H (2002) Divergence time and evolutionary rate estimation with multilocus data. Syst Biol 51:689–702

Thorne JL, Kishino H (2005) Estimation of divergence times from molecular sequence data. In: Nielsen R (ed) Statistical methods in molecular evolution. Springer, New York, pp 233–256

Thorne JL, Kishino H, Painter IS (1998) Estimating the rate of evolution of the rate of molecular evolution. Mol Biol Evol 15:1647–1657

Upham NS, Esselstyn JA, Jetz W (2019) Inferring the mammal tree: species-level sets of phylogenies for questions in ecology, evolution, and conservation. PLOS Biol 17:e3000494

Vilella AJ, Severin J, Ureta-Vidal A, Li H, Durbin R, Birney E (2009) EnsemblCompara GeneTrees: complete, duplication-aware phylogenetic trees in vertebrates. Genome Res 19:327–335

Vingron M, von Haeseler A (1997) Towards integration of multiple alignment and phylogenetic tree construction. J Comput Biol 4:23–34

Welch J, Bromham L (2005) Molecular dating when rates vary. Trends Ecol Evol 20:320–327

Wilkinson RD, Steiper ME, Soligo C, Martin RD, Yang Z, Tavaré S (2011) Dating primate divergences through an integrated analysis of palaeontological and molecular data. Syst Biol 60:16–31

Wong PB, Wiley EO, Johnson WE, Ryder OA, O'Brien SJ, Haussler D, Koepfli K-P, Houck ML, Perelman P, Mastromonaco G, Bentley AC, Venkatesh B, Zhang Y-P, Murphy RW, G10KCOS (2012) Tissue sampling methods and standards for vertebrate genomics. Gigascience 1:8

Yandell M, Ence D (2012) A beginner's guide to eukaryotic genome annotation. Nat Rev Genet 13:329–342

Yang Z (1994) Maximum likelihood phylogenetic estimation from DNA sequences with variable rates over sites: approximate methods. J Mol Evol 39:306–314

Yang Z (2004) A heuristic rate smoothing procedure for maximum likelihood estimation of species divergence times. Acta Zool Sin 50:645–646

Yang Z (2007) PAML 4: phylogenetic analysis by maximum likelihood. Mol Biol Evol 24:1586–1591

Yang Z (2014) Molecular evolution: a statistical approach. Oxford University Press, Oxford, UK

Yang Z, Rannala B (2006) Bayesian estimation of species divergence times under a molecular clock using multiple fossil calibrations with soft bounds. Mol Biol Evol 23:212–226

Yang Z, Rannala B (2010) Bayesian species delimitation using multilocus sequence data. Proc Natl Acad Sci USA 107:9264–9269

Yang Z, Rodríguez CE (2013) Searching for efficient Markov chain Monte Carlo proposal kernels. Proc Natl Acad Sci USA 110:19307–19312

Yang Z, Yoder AD (2003) Comparison of likelihood and Bayesian methods for estimating divergence times using multiple gene loci and calibration points, with application to a radiation of cute-looking mouse lemur species. Syst Biol 52:705–716

Zdobnov EM, Tegenfeldt F, Kuznetsov D, Waterhouse RM, Simão FA, Ioannidis P, Seppey M, Loetscher A, Kriventseva EV (2017) OrthoDB v9.1: cataloging evolutionary and functional annotations for animal, fungal, plant, archaeal, bacterial and viral orthologs. Nucleic Acids Res 45:D744–D749

Zerbino DR, Birney E (2008) Velvet: algorithms for de novo short read assembly using de Bruijn graphs. Genome Res 18:821–829

Zerbino DR, Achuthan P, Akanni W, Amode MR, Barrell D, Bhai J, Billis K, Cummins C, Gall A, Girón CG, Gil L, Gordon L, Haggerty L, Haskell E, Hourlier T, Izuogu OG, Janacek SH, Juettemann T, To JK, Laird MR, Lavidas I, Liu Z, Loveland JE, Maurel T, McLaren W, Moore B, Mudge J, Murphy DN, Newman V, Nuhn M, Ogeh D, Ong CK, Parker A, Patricio M, Riat HS, Schuilenburg H, Sheppard D, Sparrow H, Taylor K, Thormann A, Vullo A, Walts B, Zadissa A, Frankish A, Hunt SE, Kostadima M, Langridge N, Martin FJ, Muffato M, Perry E, Ruffier M, Staines DM, Trevanion SJ, Aken BL, Cunningham F, Yates A, Flicek P (2018) Ensembl 2018. Nucleic Acids Res 46:D754–D761

Zhang J, Chiodini R, Badr A, Zhang G (2011) The impact of next-generation sequencing on genomics. J Genet Genomics 38:95–109

Zhang C, Stadler T, Klopfstein S, Heath TA, Ronquist F (2016) Total-evidence dating under the fossilized birth-death process. Syst Biol 65:228–249

Zheng Y, Wiens JJ (2016) Combining phylogenomic and supermatrix approaches, and a time-calibrated phylogeny for squamate reptiles (lizards and snakes) based on 52 genes and 4162 species. Mol Phylogenet Evol 94:537–547

Zhu T, dos Reis M, Yang Z (2015) Characterization of the uncertainty of divergence time estimation under relaxed molecular clock models using multiple loci. Syst Biol 64:267–280

Zimin AV, Marçais G, Puiu D, Roberts M, Salzberg SL, Yorke JA (2013) The MaSuRCA genome assembler. Bioinformatics 29:2669–2677

Zuckerkandl E, Pauling L (1962) Molecular disease, evolution, and genic heterogeneity. In: Kasha M, Pullman B (eds) Horizons in biochemistry. Academic, New York, pp 189–225

MIX
Papier aus verantwortungsvollen Quellen
Paper from responsible sources
FSC® C105338

If you have any concerns about our products,
you can contact us on
ProductSafety@springernature.com

In case Publisher is established outside the EU,
the EU authorized representative is:
Springer Nature Customer Service Center GmbH
Europaplatz 3, 69115 Heidelberg, Germany

Printed by Libri Plureos GmbH
in Hamburg, Germany